BEHAVIORAL MODELS IN STRESS RESEARCH

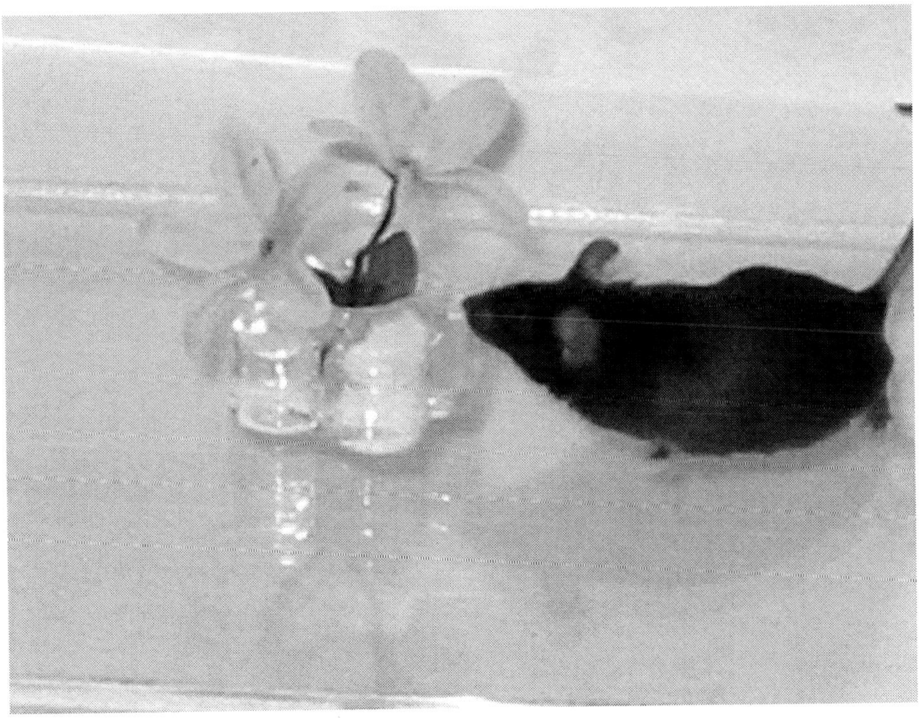

Behavioral Models in Stress Research

Allan V. Kalueff and Justin L. LaPorte
Editors

Nova Biomedical Books
New York

For permission to use material from this book please contact us:
Telephone 631-231-7269; Fax 631-231-8175
Web Site: http://www.novapublishers.com

Library of Congress Cataloging-in-Publication Data

Behavioral models in stress research / [edited by] Allan V. Kalueff and Justin L. LaPorte.
p. ; cm.
Includes bibliographical references and index.
ISBN 978-1-60456-361-0 (hardcover)
1. Stress (Psychology)--Animal models. I. Kalueff, Allan V. II. LaPorte, Justin L.
[DNLM: 1. Adaptation, Psychological--physiology. 2. Stress, Psychological--physiopathology. 3. Behavior, Animal--physiology. 4. Disease Models, Animal. 5. Models, Psychological. WM 172 B4189 2008]
BF575.S75B374 2008
156'.5--dc22

2008000043

Published by Nova Science Publishers, Inc. ✢ New York

CONTENTS

PREFACE

Stress is an adaptive response, that has developed throughout evolution, and is associated with multiple changes in the biochemistry, histology and physiology of an organism. As stress may induce or contribute to multiple neuropsychiatric disorders, the rigorous investigation of the neural substrates of stress has become a critical endeavor of contemporary biomedical science. In addition to studying physiological responses, stress researchers have attempted to assess stress-evoked behavioral changes, using both human and animal "experimental" models in clinical and non-clinical research. Though all aspects of the field have their challenges, animal experimental research of stress is a particularly difficult but meaningful task. This book provides succinct and relevant summaries of progress in the field and gives an important up-date on current methodological challenges and theoretical frameworks that exist in experimental behavioral neuroscience and genetics of stress. Each chapter is a unique contribution that showcases a vital aspect of this research.

The chapter by Czabak-Garbacz emphasizes that animal experimental studies of stress are not based on "real" human situations, and therefore may be easily influenced by additional animal-specific stressors, a factor that can complicate interpretation. Adding further complexity, the results of stress studies are greatly dependent on correct study methods, strain, sex, age and weight of subjects, their diet, experiment's time (hour, day, month, season) as well as conditions in vivarium and laboratory. This chapter outlines the importance of precise descriptions of all the above-mentioned factors in order to exclude the existing inner- and inter-laboratory variability and enhance the reproducibility of behavioral studies of stress in different scientific centers. The application of these important methodological concepts, as we will see, is crucial for the success of animal research.

Warnick and Sufka's chapter is coherent with these ideas and applies them to animal modeling as a much-utilized technique in the field of psychopharmacology. Similar to other experimental tools, animal models must be evaluated to ascertain how adequately they accomplish their research objectives. Warnick and Sufka comprehensively and critically review the existing animal models of anxiety, developing a 3-level approach to evaluating animal model simulations. According to this approach, models are assessed for their face, predictive and construct validity which can drastically increase its potential for extrapolation to clinical cases. Another feature of their approach is the ability to address utility concerns, and to be ethically justified through the 3R (reduce, refine, replace) policy. Several different experimental models were assessed by this method, and the authors compared and contrasted results of the models' evaluations, examining how each fails or succeeds in satisfying the rigorous requirements of behavioral research.

Further addressing this important issue, another chapter attends to potential solutions to the shortcomings of animal models. Feltenstein and Sufka note that most preclinical behavioral models in the field of anxiety and depression research involve the use of rodents. However, in light of issues surrounding model validity, it is important to develop new, valid, high throughput and lower cost animal models. The authors review their own recent data on the chick separation-stress paradigm, in which young domestic fowl are briefly isolated from conspecifics, leading to a number of stress behaviors, including distress vocalizations – a behavioral parameter that is highly sensitive to anxiety. This chapter shows that, in addition to construct, face and predictive validity, the chick separation-stress paradigm is more efficient than traditional rodent models because it requires smaller quantities of drugs, measures species-typical behaviors (vocalizations) that can be automatically recorded over a relatively short test period, allows for multiple animals to be tested simultaneously in the presence of a single experimenter, and because chicks are inexpensive to purchase and maintain. Importantly, this model is also relevant to depression domain, allowing the researchers to use this paradigm to sequentially model anxiety- and depressive-like states, as well as simultaneously profile anxiolytic or anxiolytic/antidepressant activity of drug compounds.

For years, single-disorder models dominated experimental neuroscience of stress, at odds with the actual complexity of clinical phenotypes. To alleviate this deficit, the field clearly needs new approaches for mimicking the anxiety-depression pathogenesis. A novel model from Sufka and colleagues not only addresses this need, but condenses the anxiety-depression continuum into a short testing period. Representing an important methodological advance, this model is highly congruent with newly appreciated concepts of commonality between stress-related brain disorders. The efficacy of this psychopharmacological approach becomes even more evident in conjunction with other - genetic and pharmacogenetic -perspectives.

An important experimental approach deals with quantitative genetics of animal emotionality, a topic that is explicated in Willis-Owen's chapter. To date, the genetic determinants of emotionality have largely resisted molecular characterization; a likely consequence of both low individual genetic effect sizes and complex gene-gene and gene–environment interactions. However, recent research has begun to overcome some of these difficulties using informative mouse models and genetic concordance between humans and mice. In this chapter, the author discusses the progress that has been made using the open-field test as a model of murine emotional reactivity, and attests to the relevance of this research towards the genetic dissection of related human phenotypes.

The chapter by Beracochea and colleagues provides further evidence of the utility of genetically-driven research of animal stress-related behaviors, and the important role of cognitive mechanisms that interplay with and influence animal emotionality. The authors explain a study involving two strains of mice that were selected based on their sensitivity or resistance to a convulsive dose of methyl ß-carboline-3-carboxylate, a benzodiazepine inverse agonist. These strains were then assessed in regard to several biochemical and behavioral properties, revealing interesting strain differences in spatial memory, endocrine stress responses, and anxiety-like behavior. Given the opposite memory- and emotional reactivity patterns, the authors showed that these selectively-bred mouse strains may be valuable tools for the study of the interaction between memory and anxiety. Along with other advances, this information will help create new animal models of neuropsychiatric diseases and disorders.

Animal models, as a research tool, are formed in a variety of ways. For example, the use of acute and chronic stress has long been used to model complex neuropsychiatric research. In her chapter, Strekalova examines the utility of chronic stress in modeling depression-like behaviors, and discusses validity and methodological limitations of anhedonia-based behavioral paradigms. The author summarizes her recent data on modifying this model [20, 22] in order to overcome the existing conceptual and methodological problems by providing an internal control for stress effects unrelated to depression. The chapter also describes a substantially modified protocol for the mouse sucrose test that aims to diminish the impact of physiological and physical artifacts, increase the model's accuracy, and detect inter-individual differences in sucrose preference within the experimental groups. As stress-induced hyperlocomotion interfered with all mouse behaviors, including depressive-like behaviors [21, 22], the use of behavioral protocols with reduced stress impact of testing conditions precluded this confounding factor, enabling a better focus on behavioral correlates of anhedonia in chronically stressed mice. Application of these methodological improvements confirmed that neurobiological correlates of stress-induced anhedonia and chronic stress are different, and that anhedonia per se (rather than chronic stress alone) is indeed associated with key pathological alterations, typical for depression. These results also show that, similar to a human population, mice display inter-individual variability in the vulnerability to a depressive-like state, which is predicted by subdominant social traits. The author further shows that pharmacological correction by antidepressant drugs reduces hedonic deficits and depressive-like syndrome, and does so selectively in anhedonic (but not the non-anhedonic) mice. Collectively, these data strongly support the validity of the chronic stress depression model in rodents, based on stress-induced anhedonic responses.

In addition to anxiety and depression, many other neuropsychiatric disorders have been associated with stress - representing a continuum or spectrum of phenomena, the integrative modeling of which is an important task *per se* [8, 9]. Among key brain molecules implicated in the co-modulation of such disorders, serotonin transporter (SERT) and brain-derived neurotrophic factor (BDNF), are particularly interesting because they enable a better understanding of molecular mechanisms of brain pathogenesis, as well as gene x gene and gene x environment interactions [1, 2, 14]. From this point of view, clinical and experimental data on genetic and environmental interactions between SERT and BDNF offer an important example of such mechanisms [10, 15, 16]. An interesting, recently developed approach focuses on mimicking a system of interacting/overlapping endophenotypes (or domains) to foster *integrative* modeling of complex behavioral and physiological phenotypes. Evaluating mutant mice with disturbed serotonergic [14] and BDNF systems [15], the authors provide evidence in favor of the utility of their approach to understanding the role of molecular and genetic mechanisms in stress responsivity, paralleling these data with clinical literature.

In their chapter, Lipina and colleagues further dissect the genetics of complex neuropsychiatric phenotypes, examining the role of a simple gene – *DISC1 (Disrupted in Schizophrenia 1)*, which encodes a large cytoplasmic protein that is expressed in the brain and modulates other brain proteins, including phosphodiesterase 4 that hydrolyzes cAMP [12]. To test the role of DISC1 in human psychiatric disorders, the authors induced mutation in Exon 2 of mouse *Disc1* near the interaction site of Disc1 with phosphodiesterase-4B [3]. Mice with mutation Q31L (Disc1-31L) showed depressive-like behavior with deficits in the forced swim test, social interactions test, sucrose preference and were sensitive to antidepressant bupropion. In addition to depressive-like changes, Disc1-31L mutants expressed mild schizophrenia-like behavioral alterations (measured by prepulse inhibition of startle response, latent inhibition, and working memory in T-maze) resistant to antipsychotic drug clozapine. Depression-like behavior in Disc1-31L mice was detected at an earlier stage than schizophrenia-like behavior, suggesting depressive phenotype as a primary core of the observed psychopathological phenotypes. Mutant Disc1 proteins exhibited reduced binding to the known Disc1 binding partner phosphodiesterase-4B. Finally, Q31L mutants had lower phosphodiesterase-4B activity, suggesting that decreased activity of this enzyme is a contributory factor in the observed psychopathological phenotypes. This study demonstrates that a *Disc1* missense mutation in mice triggers both depression-like and schizophrenia-like phenotypes, thus strongly supporting the role of genetic factors in the pathogenesis of stress-related brain disorders.

Taken together, this book represents an excellent compilation of research from a variety of disciplines pertinent to behavioral stress modeling. Anyone with an interest in neurobehavioral stress research will find that it serves as an ideal text concerning the theories, methods, and challenges in the field. The chapters provide recent, comprehensive information while remaining accessible to both established researchers and young investigators.

References

[1] A. Caspi, T.E. Moffitt, Gene-environment interactions in psychiatry: joining forces with neuroscience, *Nature reviews* 7 (2006) 583-590.

[2] A. Caspi, K. Sugden, T.E. Moffitt, A. Taylor, I.W. Craig, H. Harrington, J. McClay, J. Mill, J. Martin, A. Braithwaite, R. Poulton, Influence of life stress on depression: moderation by a polymorphism in the 5-HTT gene, *Science* 301 (2003) 386-389.

[3] S.J. Clapcote, T.V. Lipina, J.K. Millar, S. Mackie, S. Christie, F. Ogawa, J.P. Lerch, K. Trimble, M. Uchiyama, Y. Sakuraba, H. Kaneda, T. Shiroishi, M.D. Houslay, R.M. Henkelman, J.G. Sled, Y. Gondo, D.J. Porteous, J.C. Roder, Behavioral phenotypes of Disc1 missense mutations in mice, *Neuron* 54 (2007) 387-402.

[4] R. Czabak-Garbacz, B. Cygan, L. Wolanski, I. Kozlovsky, Influence of long-term treatment with tuftsin analogue TP-7 on the anxiety-phobic states and body weight, *Pharmacol Rep* 58 (2006) 562-567.

[5] M.W. Feltenstein, N.G. Ford, K.B. Freeman, K.J. Sufka, Dissociation of stress behaviors in the chick social-separation-stress procedure, *Physiol Behav* 75 (2002) 675-679.

[6] M.W. Feltenstein, K.J. Sufka, Screening antidepressants in the chick separation-stress paradigm, *Psychopharmacology (Berl)* 181 (2005) 153-159.

[7] M.W. Feltenstein, J.E. Warnick, A.N. Guth, K.J. Sufka, The chick separation stress paradigm: a validation study, *Pharmacol Biochem Behav* 77 (2004) 221-226.

[8] A.V. Kalueff, Neurobiology of memory and anxiety: from genes to behavior, *Neural plasticity* (2007) 78171.

[9] A.V. Kalueff, M. Wheaton, D.L. Murphy, What's wrong with my mouse model? Advances and strategies in animal modeling of anxiety and depression, *Behav Brain Res* 179 (2007) 1-18.

[10] J. Kaufman, B.Z. Yang, H. Douglas-Palumberi, D. Grasso, D. Lipschitz, S. Houshyar, J.H. Krystal, J. Gelernter, Brain-derived neurotrophic factor-5-HTTLPR gene interactions and environmental modifiers of depression in children, *Biol Psychiatry* 59 (2006) 673-680.

[11] N.M. Khonicheva, E.V. Loseva, R. Czabak-Garbacz, M.V. Loriia, M.G. Airapetiants, [A special case of learning disruption as a model of a disintegration phenomenon], Zhurnal vysshei nervnoi deiatelnosti imeni I P Pavlova 55 (2005) 393-399.

[12] N.C. Low, J. Hardy, What is a schizophrenic mouse?, *Neuron* 54 (2007) 348-349.

[13] W.T. McKinney, Overview of the past contributions of animal models and their changing place in psychiatry, Seminars in clinical neuropsychiatry 6 (2001) 68-78.

[14] D.L. Murphy, G.R. Uhl, A. Holmes, R. Ren-Patterson, F.S. Hall, I. Sora, S. Detera-Wadleigh, K.P. Lesch, Experimental gene interaction studies with SERT mutant mice as models for human polygenic and epistatic traits and disorders, *Genes Brain Behav* 2 (2003) 350-364.

[15] R.F. Ren-Patterson, L.W. Cochran, A. Holmes, S. Sherrill, S.J. Huang, T. Tolliver, K.P. Lesch, B. Lu, D.L. Murphy, Loss of brain-derived neurotrophic factor gene allele exacerbates brain monoamine deficiencies and increases stress abnormalities of serotonin transporter knockout mice, *J Neurosci Res* 79 (2005) 756-771.

[16] R.F. Ren-Patterson, D.K. Kim, X. Zheng, S. Sherrill, S.J. Huang, T. Tolliver, D.L. Murphy, Serotonergic-like progenitor cells propagated from neural stem cells in vitro: survival with SERT protein expression following implantation into brains of mice lacking SERT, *Faseb J* 19 (2005) 1537-1539.

[17] H. Selye, The adaptation syndrome in clinical medicine, Practitioner 172 (1954) 5-15.

[18] H. Selye, Stress and the general adaptation syndrome, *British medical journal* 1 (1950) 1383-1392.

[19] H. Selye, C. Fortier, Adaptive reaction to stress, *Psychosom Med* 12 (1950) 149-157.

[20] T. Strekalova, N. Gorenkova, E. Schunk, O. Dolgov, D. Bartsch, Selective effects of citalopram in a mouse model of stress-induced anhedonia with a control for chronic stress, *Behav Pharmacol* 17 (2006) 271-287.

[21] T. Strekalova, R. Spanagel, D. Bartsch, F.A. Henn, P. Gass, Stress-induced anhedonia in mice is associated with deficits in forced swimming and exploration, *Neuropsychopharmacology* 29 (2004) 2007-2017.

[22] T. Strekalova, R. Spanagel, O. Dolgov, D. Bartsch, Stress-induced hyperlocomotion as a confounding factor in anxiety and depression models in mice, *Behav Pharmacol* 16 (2005) 171-180.

[23] K.J. Sufka, M.W. Feltenstein, J.E. Warnick, E.O. Acevedo, H.E. Webb, C.M. Cartwright, Modeling the anxiety-depression continuum hypothesis in domestic fowl chicks, *Behav Pharmacol* 17 (2006) 681-689.

[24] P. Venault, D. Beracochea, M. Valleau, C. Joubert, G. Chapouthier, Mouse lines selected for difference in sensitivity to beta-CCM also differ in memory processes, *Behav Brain Res* 173 (2006) 282-287.

[25] J.E. Warnick, R.T. Wicks, K.J. Sufka, Modeling anxiety-like states: pharmacological characterization of the chick separation stress paradigm, *Behav Pharmacol* 17 (2006) 581-587.

[26] S.A. Willis-Owen, J. Flint, The genetic basis of emotional behaviour in mice, *Eur J Hum Genet* 14 (2006) 721-728.

[27] S.A. Willis-Owen, J. Flint, Identifying the genetic determinants of emotionality in humans; insights from rodents, *Neurosci Biobehav Rev* 31 (2007) 115-124.

[28] B. Yalcin, S.A. Willis-Owen, J. Fullerton, A. Meesaq, R.M. Deacon, J.N. Rawlins, R.R. Copley, A.P. Morris, J. Flint, R. Mott, Genetic dissection of a behavioral quantitative trait locus shows that Rgs2 modulates anxiety in mice, *Nat Genet* 36 (2004) 1197-1202.

In: Behavioral Models in Stress Research
Editors: A.V. Kalueff and J.L. LaPorte, pp. 1-22
ISBN: 978-1-60456-361-0

Chapter 1

STUDYING BEHAVIOR IN STRESS RESEARCH – NOT AS EASY AS ONE WOULD THINK

Roza Czabak-Garbacz*
Department of Human Physiology of Medical University in Lublin, Poland

INTRODUCTION: FROM HUMANS TO ANIMALS

The term "stress" was coined in the 1940's by Hans Selye [63, 169], a renowned Canadian physiologist whose prominent work continues to influence biomedical research today. Selye defined stress as a nonspecific reaction of the body after an action of harmful factors called stressors. The term, however, very quickly discontinued to be only a scientific definition and entered the common vocabulary. Currently, stress is understood as an adaptive response (caused by internal mechanisms developed throughout evolution) that allows the individual to maximize the chances of survival when confronted with a stressor. Unfortunately, the beginning of the 21st century has brought our everyday lives a great increase in the number of stressors. Among them are concerns about the natural environment, growing insecurity, globalization, economic problems and rivalry in the so-called "rat race". These concerns, in many ways, affect almost everyone, making the field of stress research evermore relevant.

Stress studies are certainly not without their difficulties, whether conducted in humans or animals. Many researchers who use humans as their study subjects recognize crucial, human-specific factors that impact the results of investigations. Among such factors are education, family, marital status, social roles, lifestyle, previous experience, mood, sex satisfaction, and work [24, 29, 79, 125, 146, 187].

The latter has a commonly known influence on human stress levels [93, 100, 149], as some professions are affected by stress more than the average [45, 73, 90, 102, 115, 165, 174, 185, 188]. Uniquely human stressors are job insecurity [46, 94, 104, 181], disparity between workers' expectations and the realities of the work environment [31, 187], and work

* E-mail address: czabak@am.lublin.pl. Phone: +48 (081) 5321733. Fax: +48 (081) 5328903. Department of Human Physiology of Medical University in Lublin, Radziwillowska 11, 20-080 Lublin, Poland

exceeding personal abilities [170]. Stress can also arise from interpersonal relationships [61, 79, 181], long working hours [79, 194], absence from home [149], and limited social life due to job pressures and excessive responsibilities [94]. Other sources of stress for humans can be quite varied [28, 33, 46, 79, 178, 181], spanning a range of stressors from political and socio-economical situations to private life [111, 177]. Previous experience and unpredictable life events, such as a dangerous situation or disaster [34, 29] can also affect a person's susceptibility to stress. So what causes someone to be more or less susceptible to stress? Though researchers have uncovered much about factors that can cause stress, it is important to understand the biological mechanisms and the levels of their actions.

In elucidating this topic, we shall first consider the influence of stress on an organism's genetics, histology and biochemistry. Stress induces a coordinated biological response from the body, including:

1. **genetic**: altered gene expression [92, 159], expression of immediate early gene products: zenk-protein, c-Fos [96, 142]; changes in RNA concentration;
2. **histological**: altered structure of tissues or organs – such as density and type of cells [148, 180], number [167, 168], size, shape and structure of cells, their organelles and molecules (e.g. receptors [70, 131]);
3. **biochemical**: cell or tissue metabolism, synthesis, release and utilization of hormones [21, 30, 37, 57, 67, 80, 101, 109, 114, 117, 145, 161, 172], mediators [1, 98, 122, 123, 131, 144, 147], peptides [99], proteins [51, 72], lipids [97, 135], carbohydrates [67, 108], as well as alterations in the activity of enzymes [48, 97, 134, 135, 136]. In parallel to the genetic, histological and biochemical alterations, stress also causes changes in the body's physiology. It influences all an organisms' important systems such as: cardiovascular (altered EEG, heart rate, blood pressure [129, 197], vessel resistance and blood flow); respiratory (frequency and deepness of respiration); gastrointestinal (motility of gastrointestinal tract [118], secretion in stomach and gut glands [179], function of interstitial barrier [119]); muscular (electromyography, electrooculography); immune [50, 128] (affecting white blood cells [162, 175], innate and acquired immunity [126, 177], production of cytokines [75, 132]); endocrine (hormone production and release [157], hypothalamic-pituitary-adrenal axis [71, 189], menstrual cycle [110]); nervous system (changes in brain electrical activity, balance control [9, 17, 58] and in integrative higher nervous activity). The latter includes altered reception and perception, cognitive abilities [97, 121] associated with synaptic plasticity [4], non-mnemonic processes (such as motivation and attention), appetitive drive [130, 133] e.g. motivation for [163, 171] and consumption of gustatory reward [76, 157] or alcohol [65, 155], and sleep patterns.

In addition to these physiological effects, stress has a well-known strong impact on emotions and behavior [20,132]. It changes, to a great degree, the behavioral rhythm of humans and animals in the duration and type of different behaviors [43, 184] like locomotion [140, 144], exploration, grooming [82, 84], eating [76, 120], freezing [95, 192] and reward [54]. Similarly, it also causes alterations in social interactions including concurrent, defensive [77] and non-defensive [86], aggressive [15, 16], sexual, maternal and reproductive [37] behaviors. Some of these behaviors will be discussed extensively later in this book.

INVESTIGATIONS OF STRESS AND BEHAVIOR

The emotional, physical and behavioral symptoms of stress are intimately linked [115]. Therefore, many researchers combine behavioral methods in their studies [43, 49] with biochemical [122], genetic, physical [47] and histological measurements to complement and parallel their findings. Behavioral endpoints are as important as genetic or neurochemical results [82], even if they are limited to a few measures.

In order to study the effects of stress on behavior, contemporary science has applied two main methods of investigation: using observations of stressful situations in normal life or in conditions created by scientists (Table 1). In general, experiments are a better tool for research because they have higher velocity, improved replication potential, lower expense and superior control over the time, place, situation, subject number and features (strain, genotype, age, sex, length and weight). Collectively, this allows researchers a greater degree of control to manipulate experimental variables according to their needs.

Table 1. The comparison of two main methods of stress investigation.

		Animal experiments	Stressful situations for humans in real life
Velocity		High	Low
Replicability		Easy	Difficult
Expense		Low	High
Time		At scientist's discretion	Accidental
Place			
Situation			
Number of subjects			
Subjects' features	Genotype		
	Age		
	Sex		
	Length		
	Weight		
	Housing		
	Handling		
	Strain		

In experimental studies, the stressor can be administered once [27, 112] or repeatedly [32, 37], can have different strengths [6, 72, 88], and may incorporate simple or complex methods. The complex methods increase the power of the statistical test, meaning that fewer subjects must be used for the study, and that the weaker effects of independent variables can be detected and measured.

Improvements in research technology have been the basis for the development of new experimental methods of stress investigation. Among these advances are automatic *in vivo* specimen data collection (often from numerous subjects simultaneously) with the use of video recorders, automated video tracking systems (such as Ethovision or Observer), infra-red photocells, ultrasonic vocalization analyzers, magnetic resonance spectroscopy, and voltammetry. There is absolutely no doubt that these new sophisticated tools provide enhanced opportunities for fast and efficient experimentation, as well as quick evaluation and analysis of the acquired data with the use of various statistical methods.

Are animals "better" subjects for stress research? It is widely accepted that most experimental procedures have been created in order to study non-human subjects. Given this fact, it is necessary to understand how animal models are relevant to humans. Importantly, the methodological basis of stress investigation is not grounded solely on the physiological and theoretical similarity between humans and animals, but also on the correspondence between their behaviors. This can be observed in social interactions such as offensive aggression. For example, in laboratory rodents, resource competition (including status and territory) is elicited by conspecific challenge situations. When the aggression is successful, the outcome of this altercation is the reduction of challenge behavior and the enhancement of resource control and status for the victor.

A similar pattern of offensive aggression is evident in human studies, although the response characteristics have been dramatically altered by human verbal, technological, and social advancements [15]. Similarities have also been shown between humans and animals in emotional responses like fear of defeat or punishment, as well as in their reactivity [8] and resistance to stress, including vulnerability to drug and alcohol abuse [2, 48]. There are also parallels between human and animal data on various types of anxiety or depressive disorders [82, 143] with an especially important role in those reactions for some mediator systems [201].

Table 2. Comparison of experiments in animals and humans

<table>
<tr><th colspan="2"></th><th>Human</th><th>Animals</th></tr>
<tr><td colspan="2">Expenses</td><td>High</td><td>Low</td></tr>
<tr><td colspan="2">Ethical concerns regarding pain
And other negative consequences</td><td>Higher</td><td>Lower</td></tr>
<tr><td rowspan="14">Individual factors</td><td>Genetic predispositions</td><td rowspan="14">Present</td><td rowspan="4">Present</td></tr>
<tr><td>Alimentation</td></tr>
<tr><td>Social role</td></tr>
<tr><td>Personal experience</td></tr>
<tr><td>Education</td><td rowspan="10">Absent</td></tr>
<tr><td>Profession</td></tr>
<tr><td>Financial situation</td></tr>
<tr><td>Marital status</td></tr>
<tr><td>Family conditions</td></tr>
<tr><td>Housework</td></tr>
<tr><td>Life rhythm and style</td></tr>
<tr><td>Leisure time</td></tr>
<tr><td>Mood</td></tr>
<tr><td>Sex satisfaction</td></tr>
</table>

Animal experiments have many other clear advantages summarized in Table 2. Studies performed on non-human subjects allow for the inexpensive and speedy selection of subjects for participation in complicated behavioral procedures, as well as quick and easy formation of experimental groups that are united by a common behavioral structure. However, in preparation for an experiment, a pretesting of the subjects may be required. In human subject studies, much time is spent on psychological tests, whereas simple pretesting is usually sufficient in animals. This ease of group formation allows the researcher to designate subjects [53, 83, 97,164].

Using animals instead of healthy human subjects can markedly foster stress research. For instance, experiments conducted on animals have fewer ethical problems (e.g. pain and other negative influences), especially in cases that involve investigations of new methods/drugs with unknown mechanisms and/or side effects, and in studies that entail obvious harmful effects. In human studies, these circumstances can additionally have grave financial consequences in the form of lawsuit liabilities and insurance costs. Examples of such procedures include X-ray radiation, magnetic field, oxidative stress [97, 135], electromagnetic waves [134, 136], ischemia and hypoxia, lesions of glands or brain structures [113, 139], high voltage electric current [113], and direct microinjection of chemicals (e.g. ibotenic acid [139]) intracerebroventricularly [183, 184] or into brain tissue [13, 105]. In addition to behavioral observation, animal studies also give the possibility to directly evaluate the processes of the central nervous system (e.g. by measuring brain mediator levels, or with the use of implanted electrodes [87]). The use of animals can also allow researchers to asses the weight and structure of particular organs such as adrenal glands [30, 103], a task that would be impossible in human subjects.

Due to the benefits of using non-human subjects in stress research, animals are often used to mimic human pathologies. Notably, many anxiolytic and antidepressant drugs that have no influence on the behavior of healthy people will affect sick humans and experimental animals. As will be seen in this book, there are many well-developed animal models and testing paradigms that enable a reliable extrapolation of results from animal research into the field of human care. The availability of these experimental animals paired with the validated paradigms alleviates the challenge of forming large enough groups of patients at an identical developmental stage of the disease or disorder.

There are, however, certain problems involved with the substitution of animals for humans in stress studies. Some researchers are skeptical of the ability to extrapolate the results derived in animal behavioral studies into meaningful data for human tests, and emphasize the difficulties of experimental interpretation [40, 198]. This criticism, along with differences in metabolism, homeostasis, topographical distribution of receptors (especially in the brain) and vulnerability to drugs point out that one must use caution when comparing different species. Animals often show over-sensitivity to external (environmental) or internal (genetic) factors, unwanted selectivity to particular neuromediatory systems [41, 85] and may display species-specific behaviors [195] in ways that humans do not.

Also, the same experimental procedure may elicit different reactions from particular animal strains (e.g. defensiveness of BALB/c, C57BL/6, CD-1 and Swiss-Webster mice in antipredator test [200], responsiveness to stress in different chicken lines [30, 48, 52]). Other scientists question the external validity of behavioral animal experimentation based on other potential problems, including difficulty in targeting behavioral vs. psychological and cognitive components [11, 60], an unclear link between behaviors and brain events, a lack of specificity [19, 160], questionable ability to detect novel compounds with unknown mechanism of action [12, 42], impossibility to reproduce complex multisyndromal human behavior and cognition, difficulty in the translation of received data into symptoms present in people [198], problems with modeling true human disorders, and variable replicability even within the same laboratory [7, 38, 141]. Nevertheless, as the reader can see in this and the subsequent chapters, the outcome of experiments carried out on animals (instead of humans) is very encouraging and clearly outweighs the shortcomings.

Other Experimental Problems with Animal Research

A growing number of experimental models involve various taxons such as mammals, birds, fishes and insects. Despite this selection, the most popular animal subjects for stress studies are mice, which account for over 80% of all experimental animals [78]. This predominance is due to low costs, the simplicity of their housing and breeding, the possibility of genetic modifications [81] and high stability in the temporal organization of their behaviors (e.g. in different novelty situations [83]). Some mouse strains are utilized more widely than others in behavioral research. The most commonly used strains are 129S1/SvImJ [84], 129/SvEv, 129S2/SvHsd [152], ABP/Le Albino, BALB/c [22, 82, 83], CBA/Lac [6], CD1(ICR)BR, C3H/HeJ, C57BL/6 [54, 83], coloboma [25], DBA/2 [54], FVB/NJ, HAP2 [32], MF1 [78], NMR1 [83, 84], St-Austin les Elbeuf, Swiss-Webster, and TNC-deficient [121] mice.

As mentioned previously, stress studies are often conducted using other species besides mice. For example, many experiments have been conducted in rats [18, 113], such as Brown Norway, Fischer [106, 157], Janvier, Lewis, Lister Hooded, Long-Evans [69, 172], PVG Black Hooded, Maudsley [99], OXYS [97], Sprague-Dawley [8, 86], Wag/Rij, Wag/GSTO and Wistar [97, 99]. Other rodents, like guinea pigs [89, 105], gerbils [147, 153], hamsters [37, 96], degus [71] and rabbits [44, 151] are used less frequently, though they still play an important role in translational research.

Non-rodent mammals, like tree shrews [70], bats [145], cats [138], dogs [36], pigs [161, 193], foxes [64], horses, sheep [21, 148], tigers [50] or non-human primates (macaque, bonobo, orangutan, rhesus, cebus paella [101]) are also used less frequently as subjects of behavioral stress studies. Similarly, birds like chickens [48, 67, 103], parrots [66], pigeons, geese [66, 164], and guinea fowl are infrequently used, as are insects (bees [127], mosquitoes [51, 92]), and fishes [56, 76, 108, 114, 133, 144]. However, the data obtained in such behavioral stress studies cannot match the effectiveness and extensiveness that may be accomplished with rodent models.

In testing rodent models it is important to be aware of the sex difference when deciding whether to use males or females for the study. Throughout evolution, males have been selected for behaviors that maximize survival in dangerous situations by overcoming the threat, whatever it may be. Self-preservation is a typical male behavioral reaction in stressful situations; it is a consequence of activating the biological mechanisms (e.g. cardiovascular performance, motor control, endocrine responses, hypothalamic-pituitary-adrenal axis) that were forged throughout evolution to assure the survival of the organism. Characteristic biological components of the male stress response are increased male sex hormone levels and the activation of the sympathetic nervous system. Additionally, the elevated release of adrenocorticoids decreases inflammation and mitigates immune response. This results in heightened physical abilities that are necessary for fight or flight mechanisms and pain inhibition, helping the organism to survive even in presence of severe injuries.

In females, there is a different set of behavioral tendencies due to their natural physiology. The life-cycle of female rodents involves pregnancy and nurturing offspring until maturity. This, of course, causes specific behaviors in the presence of a threat. Female responses are marked by pattern of “tend-and-befriend” behaviors as opposed to the risky fight or flight response that is common in males. Such reactions in females are based on the instinct to protect both the self and offspring [182], thereby maximizing the chances for survival [133]. Nurturing the young also causes changes in female social relations. For example, the formation of female networks that defend the group have been observed. Females have an inhibition of the fight or flight reactions caused, in part, by stronger activation of the parasympathetic nervous system [118], hyperactivity of corticotropin-realizing factor signaling in the brain and gut [179] and greater hormone release (e.g. opioid peptides and oxytocin [148, 181]).

Prior to 1995, most animal models of stress were validated for males. In fact, at that time they constituted more than 80% of utilized experimental animals. However, in the years since then, the imbalance involving the dominance of males as study objects and gender bias has been diminished. Although more and more studies are currently performed in both females and males [54, 175], there remain some groups who continue to perform investigations solely on males [113, 179] despite gender-specific hormonal fluctuations that underlie interesting differences in anxiety levels and stress responses [1, 32, 59, 71, 95, 133, 172, 200]. Importantly, humans have similar behavioral sex differences in the domains of stress sensitivity [35, 46, 59, 107, 156, 187, 200], coping strategies [32, 46] and stress-evoked morbidity [26, 55, 74, 178, 196]. For example, a woman's reaction in stressful situations, to some extent, results from her social roles [3, 14, 35, 137, 150, 154, 187], and physiological status (e.g. pregnancy, breastfeeding).

Differences in behavioral symptoms and reactions to stress could also be associated with age [47, 79, 115]. Enhanced stress responsiveness [101] and lower psychosocial stress occur in younger subjects [93, 187]. However, in juvenile animals, cortisol response to stress is much higher than in adults [101]. Interestingly, the fast active behavioral reactions of young subjects in stress tests are independent of circumstances while the stress-induced sickness behaviors depend on the mother's presence [75]. Stress in early life has consequences on future health [124, 199], predisposition to depression [106, 157], changes in an individual's behavior [91, 151, 172], neurochemistry [78] and hormonal and immunological functions [77]. Therefore, most researchers perform stress experiments on adult individuals and much less on the extremely young or old.

In humans, age is also an important factor which influences sensitivity, reactions and consequences of stress. Younger humans are less resistant to stress during the first few years of their life, when the brain develops. Stress may cause brain abnormalities such as reduction of hippocampus volume [100] and, in turn, may have a long-term impact on the nervous system (e.g. cognitive and behavioral deficits). Aging also causes changes in an organism's reactions to stress. Chronically stressed elderly people are at risk for stress-related pathologies like addictive behaviors or sleep disturbances because of changes in the hypothalamic-pituitary-adrenal axis, which is responsible for melatonin production [100] and alterations in glucocorticoid-immune signaling [10]. Any age group is vulnerable to stress, and some periods of life are potentially more stressful than others.

Motherhood is a prime example of such a period. In women, stress during pregnancy (gestational stress) can increase the chance of developing postpartum depression, which is observed in 10% of cases [132]. There is a strong influence of stress perceived by the mother during pregnancy on the postnatal functions of her progeny, sometimes even extending into adulthood [68, 69]. Moreover, the type and strength of effects are dependent on the timing of the maternal stress [88]. In different species, prenatal stress can impair physical [88, 114], immunological [68, 69] and hormonal development [27, 89] especially hypothalamic-pituitary-adrenal axis function [88] and behavior [5, 166]. There are suggestions that a single intense prenatal stress can enhance the learning performance and adaptive stress-related responses in the progeny [27].

An important factor in behavioral studies of stress is correct group formation according to the subject's individual social features. There exists a great difference in stress behavior between dominant and subordinate individuals [53]. In mice, dominant animals weigh less at the end of the stress period due to a decrease in the caloric efficiency and ingestion of more calories from carbohydrates, while subordinates are heavier because of increased fat intake [120]. In pigs, the larger, more dominant subjects cope better with social stressors [193]. In salmon, the fish that readily resume feeding behavior in a new environment also move less in the acute stress test [133]. Notably, people do not differ greatly from animals in this matter. For example, stress may stimulate eating disorders in predisposed humans [163], especially in perfectionist personality types [158], thus, making parallels with animal models even more important.

Housing conditions have a great impact on the results of animal stress research. Especially important are the number of animals kept in the same cage, as well as its size. It was noticed that individually housed mice spent significantly more time sleeping in the cage while groups of six mice housed together did not show such changes [78]. Fights involving previously isolated participants are significantly more intense than those involving group-housed animals, which is probably a function of established relationships between isolated subjects, heightened cortisol responsiveness, and excessive aggressive behavior [56].

Another consideration in behavioral investigations of stress is the potential role of the endogenous biological clock in the modulation of stress response [117]. Behavioral and hormonal seasonal changes are well-documented in various vertebrate species. Hens spend more time resting and eating in summer whereas more time walking, preening and drinking in winter. They also undergo prolonged stress in winter which has a strong effect on behavior [173]. In male gerbils, higher stress is present at the beginning of spring [153]. In laboratory mice, spring stress elicits a decrease in investigatory behavior while the fall season has no significant effect on behavior [117]. There are also reports of increased levels of stress in humans during the middle of the year as compared with the start and end of the year [23]. Circadian rhythms can similarly affect behavior in stress investigations. The time of day can have big impact on an animal's responses [64, 87, 197] or human behavior [176, 177, 197]. For example, people working on night shifts are particularly exposed to stress [93] because of working during the period of natural melatonin release [100]. Further modeling of such situations using animal models may be necessary to more fully understand the mechanisms of pathogenesis.

A final consideration is whether to use single or combined procedures in stress studies. There are many researchers who use multiparametrical procedures (a combination of stress situations [27, 47]) to increase the efficiency of investigations in laboratory animals. For example, prenatal stress or maternal separation procedures decrease emotionality and enhance learning performance, but the combination of both has much stronger effect than either factor alone. Likewise, maternal separation potentiates prenatal stress-induced effects by enhancing learning performance due to a dampening of hypothalamic-pituitary-adrenal axis response in the progeny to better cope with the task [27]. On the other hand, many methods reduce complex behaviors into more manageable components, a practice that may enhance the clarity of evaluation and interpretation of animal experimentation [82]. Finally, it is important to pretest, if necessary, experimental animals before the experimentation. Such procedures are not difficult to perform, and they could exclude the difficulties in interpretation of aberrant data, especially in studies with small groups that consist of individuals with high phenotypical variability.

CONCLUSION

Clearly, a thoughtful and balanced approach is needed to further advance animal experimentation in the field of stress research. Constant development of animal models and their modification is an important part of behavioral neuroscience, as it allows researchers to assess the behavioral phenotypes of mutant or transgenic animals [62, 186], find candidate genes for human behavioral disorders [39], dissect the neurobiological mechanisms of brain pathogenesis [82, 116] and test various drugs [44, 198]. It is obvious that every experimental model of stress has its own advantages and disadvantages, and may lead to new important findings. However, we must remember that the data can be influenced by many additional factors, some of which have been summarized here. Therefore, by using precise descriptions of their methods, researchers could reduce inner and inter-laboratory variability [38, 190, 191], thus increasing the overall validity of their data. Understanding the potential benefits and weaknesses of the existing animal models is crucial for obtaining valid animal data to parallel and/or complement the available clinical findings [85].

ACKNOWLEDGEMENT

This chapter is dedicated to my great friend and long-term collaborator Dr Natalia M. Khonicheva (1943-2007)

REFERENCES

[1] Andersen JP, Schwartz A, Gramsbergen JB, Loeschcke V. Dopamine levels in the mosquito Aedes aegypti during adult development, following blood feeding and in response to heat stress. *J Insect Physiol* 2006;52(11-12):1163-1170.

[2] Angst J, Gamma A, Endrass J, Rossler W, Ajdacic-Gross V, Eich D, Herrell R, Merikangas KR. Is the association of alcohol use disorders with major depressive disorder a consequence of undiagnosed bipolar-II disorder? *Eur Arch Psychiatry Clin Neurosci* 2006;256: 452-457.

[3] Artazcoz L. Artieda L. Borrell C. Cortes I. Benach J. Garcia V. Combining job and family demands and being healthy: what are the differences between men and women? *Eur J Public Health* 2004;14(1):43-48.

[4] Artola A, von Frijtag JC, Fermont PCJ, Gispen WH, Schrama LH, Kamal A, Spruijt BM. Long-lasting modulation of the induction of LTD and LTP in rat hippocampal CA1 by behavioural stress and environmental enrichment. *Eur J Neurosci* 2006;23(1):261-272.

[5] Avaliani TV, Seriakova OR, Czabak-Garbacz R, Shabaev VV. Unilateral injury of sensorimotor cortex of the dominant or subdominant hemispheres of rats-mothers and emotional behavior of their offsprings in the "open field". *Zh Vyssh Nerv Deiat Im I P Pavlova* 2002;52(5):618-625.

[6] Avgustinovich DF, Kovalenko IL, Koriakina LA. Effect of a single severe stress on behaviour of males and females in the mouse inbred strains CBA/Lac and C57BL/6JRoss Fiziol Zh Im I M Sechenova 2006;92(5):567-577.

[7] Bai F, Li X, Clay M, Lindstrom T, Skolnick P. Intra- and interstrain differences in models of "behavioural despair." *Pharmacol Biochem Behav* 2001;70:187-192.

[8] Bain MJ, Dwyer SM, Rusak B. Restraint stress affects hippocampal cell proliferation differently in rats and mice. *Neurosci Lett* 2004;368(1):7-10.

[9] Balaban CD. Neural substrates linking balance control and anxiety. *Physiol Behav* 2002;77:469-475.

[10] Bauer ME. Stress, glucocorticoids and ageing of the immune system. *Stress* 2005;8(1):69-83.

[11] Battaglia M, Ogliari A. Anxiety and panic: from human studies to animal research and back. *Neurosci Biobehav Rev* 2005;29:169-179.

[12] Belzung C. The genetic basis of the pharmacological effects of anxiolytics: a review based on rodent models. *Behav Pharmacol* 2001;12:451-460.

[13] Berretta S, Benes FM. A rat model for neural circuitry abnormalities in schizophrenia. *Nat Protocols* 2006;1:833-839.

[14] Bird CE. Gender, household labor, and psychological distress: the impact of the amount and division of housework. *J Health Soc Behav* 1999;40(1):32-45.

[15] Blanchard DC. Blanchard RJ. What can animal aggression research tell us about human aggression? *Horm Behav* 2003;44(3):171-177.

[16] Blanchard RJ, Griebel G, Farrokhi C, Markham C, Yang M., Blanchard DC. AVP V1b selective antagonist SSR149415 blocks aggressive behaviors in hamsters. *Pharmacol Biochem Behav* 2005;80(1):189-194.

[17] Bolmont B, Gangloff P, Vouriot A, Perrin PP. Mood states and anxiety influence abilities to maintain balance control in healthy human subjects. *Neurosci Lett* 2002;329:96-100.

[18] Borsini F, Brambilla A, Cesana R, Grippa N. Lack of interaction between flibanserin and antidepressants in inducing serotonergic syndrome in rats. *Int J Neuropsychopharmacol* 2001;4:9-15.

[19] Borsini F, Podhorna J, Marazziti D. Do animal models of anxiety predict anxiolytic-like effects of antidepressants? *Psychopharmacology* 2002;163:121-141.

[20] Bradley BF, Starkey NJ, Brown SL, Lea RW. Anxiolytic effects of Lavandula angustifolia odour on the Mongolian gerbil elevated plus maze. *J Ethnopharmacol* 2007;111(3):517-525.

[21] Breen KM, Oakley AE, Pytiak AV, Tilbrook AJ, Wagenmaker ER, Karsch FJ. Does cortisol acting via the type II glucocorticoid receptor mediate suppression of pulsatile luteinizing hormone secretion in response to psychosocial stress? *Endocrinology* 2007;148(4):1882-1890.

[22] Brodkin ES. BALB/c mice: Low sociability and other phenotypes that may be relevant to autism. *Behav Brain Res* 2007;176:53-65.

[23] Bruce C, Thomas PS, Yates DH. Health and stress in Australian interns. *Intern Med J* 2003;33(8):392-395.

[24] Bruce SM, Conaglen HM, Conaglen JV. Burnout in physicians: a case for peer-support. *Intern Med J* 2005;35(5):272-278.

[25] Bruno KJ, Freet CS, Twining RC, Egami K, Grigson PS, Hess EJ. Abnormal latent inhibition and impulsivity in coloboma mice, a model of ADHD. *Neurobiol Dis* 2007;25:206-216.

[26] Buxrud EG. Community health services - more stressing for female than male physicians. *Tidsskr Nor Laegeforen* 1990;110(25):3260-3264.

[27] Cannizzaro C, Plescia F, Martire M, Gagliano M, Cannizzaro G, Mantia G, Cannizzaro E. Single, intense prenatal stress decreases emotionality and enhances learning performance in the adolescent rat offspring: interaction with a brief, daily maternal separation. *Behav Brain Res* 2006;169(1):128-136.

[28] Chambers R, Sutherland VJ, Cooper CL. Identifying distress among general practitioners: predictors of psychological ill-health and job dissatisfaction. *Soc Sci Med.* 1993;37(5):575-581.

[29] Chan AO, Cheng C, Hi WM, Hu WH, Wong NY, Lam KF, Wong WM, Lai KC, Lam SK, Wong BC. Differing coping mechanisms, stress level and anorectal physiology in patients with functional constipation. *World J Gastroenterol* 2005;11(34):5362-5366.

[30] Cheng HW, Muir WM. Chronic social stress differentially regulates neuroendocrine responses in laying hens: II. Genetic basis of adrenal responses under three different social conditions. *Psychoneuroendocrinology* 2004;29(7):961-971.

[31] Cherry NM, Chen Y, McDonald JC. Reported incidence and precipitating factors of work-related stress and mental ill-health in the United Kingdom (1996-2001). *Occup Med* 2006; 56(6):414-421.

[32] Chester JA, de Paula BG, DeMaria A, Finegan A. Different effects of stress on alcohol drinking behaviour in male and female mice selectively bred for high alcohol preference. *Alcohol and Alcoholism* 2006;41(1):44-53.

[33] Chong A, Killeen O, Clarke T. Work-related stress among paediatric non-consultant hospital doctors. *Ir Med J.* 2004;97(7):203-205.

[34] Clarke D, Singh R. Life events, stress appraisals, and hospital doctors' mental health. *N Z Med J.* 2004;22;117(1204):U1121.

[35] Cohen JS, Patten S. Well-being in residency training: a survey examining resident physician satisfaction both within and outside of residency training and mental health in Alberta. *BMC Med Educ* 2005;5:21.

[36] Coppola CL, Grandin T, Enns RM. Human interaction and cortisol: can human contact reduce stress for shelter dogs? *Physiol Behav* 2006;87(3):537-541.

[37] Cordner AP, Herwood MB, Helmreich DL, Parfitt DB. Antidepressants blunt the effects of inescapable stress on male mating behaviour and decrease corticotropin-releasing hormone mRNA expression in the hypothalamic paraventricular nucleus of the Syrian hamster (Mesocricetus auratus). *J Neuroendocrinol 2004*;16(7):628-636.

[38] Crabbe JC, Wahlstein D, Dudek BC. Genetics of mouse behavior: interaction with laboratory environment. *Science* 1999;284:1670-1672.

[39] Crowley JJ, Lucki I. Opportunities to discover genes regulating depression and antidepressant response from rodent behavioural genetics. *Curr Pharm Des* 2005; 11: 157-169.

[40] Cryan JF, Holmes A. The ascent of mouse: advances in modelling human depression and anxiety. *Nat Rev Drug Discov* 2005;4:775-790.

[41] Cryan JF, Mombereau C. In search of a depressed mouse: utility of models for studying depression-related behavior in genetically modified mice. *Mol Psychiatry* 2004;9:
326-357.

[42] Cryan JF, Valentino RJ, Lucki I. Assessing substrates underlying the behavioral effects of antidepressants using the modified rat forced swimming test. *Neurosci Biobehav Rev* 2005;29:547-569.

[43] Cygan B, Wójcik G, Czabak-Garbacz R, Teter M, Kozlovsky I. Influence of the new benzodiazepine derivative BD-1158 on the behaviour of rabbits under stress conditions. *Bull of the Veterinary Institute in Pulawy* 2001;45(2):253-258.

[44] Czabak-Garbacz R, Chomicki M, Cygan B, Khonicheva NM. Influence of diazepam on the escape reaction and rabbits' behaviour under stress. *Bull of the Veterinary Institute in Pulawy* 2004; 48(4):493-497.

[45] Czabak-Garbacz R, Cygan B, Mizerska A, Mosiewicz J, Stążka J, Sekita-Krzak J, Charytanowicz M, Burghardt A. Influence of university study course on health of young people. *Polish J Environ Stud* 2006;15(2B):1051-1053.

[46] Czabak-Garbacz R, Cygan B, Sekita-Krzak J, Maziarz L, Furtak I. Work environment as a source of stress for physicians. *Polish J Environ Stud* 2006;15(2B):1054-1057.

[47] Czabak-Garbacz R, Cygan B, Wolanski L, Kozlovsky I. Influence of long-term treatment with tuftsin analogue TP-7 on the anxiety–phobic states and body weight. *Pharmacol Rep* 2006;58(4):562-567.

[48] Debut M, Berri C, Arnould C, Guemene D, Sante-Lhoutellier V, Sellier N, Baeza E, Jehl N, Jego Y, Beaumont C, Le Bihan-Duval E. Behavioural and physiological responses of three chicken breeds to pre-slaughter shackling and acute heat stress. Br *Poult Sci* 2005;46(5):527-535.

[49] Dec-Szlichtyng M, Teter M, Borowicz B. Influence of tetrapeptide Arg-Pro-Gly-Pro on the behaviour of rabbits in spontaneous and acute stress conditions. *Bull Veter Inst Pulawy* 2004; 48(3): 333-335.

[50] de la Vega E, Degnan BM, Hall MR, Cowley JA, Wilson KJ. Quantitative real-time RT-PCR demonstrates that handling stress can lead to rapid increases of gill-associated virus (GAV) infection levels in Penaeus monodon. *Dis Aquat Organ* 2004;59(3):195-203.

[51] de Meneses MD, Rebello MA. Effect of prostaglandin A1, arsenite and aspirin on stress proteins response in mosquito cells. *Dis Aquat Organ* 2001;56(3-4):298-302.

[52] Dennis R, Zhang HM, Bacon LD, Estevez I, Cheng HW. Behavioral and physiological features of chickens diversely selected for resistance to Avian Disease. 1. Selected inbred lines differ in behavioral and physical responses to social stress. *Poult Sci* 2004;83(9):1489-1496.

[53] Dennis RL, Muir WM, Cheng HW. Effects of raclopride on aggression and stress in diversely selected chicken lines. *Behav Brain Res* 2006;175(1):104-111.

[54] de Visser L, van den Bos R, Stoker AK, Kas MJH, Spruijt BM. Effects of genetic background and environmental novelty on wheel running as a rewarding behaviour in mice. *Behav Brain Res* 2007; 177(2):290-297.

[55] Dickhaus B, Mayer EA, Firooz N, Stains J, Conde F, Olivas TI, Fass R, Chang L, Mayer M, Naliboff BD. Irritable bowel syndrome patients show enhanced modulation of visceral perception by auditory stress. *Am J Gastroenterol* 2003;98(1):135-143.

[56] Earley RL, Edwards JT, Aseem O, Felton K, Blumer LS, Karom M., Grober MS. Social interactions tune aggression and stress responsiveness in a territorial cichlid fish (Archocentrus nigrofasciatus). *Physiol Behav* 2006;88(4-5):353-363.

[57] Epel E, Lapidus R, McEwen B, Brownell K. Stress may add bite to appetite in women: a laboratory study of stress-induced cortisol and eating behavior. *Psychoneuroendocrinology* 2001;26(1):37-49.

[58] Erez O, Gordon CR, Sever J, Sadeh A, Mintz M. Balance dysfunction in childhood anxiety: findings and theoretical approach. *J Anxiety Disords* 2002;461:1-16.

[59] Falconer EM, Galea LAM. Sex differences in cell proliferation, cell death and defensive behavior following acute predator odor stress in adult rats. *Brain Res* 2003;975(1-2):22-36.

[60] Finn DA, Rutledge-Gorman MT, Crabbe JC. Genetic animal models of anxiety. *Neurogenet* 2003;4:109-135.

[61] Firth-Cozens J. Individual and organizational predictors of depression in general practitioners. *Br J Gen Pract* 1998;48(435):1647-1651.

[62] Flint J. Animal models of anxiety and their molecular dissection. *Semin Cell Dev Biol* 2003;14:37-42.

[63] Fortier C, Selye H. Adrenocorticotrophic effect of stress after severance of the hypothalamo-hypophyseal pathways. *Am J Physiol* 1949;159(3):433-439.

[64] Funk D, Amir S. Circadian modulation of fos responses to odor of the red fox, a rodent predator, in the rat olfactory system. *Brain Res* 2000;866:262-267.

[65] Funk D, Vohra S, Le AD. Influence of stressors on the rewarding effects of alcohol in Wistar rats: studies with alcohol deprivation and place conditioning. *Psychopharmacology* 2004;176(1):82-87.

[66] Garner JP, Meehan CL, Mench JA. Stereotypies in caged parrots, schizophrenia and autism: evidence for a common mechanism. *Behav Brain Res* 2003;145:125-134.

[67] Garriga C, Hunter RR, Amat C, Planas JM, Mitchell MA, Moreto M. Heat stress increases apical glucose transport in the chicken jejunum. *Am J Physiol* 2006;290(1):R195-201.

[68] Gotz AA, Stefanski V. Psychosocial maternal stress during pregnancy affects serum corticosterone, blood immune parameters and anxiety behaviour in adult male rat offspring. *Physiol Behav* 2007;90(1):108-115.

[69] Gotz AA. Wittlinger S. Stefanski V. Maternal social stress during pregnancy alters immune function and immune cell numbers in adult male Long-Evans rat offspring during stressful life-events. *J Neuroimmunol* 2007;185(1-2):95-102.

[70] Gould E, McEwen BS, Tanapat P, Galea LA, Fuchs E. Neurogenesis in the dentate gyrus of the adult tree shrew is regulated by psychosocial stress and NMDA receptor activation. *J Neurosci* 1997;17(7):2492-2498.

[71] Gruss M., Westphal S, Luley C, Braun K. Endocrine and behavioural plasticity in response to juvenile stress in the semi-precocial rodent Octodon degus. *Psychoneuroendoc*r 2006;31(3):361-372.

[72] Gryzunov YuA, Koplik EV, Smolina NV, Kopaeva LB, Dobretsov GE, Sudakov KV. Conformational properties of serum albumin binding sites in rats with different behaviour in the open field test. *Stress* 2006;9(1):53-60.

[73] Gyorffy Z, Adam S, Csoboth C, Kopp M. The prevalence of suicide ideas and their psychosocial backgrounds among physicians. Psychiatr Hung 2005;20(5):370-379

[74] Heitkemper M, Jarrett M, Cain K, Shaver J, Bond E, Woods NF, Walker E. Increased urine catecholamines and cortisol in women with irritable bowel syndrome. *Am J Gastroenterol* 1996;91(5):906-913.

[75] Hennessy MB, Deak T, Schiml-Webb PA, Wilson SE, Greenlee TM, McCall E. Responses of guinea pig pups during isolation in a novel environment may represent stress-induced sickness behaviors. *Physiol Behav* 2004;81(1):5-13.

[76] Hoglund E, Sorensen C, Bakke MJ, Nilsson GE, Overli O. Attenuation of stress-induced anorexia in brown trout (Salmo trutta) by pre-treatment with dietary l-tryptophan. *Br J Nutr* 2007;97(4):786-789.

[77] Hubbard DT, Blanchard DC, Yang M, Markham CM, Gervacio A, Chun-I L, Blanchard RJ. Development of defensive behavior and conditioning to cat odor in the rat. *Physiol Behav* 2004;80(4):525-530.

[78] Hunt C, Hambly C. Faecal corticosterone concentrations indicate that separately housed male mice are not more stressed than group housed males. *Physiol Behav* 2006;87(3):519-526.

[79] Isikhan V, Comez T, Danis MZ. Job stress and coping strategies in health care professionals working with cancer patients. *Eur J Oncol Nurs* 2004;8(3):234-244.

[80] Jarvis S, Moinard C, Robson SK, Baxter E, Ormandy E, Douglas AJ, Seckl JR, Russell JA, Lawrence AB. Programming the offspring of the pig by prenatal social stress: neuroendocrine activity and behaviour. *Horm Behav* 2006;49(1):68-80.

[81] Kalueff AV, Fox MA, Gallagher PS, Murphy DL. Hypolocomotion, anxiety and serotonin syndrome-like behavior contribute to the complex phenotype of serotonin transporter knockout mice. *Genes Brain Behav* 2007;6(4):389-400.

[82] Kalueff AV, Keisala T, Minasyan A, Tuohimaa P. Influence of paternal genotypes on F1 behaviors: lessons from several mouse strains. *Behav Brain Res* 2007;177(1):45-50.

[83] Kalueff AV, Minasyan A, Keisala T, Shah ZH, Tuohimaa P. Hair barbering in mice: implications for neurobehavioural research. *Behav Processes* 2006;71(1):8-15.

[84] Kalueff AV, Tuohimaa P. Contrasting grooming phenotypes in three mouse strains markedly different in anxiety and activity (129S1, BALB/c and NMRI). *Behav Brain Res* 2005;160(1):1-10.

[85] Kalueff AV, Tuohimaa P. Experimental modeling of anxiety and depression. *Acta Neurobiol Exp* 2004;64: 439-448.

[86] Kambo JS, Galea LAM. Activational levels of androgens influence risk assessment behaviour but do not influence stress-induced suppression in hippocampal cell proliferation in adult male rats. *Behav Brain Res* 2006;175(2):263-270.

[87] Kant GJ, Bauman RA, Pastel RH, Myatt CA, Closser-Gomez E, D'Angelo CP. Effects of controllable vs. uncontrollable stress on circadian temperature rhythms. *Physiol Behav* 1991;49(3):625-630.

[88] Kapoor A, Matthews SG. Short periods of prenatal stress affect growth, behaviour and hypothalamo-pituitary-adrenal axis activity in male guinea pig offspring. *J Physiol* 2005;566(Pt 3):967-977.

[89] Kemme K, Kaiser S, Sachser N. Prenatal maternal programming determines testosterone response during social challenge. *Horm Behav* 2007;51(3):387-394.

[90] Kenny DT, Davis P, Oates J. Music performance anxiety and occupational stress amongst opera chorus artists and their relationship with state and trait anxiety and perfectionism. *J Anxiety Disord* 2004;18(6): 757-777.

[91] Khonicheva NM, Czabak-Garbacz R, Krupina NA. Behavioral consequences of isolation in early ontogeny in rats: selectivity of anxiety conditions. *Zh Vyssh Nerv Deiat Im I P Pavlova* 2002;52(6):743-749.

[92] Kim M., Robich RM, Rinehart JP, Denlinger DL. Upregulation of two actin genes and redistribution of actin during diapause and cold stress in the northern house mosquito, Culex pipiens. *J Insect Physiol* 2006;52(11-12):1226-1233.

[93] Kimak K, Kimak K, Skorek K. Stress in the nurse's environment of work. *Zdr Pub* 2000:110 (12):427-431.

[94] Kisa S, Kisa A. Job dissatisfaction among public hospital physicians is a universal problem: evidence from Turkey. *Health Care Manag* 2006;25(2):122-129.

[95] Klein LC, Popke EJ, Grunberg NE. Sex differences in effects of opioid blockade on stress-induced freezing behavior. Pharmacol Biochem Behav 1998;61(4):413-417.

[96] Kollack-Walker S, Don C, Watson SJ, Akil H. Differential expression of c-fos mRNA within neurocircuits of male hamsters exposed to acute or chronic defeat. *J Neuroendocrinol* 1999;11(7):547-559.

[97] Kolosova NG, Shcheglova TV, Sergeeva SV, Loskutova LV. Long-term antioxidant supplementation attenuates oxidative stress markers and cognitive deficits in senescent-accelerated OXYS rats. *Neurobiol Aging* 2006;27(9):1289-1297.

[98] Koopmans SJ, Ruis M, Dekker R, van Diepen H, Korte M, Mroz Z. Surplus dietary tryptophan reduces plasma cortisol and noradrenaline concentrations and enhances recovery after social stress in pigs. *Physiol Behav* 2005;85(4):469-478.

[99] Kosti O, Raven PW, Renshaw D, Hinson JP. Intra-adrenal mechanisms in the response to chronic stress: investigation in a rat model of emotionality. *J Endocrinol* 2006;189(2):211-218.

[100] Kwiatkowski F, Abrial C, Gachon F, Chevrier R, Cure H, Chollet P. Stress, cancer and circadian rhythm of melatonin. *Pathol Biol* 2005;53(5):269-272.

[101] Lahoz MM, Nagle CA, Porta M., Farinati Z, Manzur TD. Cortisol response and ovarian hormones in juvenile and cycling female Cebus monkeys: effect of stress and dexamethasone. *Am J Primatol* 2007;69(5):551-561.

[102] Laschinger HKS, Havens DS. The effect of workplace empoverment on staff nurses' occupational mental health and work effectiveness. *J Nurs Adm* 1997;26:42–50.

[103] Lay DCJr, Wilson ME. Development of the chicken as a model for prenatal stress. *J Anim Sci* 2002;80(7):1954-1961.

[104] Lee MC, Chou MC. Job and life satisfaction among remote physicians in Taiwan. *J Form Med Assoc* 1991;90(7):681-687.

[105] Leite-Panissi CR, Ferrarese AA, Terzian AL, Menescal-de-Oliveira L. Serotoninergic activation of the basolateral amygdala and modulation of tonic immobility in guinea pig. *Brain Res Bull* 2006;69(4):356-364.

[106] Leventopoulos M, Ruedi-Bettschen D, Knuesel I, Feldon J, Pryce CR, Opacka-Juffry J. Long-term effects of early life deprivation on brain glia in Fischer rats. *Brain Res* 2007;1142:119-126.

[107] Li J, Yang W, Cho SI. Gender differences in job strain, effort-reward imbalance, and health functioning among Chinese physicians. *Soc Sci Med* 2006;62(5):1066-1077.

[108] Liebert AM, Schreck CB. Effects of acute stress on osmoregulation, feed intake, IGF-1, and cortisol in yearling steelhead trout (Oncorhynchus mykiss) during seawater adaptation. *Gen Comp Endocrinol* 2006;148(2):195-202.

[109] Lobstein DD, Rasmussen CL, Dunphy GE, Dunphy MJ. Beta-endorphin and components of depression as powerful discriminators between joggers and sedentary middle-aged men. *J Psychosom Res* 1989;33(3):293-305

[110] Loucks AB, Redman LM. The effect of stress on menstrual function. *Trends Endocrinol Metab* 2004;15(10):466-471.

[111] Marco CA, Schwartz JE, Neale JM, Shiffman S, Catley D, Stone AA. Impact of gender and having children in the household on ambulatory blood pressure in work and nonwork settings: a partial replication and new findings. *Ann Behav Med* 2000;22(2):110-115.

[112] Marini F, Pozzato C, Andreetta V, Jansson B, Arban R, Domenici E, Carboni L. Single exposure to social defeat increases corticotropin-releasing factor and glucocorticoid receptor mRNA expression in rat hippocampus. *Brain Res* 2006;1067(1):25-35.

[113] Markham CM, Blanchard DC, Canteras NS, Cuyno CD, Blanchard RJ. Modulation of predatory odor processing following lesions to the dorsal premammillary nucleus. *Neurosci Lett* 2004;372(1-2):22-26.

[114] McCormick MI. Mothers matter: crowding leads to stressed mothers and smaller offspring in marine fish. *Ecology* 2006;87(5):1104-1109.

[115] McFarlane D, Duff EM, Bailey EY. Coping with occupational stress in an accident and emergency department. *West Indian Med J* 2004;53(4):242-247.

[116] McKinney WT. Overview of the past contributions in animal models and their changing place in psychiatry. *Sem Clin Psychiatry* 2004;6: 68-78.

[117] Meyer L, Caston J, Mensah-Nyagan AG. Seasonal variation of the impact of a stressful procedure on open field behaviour and blood corticosterone in laboratory mice. *Behav Brain Res* 2006;167(2):342-348.

[118] Million M, Wang L, Stenzel-Poore MP, Coste S.C., Yuan PQ, Lamy C, Rivier J, Buffington T, Tache Y. Enhanced pelvic responses to stressors in female CRF-overexpressing mice. *Am J Physiol Regul Integr Comp Physiol* 2007;292(4):R1429-1438.

[119] Moeser AJ, Klok CV, Ryan KA, Wooten JG, Little D, Cook VL, Blikslager AT. Stress signaling pathways activated by weaning mediate intestinal dysfunction in the pig. Am *J Physiol Gastrointest Liver Physiol* 2007;292(1):G173-181.

[120] Moles A, Bartolomucci A, Garbugino L, Conti R, Caprioli A, Coccurello R, Rizzi R, Ciani B, D'Amato FR. Psychosocial stress affects energy balance in mice: modulation by social status. *Psychoneuroendocrinology* 2006;31(5):623-633.

[121] Morellini F, Schachner M. Enhanced novelty-induced activity, reduced anxiety, delayed resynchronization to daylight reversal and weaker muscle strength in tenascin-C-deficient mice. *Eur J Neurosci* 2006;23(5):1255-1268.

[122] Morilak DA, Barrera G, Echeverria DJ, Garcia AS, Hernandez A, Ma S. Role of brain norepinephrine in the behavioral response to stress. *Prog Neuropsychopharmacol Biol Psychiatry* 2005;29(8):1214-1224.

[123] Morrow BA, Elsworth JD, Roth RH. Fear-like biochemical and behavioral responses in rats to the predator odor, TMT, are dependent on the exposure environment. *Synapse* 2002;46(1):11-18.

[124] Mourya DT, Yadav P, Mishra AC. Effect of temperature stress on immature stages and susceptibility of Aedes aegypti mosquitoes to chikungunya virus. *Am J Trop Med Hyg* 2004; 70(4):346-350.

[125] Murji A, Gomez M, Knighton J, Fish JS. Emotional implications of working in a burn unit. *J Burn Care Res* 2006;27(1):8-13.

[126] Nishioka K, Okano M, Ichihara Y, Ichihara N, Nishizaki K. Immunosuppressive effect of restraint stress on the initiation of allergic rhinitis in mice. *Int Arch Allergy Appl Immunol* 2005;136(2):142-147.

[127] Nunez J, Almeida L, Balderrama N, Giurfa M. Alarm pheromone induces stress analgesia via an opioid system in the honeybee. *Physiol Behav* 1997;63(1):75-80.

[128] Oarada M, Gonoi T, Tsuzuki T, Igarashi M., Hirasaka K, Nikawa T, Onishi Y, Toyotome T, Kamei K, Miyazawa T, Nakagawa K, Kashima M, Kurita N. Effect of dietary oils on lymphocyte immunological activity in psychologically stressed mice. *Biosci Biotechnol Biochem* 2007;71(1):174-182.

[129] O'Connor DB, O'Connor RC, Rory C. Perceived changes in food intake in response to stress: the role of conscientiousness. Stress and Health: *Journal of the International Society for the Investigation of Stress* 2004;20(5):279-291.

[130] O'Connor DB, O'Connor RC, White BL, Bundred PE. Are occupational stress levels predictive of ambulatory blood pressure in British GPs? An exploratory study. *Fam Pract* 2001;18(1):92-94.

[131] Ogren SO, Kuteeva E, Hokfelt T, Kehr J. Galanin receptor antagonists: a potential novel pharmacological treatment for mood disorders. *CNS Drugs* 2006;20(8):633-654.

[132] O'Mahony SM, Myint A-M, van den Hove D, Desbonnet L, Steinbusch H, Leonard BE. Gestational stress leads to depressive-like behavioural and immunological changes in the rat. *Neuroimmunomodulation* 2006;13(2):82-88.

[133] Overli O, Sorensen C, Nilsson GE. Behavioral indicators of stress-coping style in rainbow trout: Do males and females react differently to novelty? *Physiol Behav* 2006;87(3):506-512.

[134] Ozguner F, Altinbas A, Ozaydin M,.Dogan A, Vural H, Kisioglu AN, Cesur G, Yildirim NG. Mobile phone-induced myocardial oxidative stress: protection by a novel antioxidant agent caffeic acid phenethyl ester. *Toxicol Ind Health* 2005;21(9):223-230.

[135] Ozguner F, Bardak Y, Comlekci S. Protective effects of melatonin and caffeic acid phenethyl ester against retinal oxidative stress in long-term use of mobile phone: a comparative study. *Mol Cell Biochem* 2006;282(1-2):83-88.

[136] Ozguner F, Oktem F, Ayata A, Koyu A, Yilmaz HR. A novel antioxidant agent caffeic acid phenethyl ester prevents long-term mobile phone exposure-induced renal impairment in rat. Prognostic value of malondialdehyde, N-acetyl-beta-D-glucosaminidase and nitric oxide determination. *Mol Cell Biochem* 2005;277(1-2): 73-80.

[137] Pavalko EK, Woodbury S. Social roles as process: Caregiving careers and women's health. *J Health Soc Behav* 2000;41:91–105.

[138] Pavlova IV, Vanetsian GL. Evoked activity in the hypothalamus and amygdala of the cat in conditions of food-related motivation and emotional tension. *Neurosci Behav Physiol* 2006;36(2):131-138.

[139] Pentkowski NS, Blanchard DC, Lever C, Litvin Y, Blanchard RJ. Effects of lesions to the dorsal and ventral hippocampus on defensive behaviors in rats. *Eur J Neurosci* 2006;3(8):2185-2196.

[140] Perrot-Sinal T, Ossenkopp KP, Kavaliers M. Influence of a natural stressor (predator odor) on locomotor activity in the meadow vole (Microtus pennsylvanicus): modulation by sex, reproductive condition and gonadal hormones. *Psychoneuroendocrinology* 2002;25(3):259-276.

[141] Phillips TJ, Belknap JK, Hitzemann RJ, Buck KJ, Cunningham CL, Crabbe JC. Harnessing the mouse to unravel the genetics of human disease. *Genes Brain Behav* 2002;1:14-26.

[142] Pietersen CY, Bosker FJ, Postema F, Fokkema DS, Korf J, den Boer JA. Ketamine administration disturbs behavioural and distributed neural correlates of fear conditioning in the rat. *Prog Neuropsychopharmacol Biol Psychiatry* 2006;30(7):1209-1218.

[143] Prickaerts J, Moechars D, Cryns K, Lenaerts I, van Craenendonck H, Goris I, Daneels G, Bouwknecht JA, Steckler T. Transgenic mice overexpressing glycogen synthase kinase 3beta: a putative model of hyperactivity and mania. *J Neurosci* 2006;26: 9022-9029.

[144] Rao JV, Begum G, Pallela R, Usman PK, Rao RN. Changes in Behavior and Brain Acetylcholinesterase Activity in Mosquito Fish, Gambusia affinis in Response to the Sub-Lethal Exposure to Chlorpyrifos. *Int J Environ Res Public Health* 2005;2(3-4):478-483.

[145] Reeder DM, Kosteczko NS, Kunz TH, Widmaier EP. The hormonal and behavioral response to group formation, seasonal changes, and restraint stress in the highly social Malayan Flying Fox (Pteropus vampyrus) and the less social Little Golden-mantled Flying Fox (Pteropus pumilus) (Chiroptera: Pteropodidae). *Horm Behav* 2006;49(4):484-500.

[146] Reimer C, Trinkaus S, Jurkat HB. Suicidal tendencies of physicians - an overview *Psychiatr Prax.* 2005;32(8):381-385.

[147] Renoldi G, Invernizzi RW. Blockade of tachykinin NK1 receptors attenuates stress-induced rise of extracellular noradrenaline and dopamine in the rat and gerbil medial prefrontal cortex. *J Neurosci Res* 2006;84(5):961-968.

[148] Rivalland ET, Clarke IJ, Turner AI, Pompolo S, Tilbrook AJ. Isolation and restraint stress results in differential activation of corticotrophin-releasing hormone and arginine vasopressin neurons in sheep. *Neuroscience* 2007;145(3):1048-1058.

[149] Robinson GE Stresses on women physicians: consequences and coping techniques. *Depress Anxiety* 2003;17(3):180-189.

[150] Rocha LE, Debert-Ribeiro M. Work, health and gender: a comparative study on systems analysts. *Rev Saude Publica* 2001;35(6):539-547.

[151] Rodel HG, Monclus R, von Holst D. Behavioral styles in European rabbits: social interactions and responses to experimental stressors. *Physiol Behav* 2006;89(2):180-188.

[152] Rodgers RJ, Augar R, Berryman N, Hansom CJ, O'Mahony ML, Palmer RM, Stevens A, Tallett AJ. Atypical anxiolytic-like response to naloxone in benzodiazepine-resistant 129S2/SvHsd mice: role of opioid receptor subtypes. *Psychopharmacology* 2006;187(3):345-355.

[153] Rogovin KA, Tupikin AA, Randall JA, Kolosova IE, Moshkin MP. Long-term dynamic of fecal corticosterone and its ecological and social correlates in males of great gerbil (Rhombomys opimus Licht.). Non-invasive approach in studies of stress in natural populations. *Zh Obshch Biol* 2006;67(1):37-52.

[154] Ross CE, BirdCE. Sex stratification and health lifestyle: consequences for men's and women's perceived health. *J Health Soc Behav* 1994;35(2):161-178.

[155] Roxburgh S. Gender differences in the effect of job stressors on alcohol consumption. *Addict Behav* 1998;23(1):101-107.

[156] Roxburgh S. Gender differences in work and well-being: effects of exposure and vulnerability. 1996; 37(3):265-277.

[157] Ruedi-Bettschen D, Zhang W, Russig H, Ferger B, Weston A, Pedersen E-M, Feldon J, Pryce CR. Early deprivation leads to altered behavioural, autonomic and endocrine responses to environmental challenge in adult Fischer rats. *Eur J Neurosci* 2006;24(10):2879-2893.

[158] Ruggiero GM, Levi D, Ciuna A, Sassaroli S. Stress situation reveals an association between perfectionism and drive for thinness. *Int J Eat Disord* 2003;34(2):220-226.

[159] Rulten SL, Ripley TL, Manerakis E, Stephens DN, Mayne LV. Ethanol modifies the effect of handling stress on gene expression: problems in the analysis of two-way gene expression studies in mouse brain. *Brain Res* 2006;102(1):39-43.

[160] Rupniak NM. Animal models of depression: challenges from a drug development perspective. *Behav Pharmacol* 2003;14:385-390.

[161] Rutherford KM, Haskell MJ, Glasbey C, Lawrence AB. The responses of growing pigs to a chronic-intermittent stress treatment. *Physiol Behav* 2006;89(5):670-680.

[162] Saul AN, Oberyszyn TM, Daugherty C, Kusewitt D, Jones S, Jewell S, Malarkey WB, Lehman A, Lemeshow S, Dhabhar FS. Chronic Stress and Susceptibility to Skin Cancer. *J Natl Cancer Inst* 2005;97(23):1760-1767.

[163] Sassaroli S, Ruggiero GM. The role of stress in the association between low self-esteem, perfectionism, and worry, and eating disorders. *Int J Eat Disord* 2005;37(2):135-141.

[164] Satterlee DG, Marin RH. Stressor-induced changes in open-field behavior of Japanese quail selected for contrasting adrenocortical responsiveness to immobilization. *Poult Sci* 2006;85(3):404-409.

[165] Schattner P, Coman G. The stress of metropolitan general practice. *Med J Aust* 1998;169(3):133-137.

[166] Schiml-Webb PA, Deak T, Greenlee TM, Maken D, Hennessy MB. Alpha-melanocyte stimulating hormone reduces putative stress-induced sickness behaviors in isolated guinea pig pups. *Behav Brain Res* 2006;168(2):326-330.

[167] Sekita-Krzak J, Czajkowski L, Czajkowska W, Łupina T, Żebrowska-Łupina I, Danilczuk Z, Czabak-Garbacz R, Stępniewska M. Apoptosis of hepatocytes induced by chronic stress: morphological and TUNNEL studies. *Polish J Environ Stud* 2006;15(2B):1539-1542.

[168] Sekita-Krzak J, Czajkowski L, Czajkowska W, Łupina T, Żebrowska-Łupina I, Stępniewska M, Czabak-Garbacz R, Koziej J, Term-related, chronic stress-induced alterations in rat liver: histological and morphometrical studies. *Polish J Environ Stud* 2006;15(2B):1547-1550.

[169] Selye H, Fortier C. Adaptive reactions to stress. *Res Publ Assoc Res Nerv Ment Dis* 1949;29:3-18.

[170] Simonova NI, Kaibyshev VT, Galimov AR, Kaibysheva EV. Work in structure of social and psychologic factors in doctors. *Med Tr Prom Ekol* 2005;(7):25-30.

[171] Sipos ML, Bauman RA, Widholm JJ, Kant GJ. Behavioral effects of 8-OH-DPAT in chronically stressed male and female rats. *Pharmacol Biochem Behav* 2000;66(2): 403-411.

[172] Slotten HA, Kalinichev M, Hagan JJ, Marsden CA, Fone KCF. Long-lasting changes in behavioural and neuroendocrine indices in the rat following neonatal maternal separation: gender-dependent effects. *Brain Res* 2006;1097(1):123-132.

[173] Spinu M, Benveneste S, Degen AA. Effect of density and season on stress and behaviour in broiler breeder hens. *Br Poult Sci* 2003;44(2):170-174.

[174] Stacciarini J-MR, Troccoli BT. Occupational stress and constructive thinking: health and job satisfaction. *J Adv Nurs* 2004;46(5):480-487.

[175] Stefanski V, Gruner S. Gender difference in basal and stress levels of peripheral blood leukocytes in laboratory rats. *Brain Behav Immun* 2006;20(4):369-377.

[176] Stone AA. Kennedy-Moore E. Neale JM. Association between daily coping and end-of-day mood. *Health Psychol* 1995;14(4):341-349.

[177] Stone AA, Marco CA, Cruise CE, Cox DS, Neale JM. Are stress-induced immunological changes mediated by mood? A closer look at how both desirable and undesirable daily events influence sIgA antibody. *Int J Behav Med* 1996;3(1):1-13.

[178] Sutherland VJ, Cooper CL. Identifying distress among general practitioners: predictors of psychological ill-health and job dissatisfaction. *Soc Sci Med* 1993;37(5):575-581.

[179] Tache Y, Million M, Nelson AG, Lamy C, Wang L. Role of corticotropin-releasing factor pathways in stress-related alterations of colonic motor function and viscerosensibility in female rodents. *Gend Med* 2005;2(3):146-154.

[180] Tanapat P, Hastings NB, Rydel TA, Galea LAM, Gould E. Exposure to fox odor inhibits cell proliferation in the hippocampus of adult rats via an adrenal hormone-dependent mechanism. *J Comp Neurol* 2001; 437(4):496-504.

[181] Taylor B, Barling J. Identifying sources and effects of carer fatigue and burnout for mental health nurses: a qualitative approach. *Int J Ment Health Nurs* 2004;13(2):117-125.

[182] Taylor SE, Klein LC, Lewis BP, et al. Biobehavioral responses to stress in females: tend-and-befriend, not fight-or-flight. *Psychol Rev* 2000;107:411–429.

[183] Teter M, Dec-Szlichtyng M, Cygan B, Borowicz B, Kozlovska MM. The influence of the analogue 4-10 corticotropin derivative on the general structure of the rabbits behaviour in stress. Ann Univ Mariae Curie Sklodowska 2000;55:1-10.

[184] Teter M, Dec-Szlichtyng M, Cygan B, Kozlovska M.M. Comparison of the activity of ACTH 4-10 fragment and analogue and tripeptide Pro-Gly-Pro during 3-hour-observation of rabbits' behaviour under acute stress conditions. *Bull Veter Inst Pulawy* 2003;47(1):239-245.

[185] Tian J, Wang X. An epidemiological survey of job stress and health in four occupational populations in Fuzhou city of China. *Stress and Health* 2005;21(2):107-112.

[186] Urani A, Chourbaji S, Gass P. Mutant mouse models of depression: candidate genes and current mouse lines. *Neurosci Biobehav Rev* 2005;29:805-828.

[187] Vanagas G, Bihari-Axelsson S. The factors associated to psychosocial stress among general practitioners in Lithuania. Cross-sectional study. *BMC Health Serv Res* 2005;5(1):45.

[188] van der Ploeg E, Dorresteijn SM, Kleber RJ. Critical incidens and chronic stressors at work: their impact on forensic doctors. *J Occup Health Psychol* 2003;8(2):157-166.

[189] Veenema AH, Koolhaas JM, de Kloet ER. Basal and stress-induced differences in HPA axis, 5-HT responsiveness, and hippocampal cell proliferation in two mouse lines. *Ann NY Acad Sci* 2004;1018:255-265.

[190] Wahlstein D. Standardizing tests of mouse behavior: reasons, recommendations, and reality. *Physiol Behav* 2002;73:695-704.

[191] Wahlstein D, Rustay NR, Metten P, Crabbe JC. In search of a better mouse test. *Trends Neurosci* 2003;26: 132-136.

[192] Wallace KJ, Rosen JB. Predator odor as an unconditioned fear stimulus in rats: elicitation of freezing by trimethylthiazoline, a component of fox feces. *Behav Neurosci* 2000;114(5):912-922.

[193] Wellock IJ, Emmans GC, Kyriazakis I. Modeling the effects of stressors on the performance of populations of pigs. *J Anim Sci* 2004;82(8):2442-2450.

[194] Whalley D, Bojke C, Gravelle H, Sibbald B. GP job satisfaction in view of contract reform: a national survey. *Br J Gen Pract* 2006;56(523):87-92.

[195] Whishaw IQ, Gharbawie OA, Clark BJ, Lehmann H. The exploratory behavior of rats in an open environment optimizes security. *Behav Brain Res* 2006;171:230-239.

[196] Whitley TW, Allison EJJr, Gallery ME, Heyworth J, Cockington RA, Gaudry P, Revicki DA. Work-related stress and depression among physicians pursuing postgraduate training in emergency medicine: an international study. *Ann Emerg Med* 1991;20(9):992-996.

[197] Willis TA, O'Connor DB, Smith L. The influence of morningness-eveningness on anxiety and cardiovascular responses to stress. *Physiol Behav* 2005;85(2):125-133.

[198] Wong ML, Licinio J. From monoamines to genomic targets: a paradigm shift for drug discovery in depression. *Nat Rev Drug Discov* 2004;3:136-151.

[199] Yadav P, Barde PV, Gokhale MD, Vipat V, Mishra AC, Pal JK, Mourya DT. Effect of temperature and insecticide stresses on Aedes aegypti larvae and their influence on the susceptibility of mosquitoes to dengue-2 virus. *Southeast Asian J Trop Med Public Health* 2005;36(5):1139-1144.

[200] Yang M, Augustsson H, Markham CM, Hubbard DT, Webster D, Wall PM, Blanchard RJ, Blanchard DC. The rat exposure test: a model of mouse defensive behaviors. *Physiol Behav* 2004;81(3):465-473.

[201] Yoshikawa T, Watanabe A, Ishitsuka Y, Nakaya A, Nakatani N. Identification of multiple genetic loci linked to the propensity for "behavioral despair" in mice. *Genome Res* 2002; 12: 357-366.

In: Behavioral Models in Stress Research
Editors: A.V. Kalueff and J.L. LaPorte, pp. 23-54
ISBN: 978-1-60456-361-0

Chapter 2

Development and Validation of Alternative Animal Models of Anxiety and Depression Using Domestic Fowl

Matthew W. Feltenstein[1] and Kenneth J. Sufka[2,3,4*]
[1]Department of Neurosciences,
Medical University of South Carolina, Charleston, SC, USA;
Departments of [2]Psychology, [3]Pharmacology and
[4]Research Institute for Pharmaceutical Sciences,
University of Mississippi, Oxford, MS, USA

Preclinical Models of Anxiety

In behavioral neuroscience, animal models involve the use of living organisms in controlled conditions as a means to provide insight into brain-behavior relationships [1]. The goal of many behavioral models is to produce a condition in a nonhuman animal in an attempt to understand some aspect of the etiology, symptomatology or treatment of a given human syndrome or disorder [2, 3]. That is, animal models serve as the foundation for pre-clinical research and can be employed to serve as a simulation to study the underlying mechanisms of a disorder (behavioral bioassays), to understand its psychological processes (simulations) or as a screening tool to discover novel therapeutic agents (screening tests) [4-12]. Throughout animal model development, one area of research that has received considerable interest is the development of valid animal models to study anxiety disorders [13].

In the course of laboratory animal research, over 30 animal models of anxiety have been developed [14]. However, in order to truly model the human condition, three assumptions are required [8, 15]. First, the model should have sound theoretical rationale (construct validity). Secondly, there should be phenomenological similarities between the animal model and the clinical condition being simulated (face validity). Finally, performance in the animal model should predict performance in the clinical condition (predictive validity). The model should also address the adherence of responsible animal research to the "3 Rs" (replacement, reduction and refinement) [16] by replacing traditional animal models with a less sentient species, reducing the number of animals bred for research purposes and refining procedures so that pain and distress are minimized. Finally, it should be reproducible both within and across various laboratories and, while not necessary, it can also be viewed as beneficial for a model to be economical in terms of cost, resources and time (utility).

Traditionally, animal models of anxiety can be classified into two different categories depending on how the state of anxiety is induced [17]. The first category is conditioned models, including those that are based on conflict or fear such as the Geller-Seifter test and the Vogel punished drinking test. In the Geller-Seifter test, food-deprived rats receive an electrical shock for every lever press emitted for a food reward [18]. Similarly, water-deprived rats in the Vogel test receive an electric shock every time the animal takes a drink through the available water spout [19]. In these tests, it is assumed that anticipation of punishment will lead to a decrease in responding and drugs that reduce anxiety will lead to a reversal of punished responding. Although these models have predictive validity in that anxiolytics of the benzodiazepine [20, 21] and non-benzodiazepine [18-20] classes enhance punished responding, their sensitivity to the anxiolytic effects of antidepressants has been mixed [22]. The use of conflict or conditioned fear anxiety tests has also been limited due to ethical concerns involving food/water deprivation and the use of electric shock. Moreover, these animal models possess low utility in that considerable time must be invested to train subjects [14].

The second category of animal models of anxiety is unconditioned models, such as those in which an anxious state is induced through exposure of the animal to a naturally aversive situation. Compared to conditioned models, models involving unconditioned (i.e., spontaneous) behaviors are generally preferred in that they not only have a higher degree of ecological validity, but are also subject to less confounds (e.g., interference with nociceptive, hunger/thirst, and/or learning/memory mechanisms) that could affect behavioral indices of anxiety [14]. The utility of unconditioned models is also higher than conditioned models in that they do not require extensive periods of pre-training and their test phases also tend to be relatively short [23]. In one such model, the rat social interaction test, a pair of male rats are placed in a novel environment and the amount of time that they spend interacting (e.g., sniffing, grooming) is measured [17, 24, 25]. When exposed to an unfamiliar or brightly lit environment, social interaction decreases relative to a low-light, familiar environmental conditions- an effect that is attributed to anxiety [26]. Agents that possess anxiolytic properties are predicted to prevent this decline. Research has indicated that the rat social interaction test is a behaviorally valid animal model of anxiety in that other behaviors of anxiety (such as defecation, self-grooming and displacement activities) have been found to correlate with reductions in social interaction [17]. Furthermore, this paradigm has also demonstrated construct validity in that physiological measures such as adrenocorticotropic hormone and corticosterone have been found to negatively correlate with reductions in social interaction [17]. Although sensitive to a large number of benzodiazepine anxiolytics of

* E-mail address: pysufka@olemiss.edu. Phone: 1 (662) 915-7728. Department of Psychology, University of Mississippi, Oxford, MS 38677, USA.

varying potencies [25-28], the predictive validity of the rat social interaction test seems to be limited in that it appears only sensitive to anxiolytics of this class [11].

Another unconditioned animal model of anxiety is the elevated plus maze. In this model, rats are allowed to freely explore a plus-shaped apparatus consisting of two open and two closed arms. As the open arms are considered more anxiety-provoking [29, 30], animals tend to spend more time exploring the closed arms. Anxiolytic agents are predicted to increase the amount of activity (i.e., amount of time and/or number of crossings) in the open arms of the maze [31, 32]. Similar to the rat social interaction test, research has indicated that this model is a valid model of anxiety both behaviorally and physiologically [32, 33]. Furthermore, the elevated plus maze has also demonstrated predictive validity in that a variety of benzodiazepine anxiolytics have been shown to increase open arm activity, while non-anxiolytic agents have no such effect [31-36]. It is also one of few bidirectional tests in that it is also sensitive to anxiogenic agents (i.e., a reduction in time spent/number of open arm entries) [5, 33]. Although the elevated plus maze is economical in terms of time and effort, it is somewhat problematic in that it tends to produce a large amount of variance, leading to an inability to determine fine-gained drug effects (e.g., dose-dependent effects) [5]. Moreover, like the rat social interaction test, the pharmacological validity of this model may be limited, in that it only appears sensitive to anxiolytics of the benzodiazepine class [14].

A different class of unconditioned animal models of anxiety involves the separation of infant animals from their mother or littermates. This isolation manipulation can lead to a number of behaviors, including somatic reactions such as an elevation of blood pressure [37] and hyperthermia [38, 39]. Although anxiolytics have been found to antagonize these reactions in a number of models [39], the utility of somatic reactions as screening tests has been weakened in that these indices are also sensitive to non-anxiolytic agents (i.e., false positives) [38, 40]. An additional behavior that occurs when some infant animals are isolated from their mother or littermates is the elicitation of distress vocalizations (DVocs). This isolation-induced behavior occurs in a number of species and can either be ultrasonic (e.g., rats [41] and mice [42]) or audible (e.g., guinea pigs [43-45]). An anxiolytic effect is indicated by a reduction in the number of DVocs emitted by isolated animals in the absence of behavioral sedation [46-48]. Although a variety of benzodiazepine and non-benzodiazepine anxiolytics have produced positive results in these models, some false negatives as well as false positives have also been found [42, 46, 49] leading to the conclusion that DVocs in these models may lack selectivity [11].

The Chick Separation-Stress Paradigm

In response to the various limitations of existing animal models in the field of anxiety research (for review, see [50]), we have embarked on a programmatic line of research to develop and validate an alternative animal model of anxiety called the chick separation-stress paradigm. In this model, socially-raised young domestic fowl (*Gallus gallus*) are isolated from their conspecifics. Although originally developed as a model to study the neurochemical basis of social attachment (i.e., the development of social bonds between individuals) [39, 51-53], we have utilized this model in that social-separation causes an activation of the general stress response system similar to traditional rodent models, leading to stress response activity and a number of behaviors including hyperthermia [39, 54], reduced pain sensitivity (i.e., stress-induced analgesia) [39, 54-56] and an increase in audible DVocs [39, 51, 52, 57-59].

In order to evaluate the construct validity of these behaviors as indices of stress in domestic fowl, Sufka and Weed [54] conducted a series of experiments in which seven-day-old chicks received intraplantar injections of 0.05% formalin (or saline) immediately prior to being placed in a sound-attenuating apparatus either in the presence or absence of conspecifics for a 180 sec observation period (Table 1). The behaviors collected [including body temperature, footlift frequency and duration, number of pecks directed at the infected foot, DVocs and ventral recumbency latency (to index behavioral sedation)] were analyzed using principle components analysis in order to determine any grouping of behaviors and relationship between components hypothesized to reflect isolation stress and inflammatory nociception (Table 2). Relative to chicks tested in the presence of social companions, isolated chicks displayed an increase in body temperature (i.e., hyperthermia) and number of DVocs. Moreover, chicks that received intraplantar formalin injections displayed a greater number of footlights and footlift duration, as well as number of pecks at the infected foot, than saline control animals, an effect that was attenuated in animals tested in isolation (i.e., stress-induced analgesia). Finally, in addition to body temperature and DVocs converging to form a construct of isolation stress, while footlift frequency, duration and number of pecks formed an inflammatory nociception measure, the relationship between these two constructs was curvilinear. That is, the inverted U-shaped relationship between the stress and inflammatory nociception behaviors not only support the presumed effects of stress on nociceptive processing (i.e., stress-induced analgesia), but also suggests that these behavioral clusters can be utilized as valid indices of stress in the chick separation-stress paradigm.

Table 1. Indices of social-separation stress and inflammatory nociception in seven-day-old chicks.

	Non-Isolated		Isolated	
	Saline	Formalin	Saline	Formalin
Distress Vocalizations	14.31 ± 6.51	13.00 ± 5.40	105.94 ± 16.20	95.12 ± 15.49
Ventral Recumbency Latency	159.88 ± 10.14	172.94 ± 5.62	175.56 ± 3.31	168.06 ± 7.03
Footlift Frequency	0.50 ± 0.39	59.81 ± 6.17	0.19 ± 0.19	23.38 ± 5.11
Footlift Duration	0.08 ± 6.51	23.66 ± 5.40	0.09 ± 16.20	7.30 ± 15.49
Foot peck Frequency	0.00 ± 0.00	3.69 ± 0.86	0.00 ± 0.00	2.31 ± 1.25
Body Temperature	40.09± 0.11	40.48 ± 0.08	40.51 ± 0.08	40.47 ± 0.11
Body Weight	74.95 ± 1.07	74.50 ± 1.29	75.43 ± 1.08	74.21 ± 1.28

Values represent means ± SEM (n = 16). Adapted from [54].

Table 2. Sorted oblique principal components loadings.

	Component		
	1	**2**	**3**
Distress Vocalizations	-13	**58**	16
Ventral Recumbency Latency	13	**75**	-06
Footlift Frequency	**91**	-05	-08
Footlift Duration	**90**	-02	-04
Foot peck Frequency	**76**	-02	06
Body Temperature	19	58	**62**
Body Weight	-11	00	**93**

Decimal places omitted. Highest component loadings for each variable in bold (n = 64). Adapted from [54].

Following the validation of these behaviors, two studies [59, 60] examined the predictive validity of the chick separation-stress paradigm by assessing the sensitivity of the model to the benzodiazepine anxiolytic chlordiazepoxide. Similar to the Sufka and Weed [54] study, isolated chicks displayed an elevation in DVocs (Figures 1A and 2A) and fewer formalin-induced pain-related behaviors [i.e., stress-induced analgesia, as indexed by a composite pain z-score score derived from footlift frequency, duration and number of pecks (Figure 1B) or footlights alone (Figure 2B)] than non-isolated/social chicks. Additionally, chlordiazepoxide (1-5 mg/kg, intramuscular) produced a dose-dependent improvement in these behaviors in isolated chicks, as indicated by a reduction and enhancement in DVocs and pain-related behaviors, respectively (Figures 1 and 2). Moreover, the effect of chlordiazepoxide on these behaviors was blocked by co-administration of the benzodiazepine antagonist flumazenil (0.1 – 1 mg/kg, IM; Figure 2), confirming that its anxiolytic effects are mediated through benzodiazepine receptor activity.

Given the sensitivity of the model to anxiolytic effects of chlordiazepoxide, Watson et al. [60] sought to provide further predictive validity to the model by examining the potency and dose sensitivity of the chick separation-stress paradigm to various benzodiazepine agonists. In this study, multiple doses of alprazolam and lorazepam, two potent benzodiazepine agonists shown to possess anxiolytic properties [61, 62], were examined for their effects on social separation-induced DVocs (Figure 3A) and stress-induced analgesia [as indexed by footlights (Figure 3B)]. As predicted, the DVocs measure in the chick separation-stress paradigm demonstrating dose-dependent sensitivity to lorazepam (0.125 – 1 mg/kg, intramuscular), alprazolam (0.065 – 0.5 mg/kg, intramuscular) and chlordiazepoxide. Moreover, the relative difference in effectiveness of the three benzodiazepines was found to be analogous to typical daily doses used in humans, with lorazepam being less potent than alprazolam, yet substantially more potent than chlordiazepoxide. It was also interesting to note that the ED_{50} values (i.e., median effective dose) for these benzodiazepine anxiolytics also matched the typical daily doses used in humans, providing further support for the predictive validity of the chick separation-stress paradigm as a valid animal model of anxiety. While dose-dependently reversed by both drugs, the stress-induced analgesia measure was not able to differentiate the potency differences between the two compounds (Figure 3B), indicating that this index may not be a pure measure of social-separation stress and as such, not as sensitive as DVocs to the anxiolytic properties of drugs.

In considering this issue, Feltenstein et al. [63] examined whether apparatus novelty could serve as a potential confound in the model by examining the effect of habituation to the test chamber on separation-induced DVocs and stress-induced analgesia. Similar to other factors, including presence of the experimenter [64] and early handling [65], apparatus novelty has been shown to affect stress responses in animals [66]. To address this concern, eight-day-old chicks were tested in isolation following chronic habituation to the test apparatus. The habituation procedure consisted of a 180 sec exposure to the testing chamber (under social conditions to prevent habituation to isolation) once per day on Days 2 through 7 posthatch. Control groups consisted of animals tested under a social condition, as well as non-habituated groups. Similar to previous findings, isolated chicks exhibited a significantly greater number of DVocs (Figure 4A) and fewer formalin-induced pain behaviors [i.e., stress-induced analgesia, as indexed by footlifts (Figure 4B)] than social animals. Although separation-induced DVocs were unaffected, isolated chicks that were previously habituated to the test apparatus exhibited more pain-related behaviors (i.e., a reduction in stress-induced analgesia) than their non-habituated counterparts, suggesting this index is confounded by apparatus novelty and unlike DVocs, not a pure measure of social separation-stress.

Following the observation that DVocs were unaffected by habituation to the test apparatus, Feltenstein et al. [63] conducted a second experiment in which chicks (both isolated and social) were tested with or without mirrors in the test chambers. If DVocs are indeed a more pure measure of social separation-stress aimed at re-establishing social contact [67, 68], then the perceived presence of social companions would significantly attenuate DVocs, while only having a modest effect on the stress-induced analgesia measure. Consistent with this hypothesis, isolated chicks displayed a greater number of DVocs than social chicks, an effect was significantly attenuated by the presence of mirrors (Figure 5A). Moreover, this manipulation did not have any effect on the stress-induced analgesia measure (Figure 5B). Taken together, these results suggest that DVocs and stress-induced analgesia are dissociable indexes of stress in young domestic fowl and that DVocs are a more pure measure of social separation-stress than stress-induced analgesia in the chick separation-stress paradigm.

In light of these issues, our laboratory sought to enhance the utility of the model by reducing the number of behaviors recorded to DVocs and replacing conspecifics with mirrors in the social condition. That is, by eliminating the stress-induced analgesia measure, we removed an index confounded by novelty to the test apparatus [63] and the possible interference of nociception on other behavioral indexes of stress [14]. This change also refined the procedure by reducing the pain and discomfort of the test subjects and eliminated labor issues associated with collecting this measure. In that DVocs are a species-typical behavior that can be automatically recorded over a relatively short test period (180 sec), using this single index in the model enhances efficiency given that multiple animals can be tested simultaneously (i.e., 6-9) in the presence of a single experimenter. Moreover, replacing non-test conspecifics with mirrors not only reduces the number of research animals, but controls for any inadvertent behavioral effects that these animals may have on the test animal.

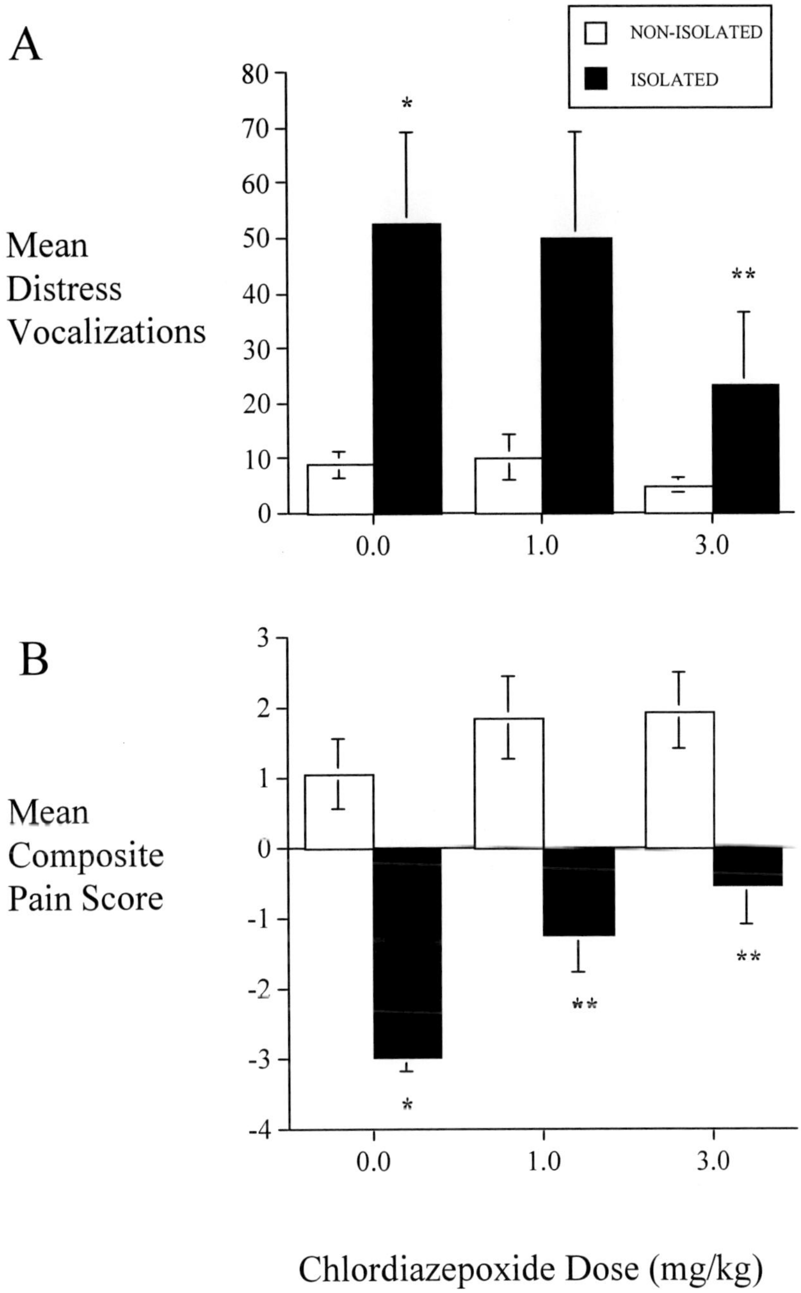

Figure 1. Mean separation distress vocalizations (A) and mean composite pain score (i.e., sum of the z-scores for lift frequency, lift duration and peck frequency) elicited by intraplantar injection of 0.10% formalin (B) during a 150-s observation session as a function of chlordiazepoxide dose in non-isolated and isolated chicks. Vertical bars represent the SEM (n = 14). * Significant isolation stress effect, ** significant attenuation of the stress effect by chlordiazepoxide. All P <0.05 (Duncan's procedures). Adapted from [59].

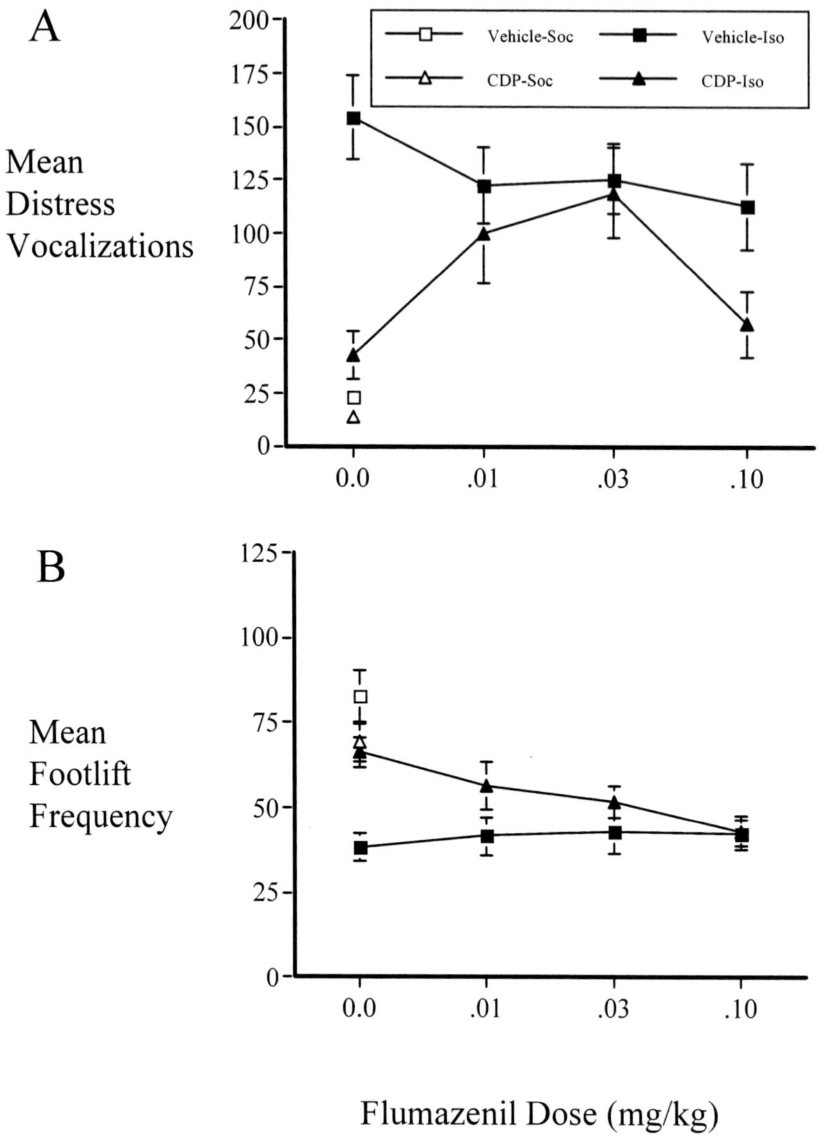

Figure 2. Mean separation distress vocalizations (A) and mean footlift frequency elicited by intraplantar injection of 0.10% formalin (B) during a 180-s observation session as a function of flumazenil dose in chicks that received either chlordiazepoxide (CDP 5 mg/kg, IM) or vehicle under a social (Soc) or isolated (Iso) test condition. Vertical bars represent the SEM (n = 15-19). Adapted from [60].

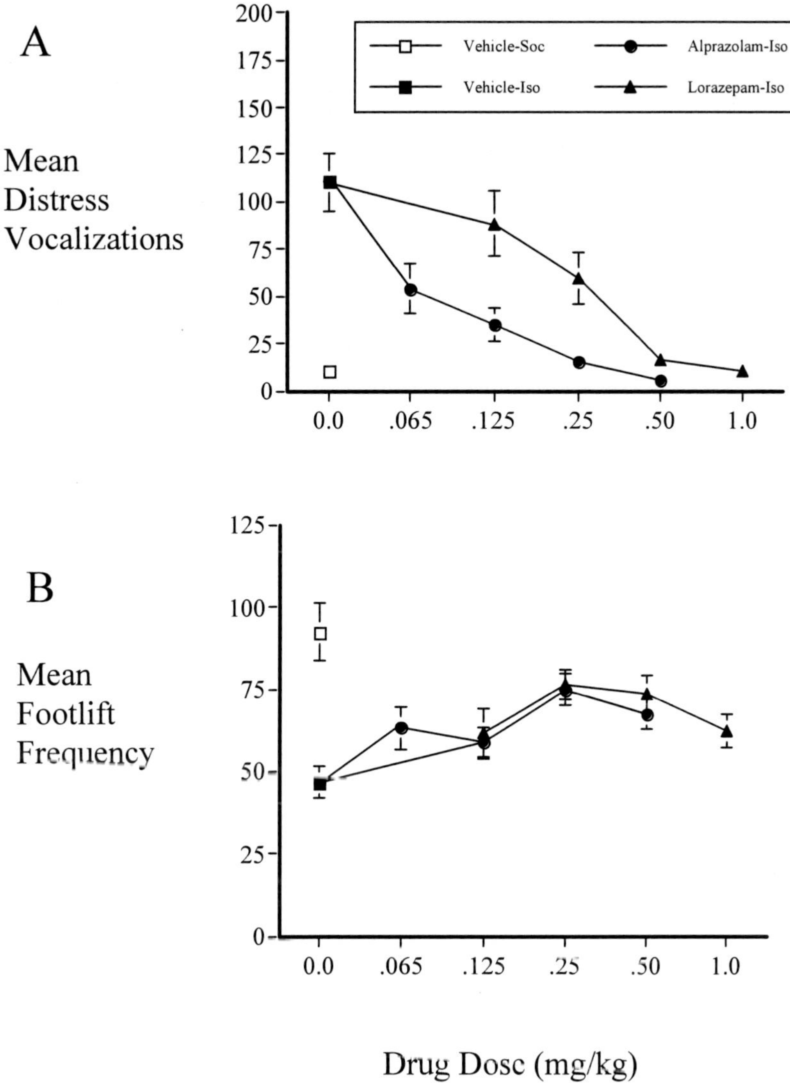

Figure 3. Mean separation distress vocalizations (A) and mean footlift frequency elicited by intraplantar injection of 0.10% formalin (B) during a 180-s observation session as a function of alprazolam and lorazepam dose or vehicle under a social (Soc) or isolated (Iso) test condition. Vertical bars represent the SEM (n= 22-25). Adapted from [60].

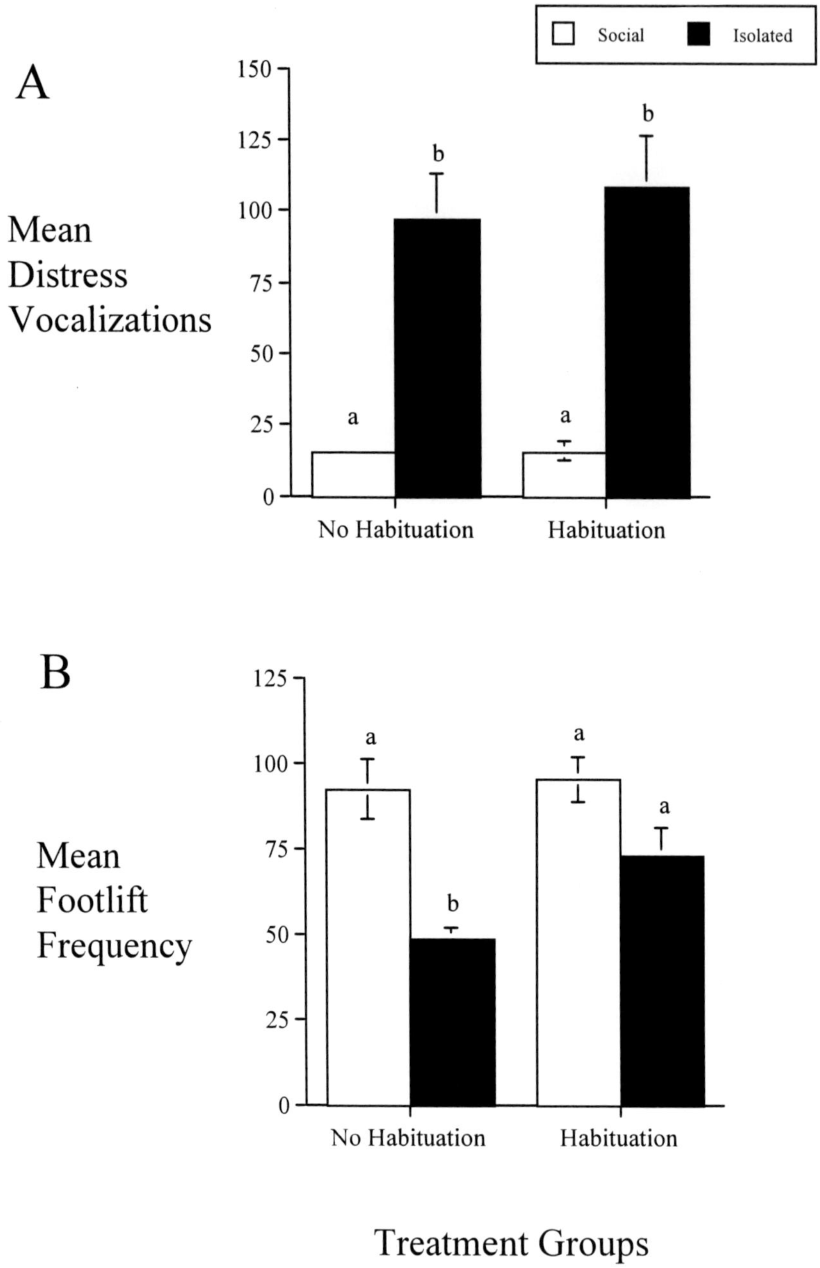

Figure 4. The effects of isolation and habituation on separation distress vocalizations (A) and footlift frequency elicited by intraplantar injection of 0.10% formalin (B) during a 180-s observation session. Vertical bars represent the SEM. Lowercase letters indicate Student-Newman-Keuls derived homogenous subsets ($p < 0.05$; n = 20-22). Adapted from [63].

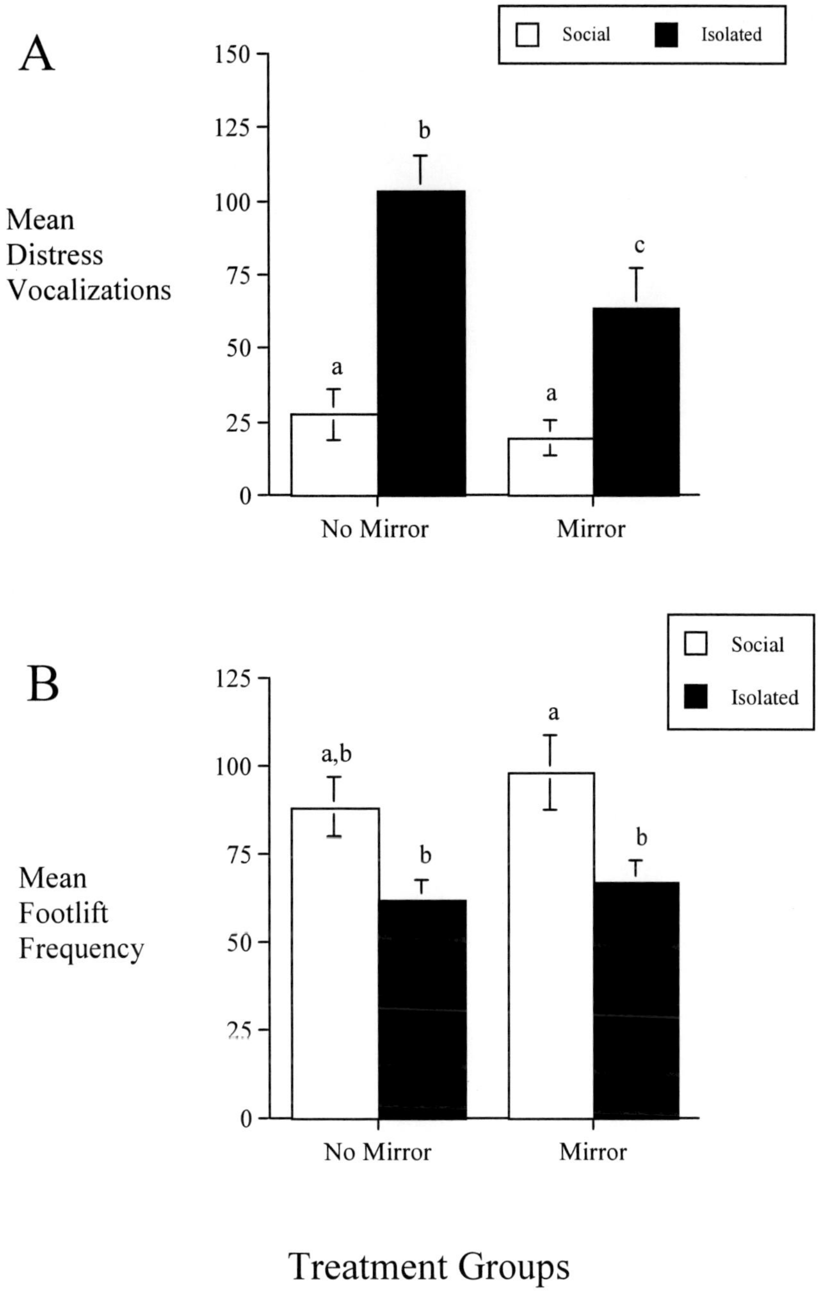

Figure 5. The effects of isolation and mirrors on separation distress vocalizations (A) and footlift frequency elicited by intraplantar injection of 0.10% formalin (B) during a 180-s observation session. Vertical bars represent the SEM. Lowercase letters indicate Student-Newman-Keuls derived homogenous subsets ($p < 0.05$; $n = 19\text{-}20$). Adapted from [63].

Subsequent to these procedural changes, Feltenstein et al. [69] examined the face and convergent validity of the model by assessing whether it demonstrates activity within the hypothalamic-pituitary adrenal axis. Similar to cortisol in humans, one biological marker of stress that has been found in nonhuman animals is the corticosteroid, corticosterone. In response to a variety of environmental stressors [70-72], corticosterone is released by the adrenal glands via hypothalamic-pituitary adrenal axis activity [70, 71, 73, 74]. In this study, seven-day-old chicks were intramuscularly pretreated with saline or chlordiazepoxide (8 mg/kg) in isolation with or without mirrors for a 15 min observation period. A higher dose of chlordiazepoxide was used in response to the prolonged isolation period that we utilized in order to ensure enhanced hypothalamic-pituitary adrenal axis and corticosterone activity. Immediately after testing, chicks were euthanized and blood samples were collected to assay plasma corticosterone.

Consistent with the Feltenstein et al. [63] study, chicks tested in the isolation condition exhibited an increase in DVocs relative to chicks in the Mirror condition, indicating an isolation-stress effect that was significantly attenuated by chlordiazepoxide (Figure 6A). Similarly, isolation condition animals had plasma corticosterone levels that were significantly higher than Mirror-tested chicks (Figure 6B). However, this corticosterone effect was only modestly attenuated by chlordiazepoxide. Interestingly, when these data were analyzed using Pearson's product-moment correlation, a positive correlation was revealed between these DVocs and plasma corticosterone levels ($r = .49$, $p < 0.01$), indicating that corticosterone levels can serve as a valid biological marker of separation-stress in this paradigm. Overall, these results provide face and convergent validity to the model by demonstrating a parallel between plasma corticosterone levels and the behavioral DVocs measure; however, use of this biological marker as an index should be limited in that it appears less sensitive than the behavioral marker to anxiolytic drug manipulations.

After demonstrating that the model possesses face and construct validity, our laboratory expanded the predictive validity of the chick separation-stress paradigm by examining whether it can discriminate between agents that are clinically effective for treating anxiety and those that are not [75]. In this study, six known anxiolytics (i.e., positive controls; meprobamate, pentobarbital, chlordiazepoxide, buspirone, imipramine and clonidine; Table 3) and five compounds that have no activity or anxiogenic properties (i.e., negative controls; amphetamine, scopolamine, caffeine, chlorpromazine, and haloperidol; Table 4) were examined in the model in chicks tested in the isolation condition. The control group was vehicle-pretreated animals tested in the low-stress Mirror condition. Vehicle or drug probe injections were administered intramuscularly 15 min prior to testing. If the chick separation-stress paradigm is a valid animal model of anxiety, then it should be sensitive to these anxiolytic compounds, while ensuring that it is not susceptible to false positives. An additional behavior (sleep onset latency) was included as an index of behavioral sedation.

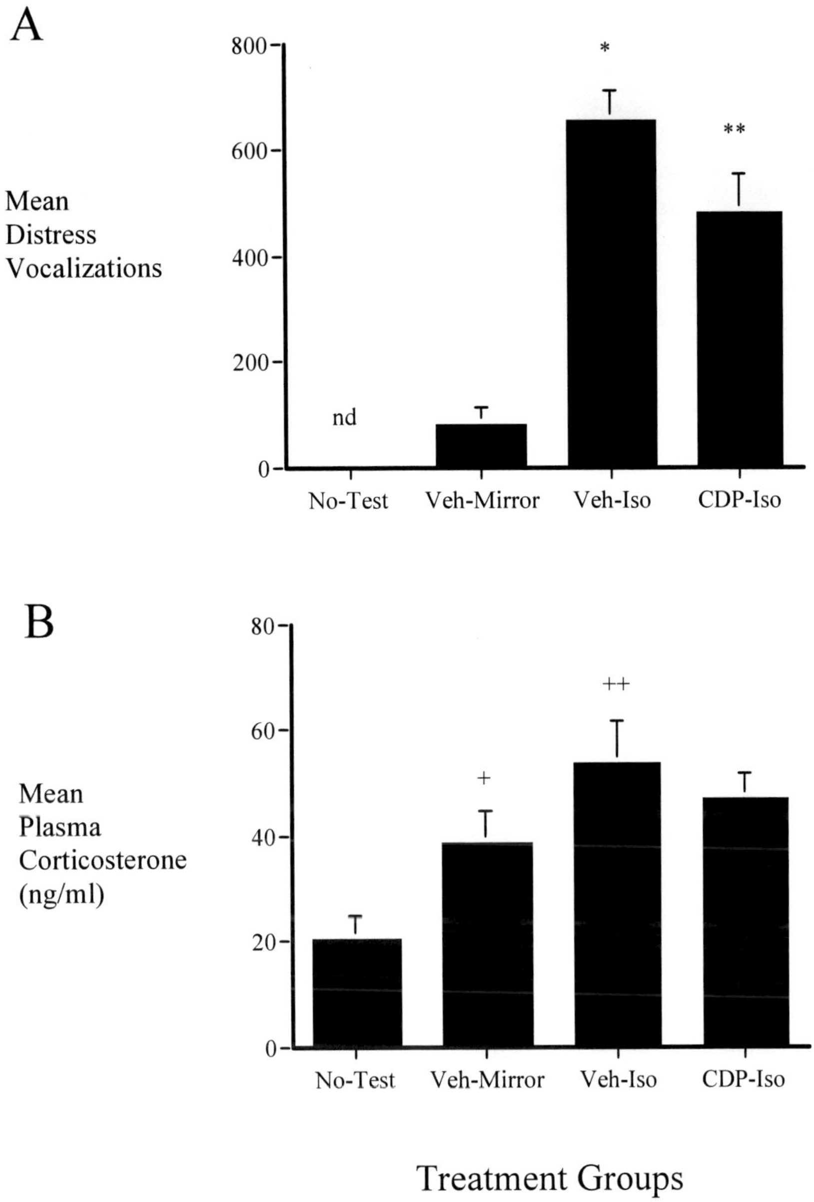

Figure 6. The effects of vehicle (Veh) and 8 mg/kg chlordiazepoxide (CDP) separation distress vocalizations (A) and plasma corticosterone (B) in chicks tested in isolation (Iso) or mirrors of a 15-min test. No-test chicks were sacrificed immediately upon removal from their home cage for collection of blood samples (nd = no data). Vertical bars represent the SEM. * Significant isolation stress effect; ** significant attenuation of vocalizations by CDP; + significant increase in corticosterone over No-Test group; ++ significant increase in corticosterone from Vehicle-Mirror group. All P <.005 (Fisher's LSD); n = 34-35 (vocalization data) and 15-18 (corticosterone data). Adapted from [69].

Table 3. Effect of anxiolytic drugs on distress vocalizations and sleep onset latency.

	Distress Vocalizations	Sleep Onset Latency
Meprobamate		
Vehicle/Mirror	91.40 ± 22.89	170.33 ± 9.67
Vehicle/Isolated	259.50 ± 10.49*	180.00 ± 0.00
15 mg/kg	236.13 ± 13.35	180.00 ± 0.00
30 mg/kg	267.20 ± 13.10	180.00 ± 0.00
60 mg/kg	241.40 ± 12.13	180.00 ± 0.00
120 mg/kg	149.69 ± 32.78 **	136.06 ± 17.41 †
Pentobarbital		
Vehicle/Mirror	1 4.69 ± 4.24	180.00 ± 0.00
Vehicle/Isolated	217.81 ± 19.99 *	180.00 ± 0.00
2.5 mg/kg	194.50 ± 16.96	171.00 ± 9.00
5.0 mg/kg	191.63 ± 19.13	162.50 ± 10.17
10.0 mg/kg	34.47 ± 17.85 **	62.40 ± 19.47 †
20.0 mg/kg	0.00 ± 0.00 **	1.75 ± 1.45 †
Chlordiazepoxide		
Vehicle/Mirror	35.19 ± 14.26	180.00 ± 0.00
Vehicle/Isolated	187.13 ± 19.82 *	171.88 ± 8.13
2.5 mg/kg	220.19 ± 17.36	180.00 ± 0.00
5.0 mg/kg	134.31 ± 25.21 **	132.81 ± 15.99 †
10.0 mg/kg	30.20 ± 14.55 **	59.93 ± 17.41 †
15.0 mg/kg	3.88 ± 3.50 **	51.88 ± 16.16 †
Buspirone		
Vehicle/Mirror	128.20 ± 16.96	180.00 ± 0.00
Vehicle/Isolated	263.20 ± 29.56*	180.00 ± 0.00
2.5 mg/kg	265.57 ± 22.80	180.00 ± 0.00
5.0 mg/kg	303.31 ± 23.71	180.00 ± 0.00
7.5 mg/kg	321.47 ± 26.21	180.00 ± 0.00
10.0 mg/kg	326.87 ± 10.96	180.00 ± 0.00
Imipramine		
Vehicle/Mirror	27.93 ± 9.88	168.00 ± 12.00
Vehicle/Isolated	156.00 ± 16.46 *	152.00 ± 15.25
1 mg/kg	110.88 ± 23.10	180.00 ± 0.00 ‡
3 mg/kg	111.53 ± 18.00	180.00 ± 0.00 ‡
10 mg/kg	89.35 ± 22.80 **	180.00 ± 0.00 ‡
15 mg/kg	47.09 ± (16.53) **	147.27 ± (21.95)
Clonidine		
Vehicle/Mirror	42.67 ± 18.00	180.00 ± 0.00
Vehicle/Isolated	186.93 ± 29.11 *	180.00 ± 0.00
0.10 mg/kg	140.47 ± 32.16	148.60 ± 14.12
0.15 mg/kg	0.00 ± 0.00 **	60.33 ± 14.12 †
0.20 mg/kg	2.14 ± 2.14 **	84.50 ± 18.53 †
0.25 mg/kg	1.07 ± 0.93 **	73.13 ± 15.86 †

Values represent means ± SEM (n = 11-18). * Significant stress effect, ** significant attenuation of the stress effect, † significant sedative effect, ‡ significant increase in sleep onset latency scores. All $p < 0.05$ (Fisher's LSD test). Adapted from [75].

Table 4. Effect of non-anxiolytic drugs on distress vocalizations and sleep onset latency.

	Distress Vocalizations	Sleep Onset Latency
Amphetamine		
Vehicle/Mirror	121.93 ± 31.30	180.00 ± 0.00
Vehicle/Isolated	231.77 ± 18.67 *	169.41 ± 10.59
0.5 mg/kg	260.50 ± 22.75	174.50 ± 5.50
1 mg/kg	235.47 ± 18.93	180.00 ± 0.00
2 mg/kg	241.86 ± 11.34	180.00 ± 0.00
4 mg/kg	25.19 ± 15.28 **	102.94 ± 21.18 †
Scopolamine		
Vehicle/Mirror	37.13 ± 14.65	180.00 ± 0.00
Vehicle/Isolated	181.75 ± 21.56 *	180.00 ± 0.00
0.2 mg/kg	199.53 ± 16.84	180.00 ± 0.00
0.4 mg/kg	162.31 ± 17.80	180.00 ± 0.00
0.8 mg/kg	188.19 ± 20.51	180.00 ± 0.00
1.6 mg/kg	189.00 ± 22.72	180.00 ± 0.00
Caffeine		
Vehicle/Mirror	63.06 ± 21.13	169.41 ± 10.59
Vehicle/Isolated	213.88 ± 16.72 *	160.00 ± 13.70
5 mg/kg	200.00 ± 16.70	180.00 ± 0.00
10 mg/kg	169.31 ± 26.67	177.50 ± 2.50
20 mg/kg	225.13 ± 25.24	180.00 ± 0.00
Chlorpromazine		
Vehicle/Mirror	90.93 ± 26.60	180.00 ± 0.00
Vehicle/Isolated	277.14 ± 18.30 *	180.00 ± 0.00
1 mg/kg	258.71 ± 17.74	180.00 ± 0.00
3 mg/kg	263.92 ± 17.10	180.00 ± 0.00
10 mg/kg	214.43 ± 20.34	180.00 ± 0.00
30 mg/kg	225.00 ± 22.11	180.00 ± 0.00
Haloperidol		
Vehicle/Mirror	43.25 ± 17.16	180.00 ± 0.00
Vehicle/Isolated	241.06 ± 17.99 *	180.00 ± 0.00
0.03 mg/kg	275.06 ± 11.33	180.00 ± 0.00
0.10 mg/kg	217.94 ± 21.45	180.00 ± 0.00
0.30 mg/kg	182.00 ± 29.87	180.00 ± 0.00
1 mg/kg	218.56 ± 21.20	173.69 ± 6.31

Values represent means ± SEM (n = 14-17). * indicates a significant stress effect. ** indicates a significant attenuation of the stress effect. † indicates a significant sedative effect. All $p < 0.05$ (Fisher's LSD test). Adapted from [54].Relative to their Mirror tested counterparts, separation-stress produced a significant increase in DVocs in vehicle-pretreated chicks tested in the isolation condition. Moreover, the results of this study supported the predictive validity of the model by demonstrating dose-dependent sensitivity to all of the positive drug probes tested (except buspirone; Table 5), whereas none of the negative drug probes significantly affected any of the measured behaviors (except for the highest dose of amphetamine, which produced a competing behavioral stereotypy response). Many of the positive drug probes also possessed sedative properties at the higher doses tested. While problematic for its validity, it was not surprising that buspirone did not possess anxiolytic properties in the model, in that previous studies have shown that chronic administration is needed to induce an anxiolytic effect [7]. However, given that its anxiolytic properties are specific for treating generalized anxiety disorder, it is also possible that the chick separation-stress paradigm is insensitive to buspirone should it model a distinct, rather than a non-specific, anxiety-like state (e.g., panic disorder).

To address this issue, Warnick et al. [76] examined the sensitivity of the model to anxiolytic agents that are specific for treating panic disorder (phenelzine), generalized anxiety disorder (buspirone and trazodone) or both conditions (alprazolam, clonidine and imipramine), as well as an agent that is capable of inducing panic symptomology in humans (yohimbine). Seven-day-old chicks received intramuscular injections of vehicle or drug probe 15 min prior to a 180 sec isolation session under the low-stress (mirror) or high-stress (isolation) test condition (Table 6).

Contrary to Mirror-tested chicks, vehicle-pretreated animals in the isolated condition displayed enhanced DVocs, an effect that was significantly attenuated by agents that demonstrate some efficacy for treating panic disorder (i.e., phenelzine, alprazolam, clonidine and imipramine), but not generalized anxiety disorder (i.e., buspirone and trazodone; Table 7). Similar to the Feltenstein et al. [75] study, many of these agents also possessed sedative properties at higher doses. Interestingly, yohimbine failed to produce the hypothesized increase in DVocs; in fact, a modest attenuation occurred. While it may be considered problematic for the validity of the model (i.e., insensitivity to anxiogenic agents) [8], this small effect ($\eta^2 = 0.11$) may have been due to a number of factors, including doses on the descending limb of the dose-response curve or a ceiling effect for DVocs in vehicle-treated animals. Nonetheless, the sensitivity of the chick separation-stress paradigm to some, but not all, of the anxiolytic agents further enhances its predictive validity and suggests that DVocs in the model may better serve as an index of panic, rather than a more generalized state of anxiety.

Table 5. Percent anxiolytic effect compared to the percent sedative effect.

	$Dose_1$	$Dose_2$	$Dose_3$	$Dose_4$
Meprobamate				
Distress Vocalizations	9.01	-2.97	6.97	42.32 **
Sleep Onset Latency	0.00	0.00	0.00	24.41 †
Pentobarbital				
Distress Vocalizations	10.70	12.02	84.17 **	100.00 **
Sleep Onset Latency	5.00	9.72	65.33 †	99.03 †
Chlordiazepoxide				
Distress Vocalizations	-17.67	28.23 **	83.86 **	97.93 **
Sleep Onset Latency	0.00	26.22 †	66.70 †	71.18 †
Buspirone				
Distress Vocalizations	-0.91	-15.24	-22.14	-24.19
Sleep Onset Latency	0.00	0.00	0.00	0.00
Imipramine				
Distress Vocalizations	28.92	28.51	42.72 **	69.81 **
Sleep Onset Latency	-18.42 ‡	-18.42 ‡	-18.42 ‡	3.11
Clonidine				
Distress Vocalizations	24.85	100.00 **	98.86 **	99.43 **
Sleep Onset Latency	17.44	66.48 †	53.06 †	59.37 †

** Significant attenuation of the stress effect, † significant sedative effect, ‡ significant increase in sleep onset latency scores. All $p < 0.05$ (Fisher's LSD test). Adapted from [54].

Table 6. Effects of drug probes on Isolation-induced distress vocalizations and sleep onset latency.

	Distress Vocalizations	Sleep Onset Latency
Phenelzine		
Vehicle/Mirror	97.0 ± 23.3	180.0 ± 0.0
Vehicle/Isolated	235.9 ± 19.8 *	180.0 ± 0.0
3.125 mg/kg	241.4 ± 11.5	180.0 ± 0.0
6.25 mg/kg	231.1 ± 21.6	180.0 ± 0.0
12.5 mg/kg	192.4 ± 21.9	180.0 ± 0.0
25 mg/kg	159.3 ± 21.0 **	180.0 ± 0.0
Alprazolam		
Vehicle/Mirror	11.1 ± 5.7	180.0 ± 0.0
Vehicle/Isolated	186.8 ± 13.2 *	180.0 ± 0.0
0.065 mg/kg	51.9 ± 15.8 **	139.9 ± 14.3 †
0.125 mg/kg	21.4 ± 15.7 **	67.9 ± 14.2 †
0.25 mg/kg	1.4 ± 0.7 **	4.4 ± 4.4 †
0.5 mg/kg	0.4 ± 0.2 **	0.0 ± 0.0 †
Imipramine		
Vehicle/Mirror	28.2 ± 16.6	180.0 ± 0.0
Vehicle/Isolated	226.5 ± 23.2 *	180.0 ± 0.0
1 mg/kg	252.6 ± 10.9	180.0 ± 0.0
3 mg/kg	218.9 ± 22.0	180.0 ± 0.0
10 mg/kg	141.9 ± 20.1 **	180.0 ± 0.0
15 mg/kg	16.0 ± 4.8 **	180.0 ± 0.0

Values represent means ± SEM (n = 13-18). * Significant isolation-stress effect, ** significant anxiolytic effect, ‡ significant anxiogenic effect, † significant sedative effect. All $p < 0.05$ (Fisher's LSD test for distress vocalizations data, and Mann-Whitney-U test for sleep onset latency data). Adapted from [76].

Table 6 (Continued).

	Distress Vocalizations	Sleep Onset Latency
Clonidine		
Vehicle/Mirror	31.8 ± 11.7	180 ± 0.0
Vehicle/Isolated	192.5 ± 19.4 *	180.0 ± 0.0
0.1 mg/kg	25.1 ± 14.0 **	159.8 ± 10.1
0.15 mg/kg	1.4 ± 1.2 **	144.7 ± 11.4
0.2 mg/kg	4.7 ± 3.7 **	112.7 ± 15.1 †
0.25 mg/kg	0.0 ± 0.0 **	110.1 ± 16.3 †
Buspirone		
Vehicle/Mirror	110.4 ± 21.1	180.0 ± 0.0
Vehicle/Isolated	219.5 ± 21.5 *	180.0 ± 0.0
2.5mg/kg	276.5 ± 13.0 ‡	180.0 ± 0.0
5.0mg/kg	319.6 ± 16.1 ‡	180.0 ± 0.0
7.5mg/kg	240.1 ± 18.4	180.0 ± 0.0
10.0mg/kg	248.7 ± 18.0	180.0 ± 0.0
Trazodone		
Vehicle/Mirror	17.3 ± 9.6	180.0 ± 0.0
Vehicle/Isolated	238.7 ± 13.6 *	180.0 ± 0.0
0.1mg/kg	190.8 ± 21.6	173.8 ± 6.2
0.3mg/kg	230.2 ± 18.3	180.0 ± 0.0
1.0mg/kg	202.4 ± 20.7	180.0 ± 0.0
3.0mg/kg	240.8 ± 11.2	180.0 ± 0.0
Yohimbine		
Vehicle/Mirror	82.3 ± 21.2	180.0 ± 0.0
Vehicle/Isolated	259.2 ± 11.5 *	180.0 ± 0.0
0.1mg/kg	215.1 ± 18.6 **	180.0 ± 0.0
0.3mg/kg	229.6 ± 9.1	180.0 ± 0.0
1.0mg/kg	220.6 ± 11.7 **	180.0 ± 0.0
3.0mg/kg	191.7 ± 15.5 **	180.0 ± 0.0

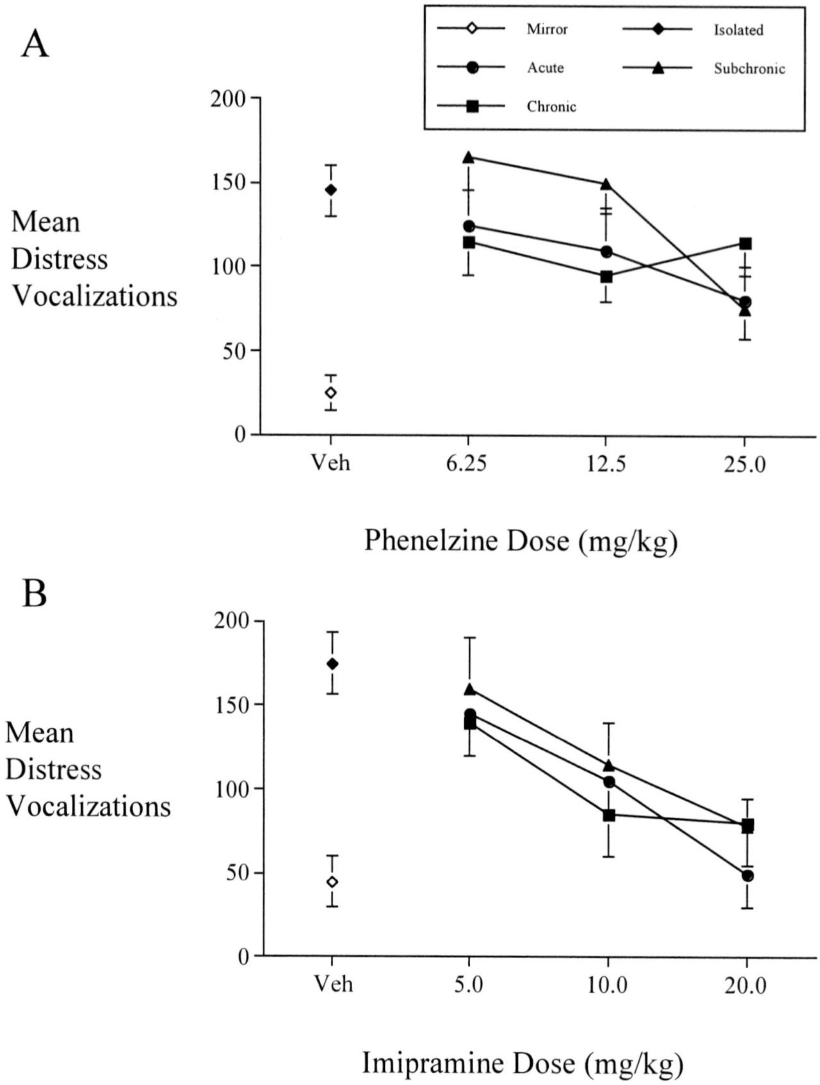

Figure 7. The effects of phenelzine (A) and imipramine (B) on separation distress vocalizations in 8-day old chicks tested in isolation (closed symbols) or with mirrors (open symbol) during 3-min test. Subchronic and chronic administration procedures involved drug injections on days 5-7 and 2-7 posthatch, respectively. Vertical bars represent SEM (n = 17-19). Adapted from [80].

Another criterion required for an animal model to demonstrate predictive validity is that it should respond to all classes of drugs that are clinically effective for treating the disorder. One drug class that has not been extensively examined in chick separation-stress paradigm is antidepressants. Clinical research indicates that antidepressants, especially when repeatedly administered, are often more robust for treating anxiety disorders than traditional anxiolytics [77-79]. Despite their clinical efficacy, few animal models are sensitive to these agents [22], a finding that may be due to procedures limited to acute administrations. To address this issue, Feltenstein and Sufka [80] examined the sensitivity of the model to the monoamine oxidase inhibitor phenelzine (Figure 7A), the tricyclic antidepressant imipramine (Figure 7B), the selective serotonin reuptake inhibitor citalopram (Figure 8A) and the norepinephrine reuptake inhibitor maprotiline (Figure 8B) under acute (no pretreatment), subchronic (3 days pretreatment) or chronic (6 days pretreatment) administration procedures. Following any pretreatment, eight-day-old chicks received their respective vehicle or drug probe injection intraperitoneally 15 min prior to a 180 sec test session. Unlike a large number of traditional rodent models of anxiety [22], the model demonstrated sensitivity to acutely administered phenelzine, imipramine and maprotiline, but not citalopram, and retained its sensitivity to these drug probes across both repeated administration procedures. Although its sensitivity to antidepressants that generally require repeated administration to demonstrate activity may raise some concerns over its validity, Willner [8] emphasizes that this issue is of no real interest in the assessment of a screening test. It is further argued that a screening test, which is solely concerned with making accurate predictions, should ideally respond to agents acutely as this increases model utility. Furthermore, Willner [8] asserts that it is necessary to show that a drug maintains its activity in a model following repeated administration. Overall, these results extend the validity and utility of the chick separation-stress paradigm as an animal model of anxiety by demonstrating its sensitivity to antidepressants under both acute and repeated administration procedures.

In that the chick separation-stress paradigm is a valid animal model of anxiety, we have utilized this procedure as a low-cost, high throughput screening tool to detect the anxiolytic properties of various botanical extracts, extract fractions and isolated constituents [55, 56, 81]. In addition to issues of validity, a good animal model should be economic, not only in terms of reliability and simplicity, but also in speed and cost [8].

Although rodent isolation-induced models of anxiety have demonstrated a degree of validity, they are not very economic. However, the chick separation-stress paradigm is a more cost-efficient animal model of anxiety than traditional rodent models in that chicks are relatively inexpensive to purchase and maintain and require small quantities of drug probes [55, 56, 81]. Moreover, we have simplified the model by reducing the number of behaviors recorded to DVocs. Finally, the procedure is exceptionally efficient in that it measures a species-typical behavior (i.e., DVocs) that can be automatically recorded, involves a relatively short test period (180 sec) and allows for multiple animals to be tested simultaneously (i.e., 6-9) in the presence of a single experimenter.

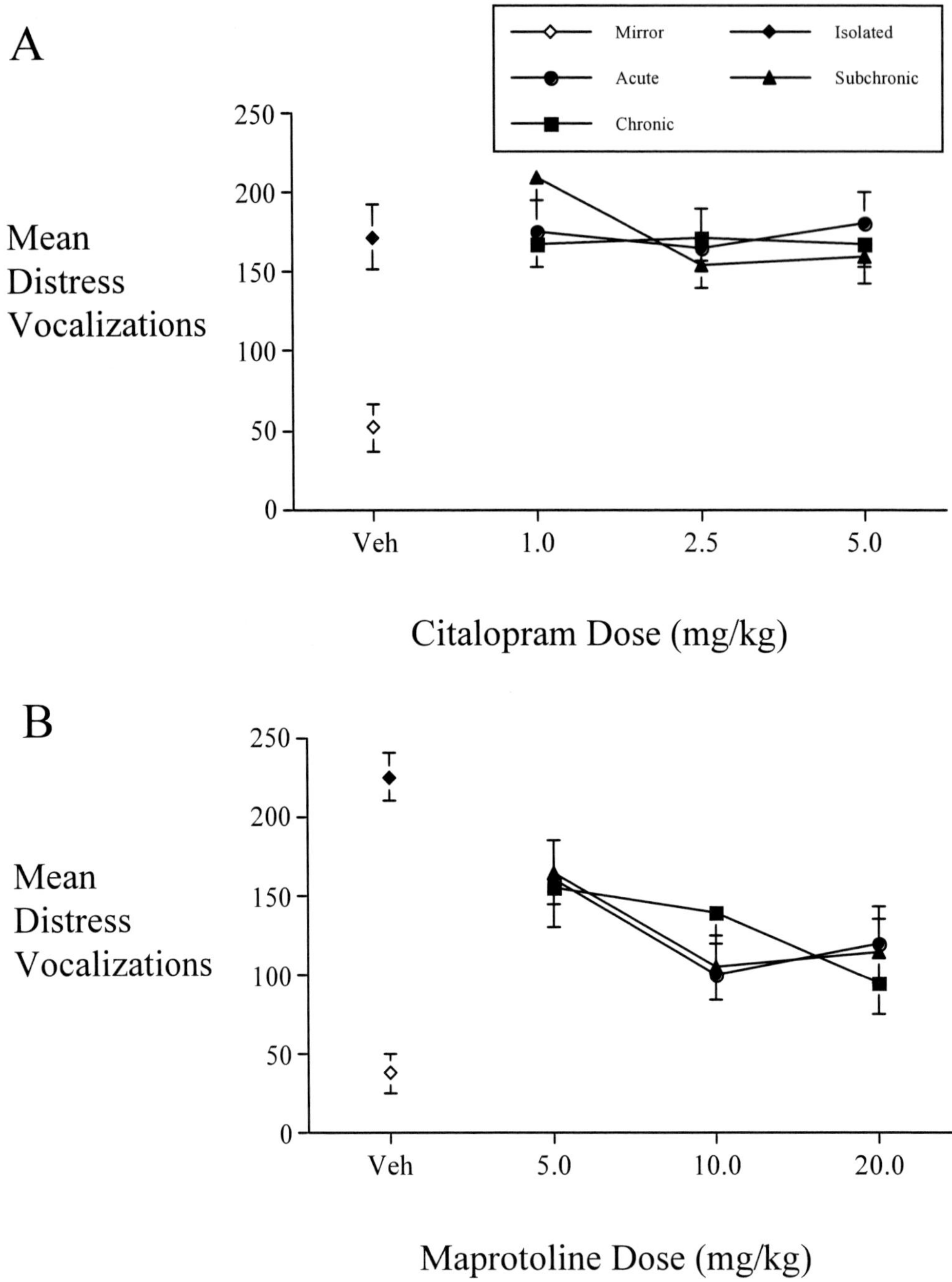

Figure 8. The effects of citalopram (A) and maprotiline (B) on separation distress vocalizations in 8-day old chicks tested in isolation (closed symbols) or with mirrors (open symbol) during 3-min test. Subchronic and chronic administration procedures involved drug injections on days 5-7 and 2-7 posthatch, respectively. Vertical bars represent SEM (n = 17-19). Adapted from [80].

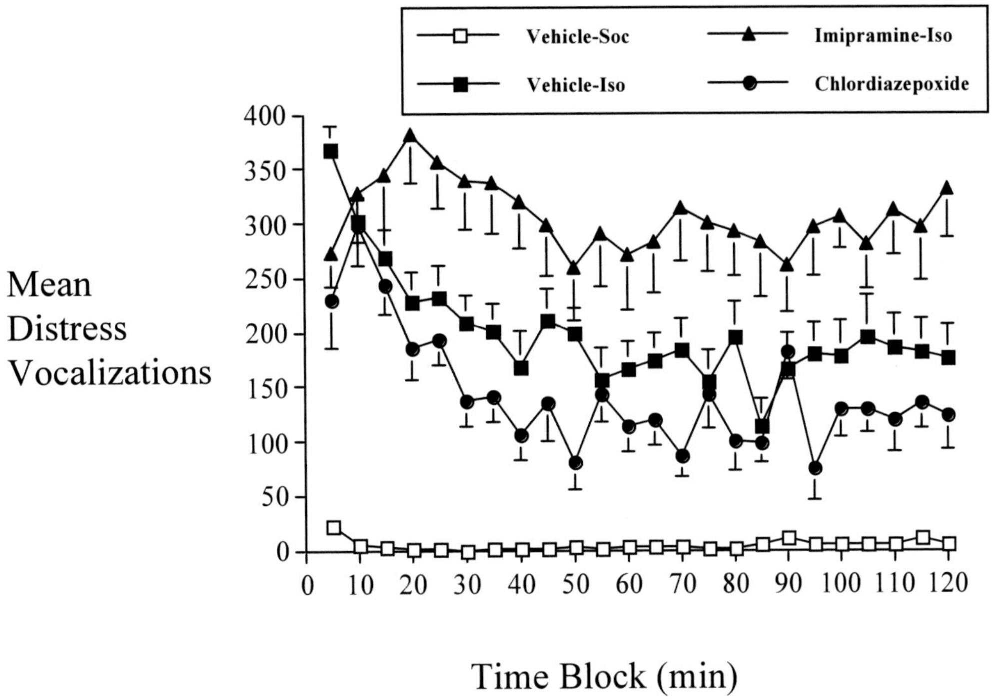

Figure 9. Behavioral characterization and pharmacological validation of the chick anxiety-depression continuum model. Data points represent means ± SEM (n = 12). Compared to chicks tested in the social condition (open symbol), isolated chicks (filled symbols) displayed a significant increase in DVocs which was maximal during the first 5 min block (anxiety-like state), significantly declined over the next 20 min (transitional phase) and stabilized at approximately 50% the initial rate for the remainder of the session (depressive-like state). The benzodiazepine anxiolytic chlordiazepoxide significantly attenuated DVocs during the anxiety phase but did not generally affect DVoc rates thereafter. The tricyclic antidepressant imipramine significantly decreased DVocs during the anxiety phase and significantly increased DVocs during the depression phase. Adapted from [101].

THE CHICK SEPARATION-STRESS PARADIGM AS AN ANIMAL MODEL OF DEPRESSION

Although traditionally used as an animal model of anxiety, more recent research from our laboratory has expanded the use of the chick separation-stress paradigm as a putative model of depression. While currently categorized as separate Axis I clinical syndromes [82], anxiety and depressive disorders share many signs and symptoms [83], including common biological markers such as glucocorticoid [84], monoamine [85] and neurotrophic factor [86] dysregulation. In addition to comorbidity rates ranging from 50 to 90% [87-89], anxiety and depression respond similarly to cognitive-behavioral therapy [90] and certain classes of pharmacological agents (e.g., antidepressants) [91]. In light of evidence from structural equation modeling indicating a close relationship between these two disorders [92], it has been suggested that anxiety and depression may be better served by a single overarching construct [83] and could represent different temporal facets of a single syndrome in which depression follows the anxiety state in the presence of an unresolved stressor [93].

Similar to animal models of anxiety, there have been a number of concerns regarding the validity of existing preclinical models of depression [50, 94-96]. Moreover, little attention has been given to simulating a hypothesized anxiety-depression continuum in a single paradigm. Development of such a model would not only add empirical evidence to support a new taxonomy of mental disorders, but it may also shed light on the course, pathology and treatment of such a clinical syndrome.

Like the chick separation-stress paradigm, chick DVocs have also been used as an index of depression [97, 98]. In the Lehr [97] study, chicks isolated for 2 h from conspecifics displayed a pronounced decline in rates of DVocs that appear to resemble a depression-like "learned helplessness" or "behavioral despair" response [99, 100]. Moreover, this modeling possessed predictive validity, in that a wide variety of antidepressant drugs reversed this state (i.e., an increase in DVocs), whereas compounds lacking antidepressant activity did not [97]. These results, along with previous research utilizing the chick separation-stress paradigm as an animal model of anxiety, suggests that the isolation of domestic fowl may sequentially model the two clinical states of anxiety and depression vis-à-vis the anxiety-depression continuum hypothesis. To test this hypothesis, DVocs from isolated chicks were collected in 5 min blocks across a 2 h test session [101]. It was hypothesized that DVocs would be highest early in the test session, decline and then stabilize at a much lower rate in the latter half of the test session. To further dissociate putative anxiety- and depressive-like phases of the DVocs response, separate groups of isolated chicks were given either the benzodiazepine anxiolytic chlordiazepoxide (8 mg/kg, intraperitoneal) or the tricyclic antidepressant imipramine (15 mg/kg, intraperitoneal) 15 min prior to testing. The control groups were vehicle pretreated chicks tested socially or in isolation.

As predicted, separation from conspecifics elicited initial high rates of DVocs (anxiety-like state) that declined and stabilized to 40-50% of the initial rate over the course of the 2 h isolation experience (depressive-like state; Figure 9). Interestingly, both chlordiazepoxide and imipramine (which both possess anxiolytic activity) were effective at modulating the early anxiety-like phase (i.e., a reduction in DVocs), while only imipramine prevented the onset of the depressive-like phase (i.e., an increase in DVocs) of the chick isolation stress response. Relative to the no test and socially-tested control groups, the convergent validity of the paradigm as a stress model was enhanced in a second experiment by demonstrating an enhancement in plasma corticosterone levels in isolated chicks (Figure 10).

Moreover, this elevation in plasma corticosterone declined over the course of isolation duration, suggesting a negative feedback process on hypothalamic-pituitary adrenal axis activity. Taken together, these results imply that anxiety and depression, while being temporally related, may be separable facets of a single disorder. Rather than viewing the inability of drug probes to either prevent or completely reverse different phases of the model as a weakness, it can be viewed as one of its strengths. That is, the model not only defines and distinguishes pharmacological activity of different classes of drug compounds, but it can also differentiate multiple clinical actions of a single drug compound. For a screening assay, the ability of the model to screen simultaneously for efficacy in two clinical activities greatly enhances utility.

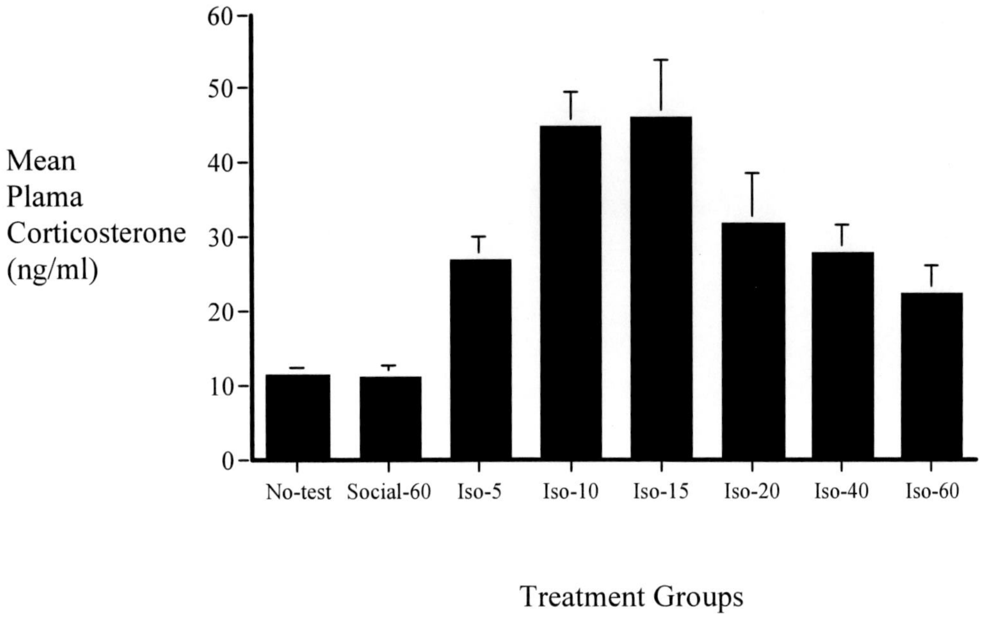

Figure 10. Mean plasma corticosterone levels (ng/ml) for no test, socially tested and isolated groups (n=6). Corticosterone levels were significantly higher for all the isolation conditions compared to the no-test and social-60 groups. Among the isolated groups, corticosterone levels were the highest in the 10 and 15 min groups compared to the 5, 40, and 60 min groups. Adapted from [101].

As a simulation, the chick anxiety-depression continuum model appears to possess face, construct and predictive validity in that i) the procedure involves a potent identifiable stressor, ii) the pattern of DVocs sequentially models an anxiety-like state followed by a depressive-like state and iii) the dependent measure shows the appropriate sensitivity to anxiolytic and antidepressant probes. However, because depression is considered to be a product of persistent stress [102] and requires chronic administration of agents to yield antidepressant effects, future studies should incorporate a chronic mild stressor and prolonged drug administration procedures to provide further homology of the model to the clinical condition. The utilization of different domestic fowl strains could also aid in examining possible gentic vulnerabilities and/or target genes that may be involved in susceptibility to depression, as well as provide insight into individual differences in medication sensitivity. Despite some limitations, these initial findings provide empirical support of the chick anxiety-depression continuum model as a simulation and fit well with human clinical data that argues adopting a new taxonomic structure highlighting the inter-relatedness of these clinical syndromes rather than their distinct boundaries.

Table 7. Comparative activity of drug probes in clinical cases of panic disorder, generalized anxiety disorder and in the chick separation stress paradigm.

Drug Probe	Effect in Panic Disorder	Effect in Generalized Anxiety Disorder	Effect in the Chick Separation-Stress Paradigm
Phenelzine	+	-	+
Alprazolam	+	+	+
Imipramine	+	+	+
Clonidine	+	+	+
Buspirone	-	+	-
Trazodone	-	+	-

+ Anxiolytic activity, - the absence of anxiolytic activity. Adapted from [76].

Conclusions

The use of domestic fowl in screening assays has numerous advantages over traditional rodent-based anxiety or depression models. Using the criteria outlined by Willner [8, 103], the chick separation-stress and anxiety-depression continuum models possess high utility and high-throughput as *in vivo* screens in that they both:

i) use a low-cost animal ($0.50 a chick),
ii) test animals at a young age,
iii) use a single relatively short test session (< 1h),
iv) measure a species typical response that is easily recorded, and
v) can screen for two drug properties in a single test.

Furthermore, both chick models address the National Institutes of Health 3R policy to Reduce, Refine and Replace animals in research [104]. That is, the use of domestic fowl reduces the number of purpose-bred research animals as male chicks are a by-product of the commercial egg-laying industry and discarded at hatch. The models also possess a refined methodology in that pain and distress associated with stress-provoking stimuli are minimized (i.e., no stress-induced analgesia measure and the single test session is relatively short). Finally, the models replace standard rodent-based models of anxiety and depression with a phylogenetically lower, and perhaps, less sentient species. While rodent models will likely continue to be the mainstay in biomedical research, the attributes outlined strongly argue for the adoption of domestic fowl models as a supplement to traditional rodent models. Thus, the chick separation-stress and anxiety-depression continuum models should continue to provide a means by which new drug candidates may be evaluated for the treatment of anxiety and depression.

REFERENCES

[1] van der Staay FJ: Animal models of behavioral dysfunctions: basic concepts and classifications, and an evaluation strategy. *Brain Res Rev* 2006; 52(1):131-159.

[2] Mineka S: *Animal models of anxiety-based disorders: their usefulness and limitations, in Anxiety and Anxiety Disorders.* Edited by Tuma AH, Maser J. Hilldale, Lawrence Erlbaum Associates Inc., 1985, pp 199-244.

[3] Marks IM: *Fears, phobias, and rituals: Panic, anxiety, and their disorders.* New York, Oxford University Press, 1987.

[4] Lister RG: Ethologically-based animal models of anxiety disorders. *Pharmacol Ther* 1990; 46(3):321-340.

[5] Green S, Hodges H: Animal models of anxiety, in *Behavioral Models in Psychopharmacology: Theoretical, Industrial, and Clinical Perspectives.* Edited by Willner P. Cambridge, Cambridge University Press, 1991, pp 21-49.

[6] Handley SL: Serotonin in animal models of anxiety: the importance of stimulus and response, in *Serotonin, Sleep, and Mental Disorders.* Edited by Idzikowski C, Cowen P. Petersfield, Wrightson Biomedical Publishing Ltd., 1991, pp 89-115.

[7] Stephens DN, Andrews JS: Screening for anxiolytic drugs, in *Behavioural models in psychopharmacology: theoretical, industrial, and clinical perspectives.* Edited by Willner P. Cambridge, Cambridge University Press, 1991, pp 50-75.

[8] Willner P: Behavioural models in psychopharmacology, in *Behavioural models in psychopharmacology: theoretical, industrial and clinical perspectives.* Edited by Willner P. Cambridge, Cambridge University Press, 1991, pp 3-18.

[9] File SE: Behavioral detection of anxiolytic action, in *Experimental Approaches to Anxiety and Depression.* Edited by Elliott JM, Heal DJ, Marsden CA. Chichester, Wiley, 1992, pp 25-44.

[10] Rodgers RJ, Cole JC: The elevated plus maze: pharmacology, methodology, and ethology, in *Ethology and Psychopharmacology.* Edited by Cooper SJ, Hendrie CA. Chichester, Wiley, 1994, pp 9-44.

[11] Treit D: Animal models of anxiety and anxiolytic drug action, in *Handbook of Depression and Anxiety.* Edited by den Boer JA, Sitsen JMA. New York, Marcel Dekker, 1994, pp 201-224.

[12] Griebel G: 5-Hydroxytryptamine-interacting drugs in animal models of anxiety disorders: more than 30 years of research. *Pharmacol Ther* 1995; 65(3):319-395.

[13] Tuma AH, Maser JD: *Anxiety and Anxiety Disorders.* Hilldale, Lawrence Erlbaum Associates, 1985.

[14] Rodgers RJ, Cao BJ, Dalvi A, Holmes A: Animal models of anxiety: an ethological perspective. *Braz J Med Biol Res* 1997; 30(3):289-304.

[15] Willner P: Validation criteria for animal models of human mental disorders: learned helplessness as a paradigm case. *Prog Neuropsychopharmacol Biol Psychiatry* 1986; 10(6):677-690.

[16] Flecknell P: Replacement, reduction and refinement. Altex 2002; 19(2):73-78.

[17] File SE: The contribution of behavioural studies to the neuropharmacology of anxiety. *Neuropharmacology* 1987; 26(7B):877-886.

[18] Geller I, Seifter J: The effects of meprobamate, barbiturates, d-amphetamine and promazine on experimentally induced conflict in the rat. *Psychopharmacologia* 1960; 1:482-492.

[19] Vogel JR, Beer B, Clody DE: A simple and reliable conflict procedure for testing anti-anxiety agents. *Psychopharmacologia* 1971; 21(1):1-7.

[20] Howard JL, Rohrbach KW, Pollard GT: Cumulative dose-effect curves in a conflict test with incremental shock. *Psychopharmacology* (Berl) 1982; 78(2):195-196.

[21] Brocco MJ, Koek W, Degryse AD, Colpaert FC: Comparative studies on the anti-punishment effects of chlordiazepoxide, buspirone and ritanserin in the pigeon, Geller-Seifter and Vogel conflict procedures. *Behav Pharmacol* 1990; 1(5):403-418.

[22] Borsini F, Podhorna J, Marazziti D: Do animal models of anxiety predict anxiolytic-like effects of antidepressants? *Psychopharmacology* (Berl) 2002; 163(2):121-141.

[23] Treit D: Animal models for the study of anti-anxiety agents: a review. *Neurosci Biobehav Rev* 1985; 9(2):203-222.

[24] File SE: Neurochemistry of anxiety, in *Drugs in Psychiatry*, vol 2. Edited by Burrows GD, Norman T, Davies B. Amsterdam, Elsevier, 1984.

[25] File SE: Animal models for predicting clinical efficacy of anxiolytic drugs: social behaviour. *Neuropsychobiology* 1985; 13(1-2):55-62.

[26] File SE, Hyde JR: Can social interaction be used to measure anxiety? *Br J Pharmacol* 1978; 62(1):19-24.

[27] File SE, Hyde JR: A test of anxiety that distinguishes between the actions of benzodiazepines and those of other minor tranquilisers and of stimulants. *Pharmacol Biochem Behav* 1979; 11(1):65-69.

[28] File SE: The use of social interaction as a method for detecting anxiolytic activity of chlordiazepoxide-like drugs. *J Neurosci Methods* 1980; 2(3):219-238.

[29] Montgomery KC: The relation between fear induced by novel stimulation and exploratory behaviour. *J Comp Physiol Psychol* 1958; 48:254-260.

[30] Halliday MS: Exploration and fear in the rat. *Symposia of the Zoological Society of London* 1966; 18:45-59.

[31] Pellow S: Anxiolytic and anxiogenic drug effects in a novel test of anxiety: are exploratory models of anxiety in rodents valid? *Methods Find Exp Clin Pharmacol* 1986; 8(9):557-565.

[32] Pellow S, File SE: Anxiolytic and anxiogenic drug effects on exploratory activity in an elevated plus-maze: a novel test of anxiety in the rat. *Pharmacol Biochem Behav* 1986; 24(3):525-529.

[33] Pellow S, Chopin P, File SE, Briley M: Validation of open:closed arm entries in an elevated plus-maze as a measure of anxiety in the rat. *J Neurosci Methods* 1985; 14(3):149-167.

[34] Handley SL, Mithani S: Effects of alpha-adrenoceptor agonists and antagonists in a maze-exploration model of 'fear'-motivated behaviour. *Naunyn Schmiedebergs Arch Pharmacol* 1984; 327(1):1-5.

[35] Lister RG: The use of a plus-maze to measure anxiety in the mouse. *Psychopharmacology* (Berl) 1987; 92(2):180-185.

[36] Soderpalm B, Eriksson E, Engel JA: Anticonflict and rotarod impairing effects of alprazolam and diazepam in rat after acute and subchronic administration. *Prog Neuropsychopharmacol Biol Psychiatry* 1989; 13(1-2):269-283.

[37] Benson H, Herd JA, Morse WH, Kelleher RT: Behaviorally induced hypertension in the squirrel monkey. *Circ Res* 1970; 27(1 Suppl 1):21-26.

[38] Delini-Stula A: Drug-induced suppression of conditioned hyperthermic and conditioned avoidance behavior response in rats. *Psychopharmacologia* 1971; 20(2):153-159.

[39] Sufka KJ, Hughes RA, McCormick TM, Borland JL: Opiate effects on isolation stress in domestic fowl. *Pharmacol Biochem Behav* 1994; 49(4):1011-1015.

[40] Benson H, Herd JA, Morse WH, Kelleher RT: Hypotensive effects of chlordiazepoxide, amobarbital and chlorpromazine on behaviorally induced elevated arterial blood pressure in the squirrel monkey. *J Pharmacol Exp Ther* 1970; 173(2):399-406.

[41] Noirot E: Ultrasounds and maternal behavior in small rodents. *Dev Psychobiol* 1972; 5(4):371-387.

[42] Benton D, Nastiti K: The influence of psychotropic drugs on the ultrasonic calling of mouse pups. *Psychopharmacology* (Berl) 1988; 95(1):99-102.

[43] Herman BH, Panksepp J: Effects of morphine and naloxone on separation distress and approach attachment: evidence for opiate mediation of social affect. *Pharmacol Biochem Behav* 1978; 9(2):213-220.

[44] Hennessy MB, Ritchey RL: Hormonal and behavioral attachment responses in infant guinea pigs. *Dev Psychobiol* 1987; 20(6):613-625.

[45] Ritchey RL, Hennessy MB: Cortisol and behavioral responses to separation in mother and infant guinea pigs. *Behav Neural Biol* 1987; 48(1):1-12.

[46] Gardner CR: Distress vocalization in rat pups. A simple screening method for anxiolytic drugs. *J Pharmacol Methods* 1985; 14(3):181-187.

[47] Insel TR, Hill JL, Mayor RB: Rat pup ultrasonic isolation calls: possible mediation by the benzodiazepine receptor complex. *Pharmacol Biochem Behav* 1986; 24(5):1263-1267.

[48] Engel J, Hard E: Effects of diazepam, ethanol and Ro 15-1788 on ultrasonic vocalization, locomotor activity and body righting in the neonatal rat. *Alcohol Alcohol Suppl* 1987; 1:709-712.

[49] Carden SE, Hofer MA: The effects of opioid and benzodiazepine antagonists on dam-induced reductions in rat pup isolation distress. *Dev Psychobiol* 1990; 23(8):797-808.

[50] Kalueff AV, Wheaton M, Murphy DL: What's wrong with my mouse model? Advances and strategies in animal modeling of anxiety and depression. *Behav Brain Res* 2007; 179(1):1-18.

[51] Panksepp J, Bean NJ, Bishop P, Vilberg T, Sahley TL: Opioid blockade and social comfort in chicks. *Pharmacol Biochem Behav* 1980; 13(5):673-683.

[52] Panksepp J, Vilberg T, Bean NJ, Coy DH, Kastin AJ: Reduction of distress vocalization in chicks by opiate-like peptides. *Brain Res Bull* 1978; 3(6):663-667.

[53] Warnick JE, McCurdy CR, Sufka KJ: Opioid receptor function in social attachment in young domestic fowl. *Behav Brain Res* 2005; 160(2):277-285.

[54] Sufka KJ, Weed NC: Construct validation of behavioral indices of isolation stress and inflammatory nociception in young domestic fowl. *Physiol Behav* 1994; 55(4):741-746.

[55] Smith KK, Dharmaratne HR, Feltenstein MW, Broom SL, Roach JT, Nanayakkara NP, Khan IA, Sufka KJ: Anxiolytic effects of kava extract and kavalactones in the chick social separation-stress paradigm. *Psychopharmacology* (Berl) 2001; 155(1):86-90.

[56] Sufka KJ, Roach JT, Chambliss WG, Jr., Broom SL, Feltenstein MW, Wyandt CM, Zeng L: Anxiolytic properties of botanical extracts in the chick social separation-stress procedure. *Psychopharmacology* (Berl) 2001; 153(2):219-224.

[57] Panksepp J, Meeker R, Bean NJ: The neurochemical control of crying. *Pharmacol Biochem Behav* 1980; 12(3):437-443.

[58] Panksepp J: Can anthropomorphic analyses of separation cries in other animals inform us about the emotional nature of social loss in humans? Comment on Blumberg and Sokoloff (2001). *Psychol Rev* 2003; 110(2):376-388; discussion 389-396.

[59] Watson GS, Sufka KJ: Chlordiazepoxide reverses social-separation-induced distress vocalizations and analgesia in young domestic fowl. *Exp Clin Psychopharmacol* 1996; 4:347-353.

[60] Watson GS, Roach JT, Sufka KJ: Benzodiazepine receptor function in the chick social separation-stress procedure. *Exp Clin Psychopharmacol* 1999; 7(2):83-89.

[61] Baldessarini RJ: Drugs and the treatment of psychiatric disorders: depression and anxiety disorders, in *Goodman and Gilman's The Pharmacological Basis of Therapeutics.* Edited by Hardman JG, Limbird LE. New York, NY, McGraw-Hill, 2001, pp 447-483.

[62] Feely M, Pullar T: Pharmacokinetic differences between benzodiazepines, in *Benzodiazepines: Current Concepts.* Edited by Hindmarch I, Beauniont G, Brandon S, Leonard B. London, Wiley and Sons, 1990, pp 61-74.

[63] Feltenstein MW, Ford NG, Freeman KB, Sufka KJ: Dissociation of stress behaviors in the chick social-separation-stress procedure. *Physiol Behav* 2002; 75(5):675-679.

[64] Jones RB, Harvey S: Behavioural and adrenocortical responses of domestic chicks to systematic reductions in group size and to sequential disturbance of companions by the experimenter. *Behavioural Processes* 1987; 14:291-303.

[65] Sufka KJ, Hughes RA: Differential effects of handling on isolation-induced vocalizations, hypoalgesia, and hyperthermia in domestic fowl. *Physiol Behav* 1991; 50(1):129-133.

[66] Abbott FV, Franklin KB, Connell B: The stress of a novel environment reduces formalin pain: possible role of serotonin. *Eur J Pharmacol* 1986; 126(1-2):141-144.

[67] Gallup G, Suarez S: An ethological analysis of open-field behaviour in chickens. *Anim Behav* 1980; 28:368-378.

[68] Faure J, Jones R, Bessei W: Fear and social motivation as factors in open-field behaviour of the domestic chick. *Biol Behav* 1983; 8:103-116.

[69] Feltenstein MW, Lambdin LC, Webb HE, Warnick JE, Khan SI, Khan IA, Acevedo EO, Sufka KJ: Corticosterone response in the chick separation-stress paradigm. *Physiol Behav* 2003; 78(3):489-493.

[70] Siegal H: Physiological stress in birds. *Bioscience* 1980; 30:529-533.

[71] Harvey S, Phillips JG, Rees A, Hall TR: Stress and adrenal function. *J Exp Zool* 1984; 232(3):633-645.

[72] Kim JW, Kirkpatrick B: Social isolation in animal models of relevance to neuropsychiatric disorders. *Biol Psychiatry* 1996; 40(9):918-922.

[73] Holmes W, Phillips J: The adrenal cortex of birds, in *General, Comparative and Clinical Endocrinology of the Adrenal Cortex*. Edited by Chester-Jones I, Henderson I. London, Academic Press, 1976, pp 293-420.

[74] Holberton R, Kelmuth B, Wingfield J: The corticosterone stress response in Gentoo and King Penguins during the non-fasting period. *Condor* 1996; 98:850-854.

[75] Feltenstein MW, Warnick JE, Guth AN, Sufka KJ: The chick separation stress paradigm: a validation study. *Pharmacol Biochem Behav* 2004; 77(2):221-226.

[76] Warnick JE, Wicks RT, Sufka KJ: Modeling anxiety-like states: pharmacological characterization of the chick separation stress paradigm. *Behav Pharmacol* 2006; 17(7):581-587.

[77] Nutt DJ: The pharmacology of human anxiety. *Pharmacol Ther* 1990; 47(2):233-266.

[78] Balestrieri M, Ruggeri M, Bellantuono C: Drug treatment of panic disorder--a critical review of controlled clinical trials. *Psychiatr Dev* 1989; 7(4):337-350.

[79] Boyer W: Serotonin uptake inhibitors are superior to imipramine and alprazolam in alleviating panic attacks: a meta-analysis. *Int Clin Psychopharmacol* 1995; 10(1):45-49.

[80] Feltenstein MW, Sufka KJ: Screening antidepressants in the chick separation-stress paradigm. *Psychopharmacology* (Berl) 2005; 181(1):153-159.

[81] Feltenstein MW, Lambdin LC, Ganzera M, Ranjith H, Dharmaratne W, Nanayakkara NP, Khan IA, Sufka KJ: Anxiolytic properties of Piper methysticum extract samples and fractions in the chick social-separation-stress procedure. *Phytother Res* 2003; 17(3):210-216.

[82] American Psychiatric Association: *Diagnostic and statistical manual of mental disorders*. Washington, DC, American Psychiatric Press, 1994.

[83] Watson D: Rethinking the mood and anxiety disorders: a quantitative hierarchical model for DSM-V. *J Abnorm Psychol* 2005; 114(4):522-536.

[84] Arborelius L, Owens MJ, Plotsky PM, Nemeroff CB: The role of corticotropin-releasing factor in depression and anxiety disorders. *J Endocrinol* 1999; 160(1):1-12.

[85] Ressler KJ, Nemeroff CB: Role of serotonergic and noradrenergic systems in the pathophysiology of depression and anxiety disorders. *Depress Anxiety* 2000; 12 Suppl 1:2-19.

[86] Jiang X, Xu K, Hoberman J, Tian F, Marko AJ, Waheed JF, Harris CR, Marini AM, Enoch MA, Lipsky RH: BDNF variation and mood disorders: a novel functional promoter polymorphism and Val66Met are associated with anxiety but have opposing effects. *Neuropsychopharmacology* 2005; 30(7):1353-1361.

[87] Kessler RC, Berglund P, Demleer O, Jin R, Merikangas KR, Walters EE: Lifetime Prevalence and Age-of-Onset Distributions of DSM-IV Disorders in the National Comorbidity Survey Replication. *Arch Gen Psychiatry* 2005; 62:593-602.

[88] Kessler RC, McGonagle KA, Zhao S, Nelson CB, Hughes M, Eshleman S, Wittchen HU, Kendler KS: Lifetime and 12-month prevalence of DSM-III-R psychiatric disorders in the United States. Results from the National Comorbidity Survey. *Arch Gen Psychiatry* 1994; 51(1):8-19.

[89] Rivas-Vazquez RA, Saff-Biller D, Ruiz I, Blais MA, Rivas-Vazquez A: Currrent issues in anxiety and depression: comorbid, mixed, and subthreshold disorders. *Prof Psychol Res Pr* 2004; 35:74-83.

[90] Hollon SD, Stewart MO, Strunk D: Enduring Effects for Cognitive Behavior Therapy in the Treatment of Depression and Anxiety. *Annu Rev Psychol* 2006; 57:285-315.

[91] Feighner JP: Overview of antidepressants currently used to treat anxiety disorders. *J Clin Psychiatry* 1999; 60 Suppl 22:18-22.

[92] Brown TA, Chorpita BF, Barlow DH: Structural relationships among dimensions of the DSM-IV anxiety and mood disorders and dimensions of negative affect, positive affect, and autonomic arousal. *J Abnorm Psychol* 1998; 107:179-192.

[93] Kasper S: Depression and anxiety--separate or continuum? *World J Biol Psychiatry* 2001; 2(4):162-163.

[94] Frazer A, Morilak DA: What should animal models of depression model? *Neurosci Biobehav Rev* 2005; 29(4-5):515-523.

[95] Matthews K, Christmas D, Swan J, Sorrell E: Animal models of depression: navigating through the clinical fog. *Neurosci Biobehav Rev* 2005; 29(4-5):503-513.

[96] McArthur R, Borsini F: Animal models of depression in drug discovery: a historical perspective. *Pharmacol Biochem Behav* 2006; 84(3):436-452.

[97] Lehr E: Distress call reactivation in isolated chicks: a behavioral indicator with high selectivity for antidepressants. *Psychopharmacology* (Berl) 1989; 97(2):145-146.

[98] Panksepp J, Yates G, Ikemoto S, Nelson E: Simple ethological models of depression: Social-isolation-induced "despair" in chicks and mice, in *Animal models in psychopharmacology*. Edited by Olivier B, Moss J, Slangen JL. Basel, Birkhauser, 1991, pp 61-181.

[99] Katz RJ: Animal models and human depressive disorders. *Neurosci Biobehav Rev* 1981; 5(2):231-246.

[100] Seligman ME, Maier SF, Geer JH: Alleviation of learned helplessness in the dog. *J Abnorm Psychol* 1968; 73(3):256-262.

[101] Sufka KJ, Feltenstein MW, Warnick JE, Acevedo EO, Webb HE, Cartwright CM: Modeling the anxiety-depression continuum in domestic fowl chicks. *Behav Pharmacol* 2006; 17(8) 681-689.

[102] Rubin RT, Poland RE, Lesser IM, Winston RA, Blodgett AL: Neuroendocrine aspects of primary endogenous depression. I. Cortisol secretory dynamics in patients and matched controls. *Arch Gen Psychiatry* 1987; 44(4):328-336.

[103] Willner P: Animal models of depression, in *Behavioural models in psychopharmacology: theoretical, industrial and clinical perspectives*. Edited by Willner P. Cambridge, Cambridge University Press, 1991, pp 91-125.

[104] Russell W, Burch R: *The principles of humane experimental technique*. London, Methuen, 1959.

In: Behavioral Models in Stress Research
Editors: A.V. Kalueff and J.L. LaPorte, pp. 55-71
ISBN: 978-1-60456-361-0

Chapter 3

ANIMAL MODELS OF ANXIETY: EXAMINING THEIR VALIDITY, UTILITY AND ETHICAL CHARACTERISTICS

Jason E. Warnick[1]* and Kenneth J. Sufka[2]
[1] Department of Behavioral Sciences,
Arkansas Tech University, Russellville,AR, USA,
[2] Department of Psychology, Department of Pharmacology, Research Institute of Pharmaceutical Sciences, University of Mississippi, Oxford, MS, USA

INTRODUCTION

Psychopharmacology is a multi-disciplinary field that relies on the utilization of many diverse experimental techniques [65]. However, throughout the discipline's history, the predominant experimental technique has been the employment of non-human animal models of psychopathology [27, 80]. While animal models have been used to study many types of psychopathological conditions, a large portion have been designed to study anxiety disorders [58]. Among the many potential explanations for this research focus, two reasons appear to be most likely. First and foremost is the need to better understand and treat human anxiety disorders [58]. Anxiety disorders affect at least nine percent of the general population [54], and have an annual cost of 42.3 billion dollars in health care costs and productivity losses in the United States alone [28]. The second reason is a common belief that some psychopathological disorders, namely anxiety disorders, are easier to model than others (e.g., schizophrenia, [39]). While it is likely that phenotypic complexity can limit the successful modeling of some psychological disorders, it is probable that animal models of anxiety have many qualities that models of other psychopathologies should strive to attain, such as, refined methodology and theoretical soundness. Thus, a comparison of a variety of anxiety models, highlighting areas of both success and failure, could be of pedagogical value for researchers using animal models of all types of psychopathological disorders.

* E-mail address: jwarnick@atu.edu. Phone: 1 (479) 356-2005. Department of Behavioral Sciences, Arkansas Tech University, Russellville, AR 72801, USA

The goals of this chapter are three-fold. First, a method of evaluating animal models of psychopathology will be described. Second, five animal models of anxiety, four widely utilized and one developed in this laboratory, will be assessed. And third, the animal models will be compared and contrasted.

METHOD OF EVALUATION

McKinney and Bunney's Perspective

Throughout the history of psychopharmacology, animal models have been primarily evaluated in two ways; their ability to simulate psychopathology (e.g., [1, 3, 29, 42]) or their ability to detect the activity of certain drug classes (e.g., [60, 71]). As drug screening was primarily conducted as an industrial enterprise peripheral to academia, most animal model critiques have centered around simulation criteria [34]. The first, and most influential, of these animal model critiques was developed by McKinney and Bunney [42]. They called for a neo-Kraepelian styled approach to the evaluation of how effectively animal models represent the: a) etiology, b) symptomatology, c) physiological basis, and d) treatment of a human psychopathological condition [42]. This approach fared well in the literature as many subsequent animal model critiques have used some variation of these evaluation criteria (e.g., [1, 3, 21, 29, 47]).

Willner's Perspective

Building on the foundation of McKinney and Bunney [42], Willner [79-81] presented a critique on animal models that introduced two important concepts. First, Willner argued that McKinney and Bunney's evaluation criteria only assessed a paradigm's face validity (i.e., the surface-level similarities between the animal model and the human psychopathology). In addition to face validity, Willner contended that models of psychopathology can also be evaluated on two other dimensions: predictive validity and construct validity. Predictive validity assesses the performance of a model, which is typically accomplished by demonstrating a similarity in treatment efficacy between an animal model and the human clinical condition (e.g., examine for false positives and negatives in the model and similarity in pharmacological treatment potencies). Construct validity assesses the theoretical rationale of the animal model. Willner argued that for an animal model to have a construct validity it must be evaluated on two criteria. "Firstly it must be established that homologous constructs are being studied in animals and people. ...Secondly, it must be shown that a change at the level of the construct being modeled is in fact central to the disorder" (p. 684, [79]). Thus, construct validity is established by the demonstration of similar theoretical underpinnings between the paradigm and the clinical condition. A classic example of a clinical syndrome's theoretical foundation that has been used to establish a paradigm's construct validity is the learned helplessness theory of depression [75].

The second concept Willner [80, 81] introduced was that the various uses of animal models can be categorized into a three-level taxonomy consisting of 1) pharmacological screening assays, 2) biobehavioral assays, and 3) simulations. Willner argued that the manner in which an animal model is employed necessitates which type of validity is important. Pharmacological screens are characterized as industrial models with the sole purpose of determining whether a drug produces an effect. Due to this singular purpose, neither the behavioral changes nor the experimental paradigm necessarily need to mimic any aspect of the psychopathological condition (e.g., muricide antidepressant model, [80]). Thus, for a pharmacological screening assay, predictive validity is the only type of validity that matters (i.e., the ability to test a specific structurally- or functionally-based drug class without false positives or negatives). Biobehavioral assays utilize the behavior of an animal as information about an underlying biological event (e.g., brain functioning, neurochemical processes, etc.). The lone evaluative criteria a biobehavioral assay should seek is to attain its accuracy in reporting of underlying physiological events. A simulation is the complete model of a psychopathological condition [80, 81]. This is the only type of animal model that requires the evaluation of face, predictive and construct validity.

Van der Staay's Perspective

Following Willner's [79-81] critique, theorists began to argue about the importance of each type of validity for simulations of psychopathology [14, 36, 61, 73, 79-81]. Most authors have agreed that each validity type is not of equal value [14, 36, 61, 73]. For example, face validity is often characterized as being too superficial and limited due to species-specific behavioral repertoires [61]. Some authors have proposed hierarchies that list the three validity types in order of importance; typically in the order of face, predictive and construct validity with the latter being considered of primary importance [36, 73].

Van der Staay [73] agrees that construct validity is the primary criteria in which an animal model should be judged. However, instead of advocating Willner's concept of construct validity, van der Staay argues that both face and predictive validity can play an important role in establishing a model's construct validity. That is, construct validity is established by the network of associations of similar neuropathological signs and behavioral symptoms, etiology and drug effects in the human clinical syndrome and animal model.

Current Perspective

The focus of this chapter is to evaluate the ability of an animal model to simulate an anxiety disorder. We believe that a good simulation of a clinical syndrome should model the etiology, pathological features and behavioral expression, and treatment profile (i.e., McKinney and Bunney's [42] criteria and Willner's face and predictive validity criteria [79-81]). Further, the more of these features that are modeled the better the construct validity (viz., van der Staay, [73]). Thus, we will evaluate each animal model on the following three criteria: neuropathological sign and behavioral symptom homology, etiological homology and pharmacological sensitivity (i.e., treatment similarity).

The evaluation criterion of etiological homology is utilized to compare the origins of the modeled behavior and the human psychopathology [80]. Frequently, animal models are only capable of modeling one aspect of etiology, while human psychopathologies often result from multiple causative factors [6]. In spite of this potential limitation, a model's limited etiological similarity can still be important in the establishment of theory and the evaluation of putative treatments [6]. In this chapter, etiological homology will be assessed as to whether the modeled psychopathological behavior is produced in a manner similar to either theoretical or empirically established origins.

In this chapter, symptom homology will be evaluated as it has been in the past [80]; specifically, the measured behavior will be qualitatively assessed for its similarity to DSM-IV-TR diagnostic criteria essential to a specific anxiety disorder [2]. The criteria of neuropathological signs will assess the homologous biological responses that occur in the model and the human condition. The physiological responses that will be assessed in this chapter are those that are consistent across the anxiety disorder spectrum, including increased heart rate, increased blood pressure, and corticosteroid release.

Pharmacological sensitivity will be evaluated by the demonstration that all clinically effective drug classes for a specific human anxiety disorder are efficacious within the model. The pharmacological treatment of anxiety disorders has gone through major changes over the last 20 years [70]. Many of the compounds previously utilized to validate simulations and pharmacological screens (e.g., barbiturates and meprobamate) are no longer used in the treatment of anxiety disorders [55]. Therefore, even though many of the reviewed models can screen these drugs, they will not be considered in the chapter. Table 1 displays the clinically-utilized drugs that are considered to be effective in treating generalized anxiety disorder, panic disorder, post-traumatic stress disorder, obsessive-compulsive disorder, and phobic disorder [13, 55, 70]. Prior to assessing pharmacological sensitivity, animal models will be judged as to what type of anxiety disorder is likely being modeled via the criterion of symptom homology. We will evaluate animal models on their ability to detect the drugs listed for the specific modeled anxiety disorder at doses comparable to that which is efficacious for the human disorder.

Table 1. Pharmacological treatments for specific anxiety disorders.

Drug Classes	GAD	PD	PTSD	OCD	Phobia
Benzodiazepines (e.g., chlordiazepoxide)		+			+
High-Potency Benzodiazepines			+		+
Monoamine Oxidase Inhibitors			+	+	
Tricyclic Antidepressants	+	+		+	
Selective Serotonin Reuptake Inhibitors (e.g., fluoxetine, fluvoxamine)	+	+		+	
Atypical Antidepressants (e.g., venlafaxine)		+			
Serotonergic Anxiolytics (e.g., buspirone)		+			

GAD - generalized anxiety disorder; PD - panic disorder; PTSD - post-traumatic stress disorder; OCD - obsessive-compulsive disorder; Phobia - phobic disorder, + established pharmacological treatment.

Another issue that was offered by Willner that will be considered is a model's utility. Willner [80] contends that animal models, in addition to having scientific integrity, should be "cheap, simple and reliable" (p. 6). If a model accomplishes all of these goals, it is said to have high utility. Thus, animals should have low purchase costs, not incur high per diem charges due to long required time in laboratory housing, and the paradigm should not be labor intensive. As judgments of a paradigm's utility must be based on comparative assessments, evidence of a paradigm's utility will rely on comparisons to the historical mainstay of animal research which is conditioning experiments using rats. Thus, if a paradigm allows the use of mice, or other species less expensive to purchase, does not require time-intensive animal training prior to testing, and can employ automated testing procedures it would be said to have high utility.

Table 2. Animal model simulation evaluation outcome.

	Animal Models				
Evaluation Criteria	Social Interaction	Elevated-Plus Maze	Vogel Response	Startle Conflict	Chick SSP[1]
Etiological Homology	+	+	+	+	+
Sign and Symptom Similarity	+	+	+	+	+
Pharmacological Sensitivity	+/-	+	-	-	+
Utility	+	+	-	+/-	+
Three R's	+/-	+/-	-	-	+

Symbols: + = criterion fully attained; +/- = criterion partially attained; - = criterion not attained

[1]Chick SSP = Chick Separation Stress Paradigm

In recent years, the welfare of animals used in research has been the focus of increased public scrutiny and legislative oversight for certain animal research practices. In order to address these concerns, animal researchers have been expected to formally address what has come to be called "the 3 R's" in Institutional Animal Care and Use Committee protocols and grant applications to major public sources of funding (e.g., National Institutes of Health in the U.S., the Medical Research Council in the U.K.; [48]). The 3 R's is an abbreviation for the consideration researchers should make to *reduce* the number of purpose-bred animals used in an experiment, *refine* the experiment to minimize the pain and distress an animal experiences, and *replace* an animal with either a phylogenetically lower species or non-animal alternative [48]. Similar to utility, judgments of how well a paradigm achieves the 3 R's is based on comparative assessments. Thus, like utility, the 3 R's will be assessed on the comparison of a paradigm to the mainstay of biomedical animal research which is typically invasive experiments using rats.

Anxiety Model Evaluation

As stated above, two of the purposes of this chapter are the evaluation of five animal models of anxiety, four widely utilized and one developed in one of the author's affiliated laboratory, and comparing and contrasting them to each other. The choice of the four established animal models consisted of two criteria; 1) being well established in the research literature, and 2) being differentiated enough experimentally to necessitate comparison. Nineteen animal models were chosen from a personal library of research articles on animal modeling. These models were then subjected to a quantitative assessment of their citation count (i.e., the model name combined with the word anxiety) on the National Library of Medicine Gateway literature database (conducted on April 29, 2007). The social interaction and elevated-plus maze models were chosen based on their overwhelming citation rates. While the open field model was the third most cited, it was not included in this chapter due to it and the elevated-plus maze being tests of anxiety to novelty [57]. Both of these models are designed to manipulate a rodent's fear of open areas [57]. Because of this exclusion, the third and fourth models chosen were the startle response and the Vogel conflict model, respectively. The animal model developed in this lab was the chick separation stress paradigm.

Social Interaction

The social interaction model was developed by File and Hyde [22] as a procedurally unique anxiety model. Prior to this model, researchers primarily depended on conditioning to develop a state of fear, or anxiety, in an organism. However, this model was different because it used the natural behavior of an animal and considered the two interacting animals as a single unit, or system. This model uses four test conditions that range in order from a low anxiety situation to a high anxiety situation, which includes, 1) familiar test apparatus and low apparatus illumination, 2) familiar test apparatus and high apparatus illumination, 3) unfamiliar test apparatus and low apparatus illumination, and 4) unfamiliar test apparatus and high apparatus illumination [22]. During the ten minute test session, the amount of time in social contact between two rodents, rats or mice, is measured [22]. The amount of time in social contact is inversely proportional to the anxiety level provoked by a test condition (e.g., high anxiety, low social contact). This graded response allowed the social interaction model to be the first model capable of screening both anxiolytics (i.e., increase the social contact in the high anxiety condition) and anxiogenics (i.e., decrease the social contact in the low anxiety condition, [25]). Due to this, the social interaction model has predominantly been used as pharmacological screen [20, 25].

The reduced amount of time that an "anxious" subject spends in social contact appears to best represent situational avoidance, a key diagnostic feature of phobic disorder [2]. The reason for only one analogous symptom is because the social interaction model only analyzes one primary behavior. Thus, the absence of further symptom similarities is due to the model's design, not to behavioral dissimilarities. Also, subjects have been shown to have increased corticosterone release in the higher stress conditions [24]. Based on this evidence, this model achieves the criterion of sign and symptom homology. Additionally, as subjects show signs and symptoms in the presence of a potentially threatening stimulus, the model achieves the criteria of etiological homology.

Unlike sign and symptom homology and etiological homology, the criterion of pharmacological sensitivity is not as easily attained in the model. This criterion is not fully attained due to the model's unreliable screening of one drug class used to treat phobic disorder. Benzodiazepines have been screened extensively within the model with good success [12, 22]. Additionally, these compounds are efficacious at doses similar to human pharmacotherapy. However, the efficacy of high-potency benzodiazepines has been equivocal; some have screened successfully (e.g., flunitazepam, [32]) and some have failed to produce an effect (e.g., triazolam, [20]). The potential for screening benzodiazepines is enhanced by the model demonstrating some pharmacological specificity as it demonstrates anxiogenic effects to anxiety producing compounds (e.g., amphetamine and caffeine, [23]). Thus, the criterion of pharmacological sensitivity are only partially met for simulation models.

The social interaction model is a high utility paradigm as it possesses attributes that reduce the economic and labor cost of conducting an experiment. Specifically, the model allows the use of mice instead of rats, requires one short (i.e., 10 minute) test session, uses a simple non-technological apparatus, and employs a simple observational measurements. The paradigm only partially achieves the 3 R criteria as the model only possesses a *refined* methodology. That is, animals are only tested once for a short test session, thus minimizing the amount of pain and distress a subject experiences compared to anxiety models that use several conditioning sessions. The other two R's, *reduce* and *replace*, are not met because the model uses mammalian subjects in sample sizes similar to that of other anxiety models.

Elevated-Plus Maze

In 1955, Montgomery [46] conducted a series of studies demonstrating that rats spend more time exploring enclosed alleyways than open-sided alleyways. Additionally, when given a choice between enclosed and open alleyways, rats overwhelmingly choose enclosed alleyways [46]. This avoidance of open arms was explained as a motivational conflict between the desire to explore new locations and a rats' fear of open areas [33, 46]. In 1984, Handley and Mithani [33] used Montgomery's work to develop a new animal model of anxiety, now called the elevated-plus maze.

The elevated-plus maze consists of four runways, two enclosed facing each other and two open facing each other, constructed in the shape of a X or a + and elevated approximately two to three feet above the ground [59]. A rat or mouse is placed in the center of the apparatus and the amount of time spent in the runways and the number of runway entries is recorded over a five to ten minute test session [52]. The level of anxiety is measured as a function of both the number of entries and the amount of time spent in the open runways [59]. Administration of an anxiolytic is predicted to increase the number of entries and time spent in the open runways, while an anxiogenic will decrease the number of entries and time spent in the open runways [59].

All validity criteria are fully achieved by model. Similar to the social interaction model, this model achieves sign and symptom homology as its behavioral index of anxiety consists of avoidance, a symptom of phobic disorder [2], and the demonstration that open runways produce increases in heart rate, blood pressure, and corticosteroid release [52, 62]. The reason for only one analogous symptom results from limited behavioral measurements, not from behavioral dissimilarities. Also, the model achieves etiological homology because subjects show signs and symptoms in the presence of a potentially threatening stimulus. The evaluation criterion of pharmacological sensitivity has been achieved fairly conclusively. As would be expected if the elevated-plus maze is a model of phobia, the drug classes that consistently produce anxiolysis are regular and high-potency benzodiazepines [59]. Additionally, these compounds are efficacious at doses similar to human treatment [59]. Further, treatments for other psychopathologies (e.g., haloperidol, indomethacin, and apomorphine) are ineffective and caffeine, cocaine, and FG7142 exert an anxiogenic response [59].

The elevated plus maze model is a high utility paradigm as it possesses attributes that reduce the economic and labor cost of conducting an experiment. Specifically, the model allows the use of mice instead of rats, requires one short (i.e., 5-10 minute) test session, uses a simple non-technological apparatus, and employs a simple observational measurements. The paradigm only partially achieves the 3 R criteria as the model only possesses a *refined* methodology. That is, animals are only tested once for a short test session, thus minimizing the amount of pain and distress a subject experiences compared to anxiety models that use several conditioning sessions. The other two R's, *reduce* and *replace*, are not met because the model uses mammalian subjects in sample sizes similar to that of other anxiety models.

Startle Response

The startle response is one of the oldest animal models of anxiety [7]. This model was designed from the "clinical observation that anxious patients exhibit exaggerated startle reactions to sudden loud sounds" (p. 318, [7]). In the model, conditioning trials are conducted in which a conditioned stimulus, usually a light, is paired a footshock. In separate trials, a startle stimulus, usually a loud noise or an air puff, is presented alone and the animal's response, a jumping motion, is recorded. In testing trials, the conditioned stimulus is presented immediately prior to the startle stimulus. During these trials, animals have a greater response to the startle stimulus than when it is presented alone. This exacerbated reaction is believed to be demonstrative of the heightened state of anxiety that the conditioned stimulus elicits.

As mentioned above, the model is based on the behavior of patients with anxiety disorders [7]. And indeed, the dependent measure is representative of two diagnostic criteria for post-traumatic stress disorder (PTSD; i.e., physiological response to a cue that resembles the traumatic event & exaggerated startle response; 2). Additionally, the model's "anxious" behavior originates in a manner similar to that of PTSD; specifically, it begins from the experience of an event that causes, or threatens, physical harm [2]. Further, the conditioned stimulus causes the physiological reaction of increased corticosterone release, which parallels patients with PTSD [8]. Collectively, these parallels help the startle response attain the criteria of sign and symptom homology and etiological homology.

However, unlike the other validity criteria, pharmacological sensitivity is not established in this model. The most efficacious pharmacological treatments for PTSD are monoamine oxidase inhibitor's (MAO-I's) and tricyclic antidepressants (see Table 1). However, in this model, imipramine (acute and chronic administration) has shown not to be effective [10], while studies using MAO-I's have not been published [4]. The drug classes that have been effective in the model have been benzodiazepines (e.g., chlordiazepoxide, [35]), high-potency benzodiazepines (e.g., alprazolam, [35]), and serotonergic anxiolytics (e.g., buspirone, [37]). This specific cluster of effective pharmacological treatments is not found across the major anxiety disorders (see Table 1).

Similar to the model's performance on the criteria of pharmacological sensitivity, the startle response does not possess high utility. There are two primary features of this paradigm that cause problems with utility. First, the model requires multiple training and testing sessions that occur over multiple days. This leads to additional per diem charges for the animals and more labor needed to complete a study. Second, it requires a technological apparatus that delivers stimuli and measures responses accurately which also increases the cost to conduct an experiment.

In addition to utility, the startle response model also performs poorly on the 3 R's criteria. The model makes no attempt to *reduce*, or *replace*, the number of purpose-bred rodents tested. And, compared to other models reviewed in this chapter, it clearly does not possess *refined* methodology that minimizes the pain and distress experienced by an animal. In fact, the model depends on multiple training sessions in which a subject is electrically shocked.

Vogel Conflict

The Vogel conflict model was developed as a simple alternative to previously established conditioning based anxiety models [74]. In this model, animals are deprived of water 18-48 hours prior to testing [45, 74]. During the three minute test session, animals are given free access to water. When the test subject licks the water tube twenty times, an electric shock is delivered through the tube. A measure of how many times the water tube is licked is used as the dependent variable. The conflict between the motivation provided by water-deprivation and the aversion provided by the electric shock produce a hypothesized state of anxiety [45].

As seen above with the social interaction and elevated-plus maze models, this model's primary behavioral measure can be characterized as an avoidance response, which is a diagnostic criteria for phobic disorder. Also, this behavioral pattern occurs as the result of a traumatic event (i.e., being shocked), which parallels a common origin of phobic disorders [2]. And additionally, electrically-shocked animals display increased corticosterone when in proximity to a feared stimulus, which is a common physiological response of phobic patients [53]. Together, these three commonalities allow the model to attain the criteria of sign and symptom homology and etiological homology.

As a simulation, this model does not fare as well on the criterion of pharmacological sensitivity. Unlike other animal models, the Vogel conflict model has problems with this criteria due to its ability to screen too many drug classes. As expected of a phobia model, benzodiazepines and high-potency benzodiazepines are reliably screened at doses similar to human treatment [64]. However, the model also screens chronically administered tricyclic antidepressants at doses similar to human treatment [11]. In addition to not being used clinically, these compounds have been shown experimentally to be non-efficacious for phobic disorders [41]. Due to this, the criterion of pharmacological sensitivity is not achieved.

In addition to validity criteria, the Vogel conflict model performs fairly well in regards to utility. The main concern about the model's utility is its expensive and technology-dependent apparatus. However, despite this concern, the model possesses many features that make it have partially high utility. Specifically, the model allows the use of mice instead of rats, requires one short (i.e., 3 minute) test session, and employs a simple observational measurements. In contrast to utility, the Vogel conflict model fails to meet the 3 R's criteria. The model does not possess *refined* methodology as it relies on electrical shock and water deprivation. Additionally, the model fails to *replace* or *reduce* the number of animals that are used in the experimental methodology.

Chick Separation Stress Paradigm

Infants of many species display a distress response when separated from whom they are attached (e.g., mother, parents, or conspecifics; [9, 31, 40, 67]). Historically, research on the separation distress response has led to many theories about psychopathology, development, and personality (e.g., [5]). One of the species in which the separation distress response has been widely studied has been the young domestic fowl (e.g., [38, 50, 51, 66, 76]). Similar to other species, the domestic fowl's separation distress response is characterized by: increased vocalizations (or distress vocalizations; DVocs), hyperthermia, stress induced analgesia, and increased corticosterone [17, 69]. Sufka and colleagues utilized this species' separation response to develop an anxiety model named the chick separation-stress paradigm (SSP). The chick SSP has been primarily used in this laboratory as a pharmacological screen [16, 19, 63, 68] and a biobehavioral assay of opiate functioning [67, 76].

While the chick SSP has undergone some experimental modifications since its development (see [15]), the core set of procedures have remained the same. In the chick SSP, seven-day post-hatch cockerels are isolated from their group-house conspecifics for a three-minute test session in one of two experimental conditions. The first experimental condition tests chicks in a low-stress social condition. Formerly, the social condition consisted of the experimental subject being tested with two social companions in the test chamber (e.g., [78]). Recently, the two social companions have been replaced by two mirrors placed along the sides of the test chamber [15]. The mirrors mimic social companions, thus creating a low-stress condition that employs fewer animals [15]. The second experimental condition tests isolated chicks without mirrors, which creates a high-stress social separation condition [15]. While the chick SSP has employed different behavioral measures (e.g., stress-induced analgesia), DVocs have been shown to be a pure measure of separation distress and constitutes the current dependent measure [15].

The chick SSP performs well on all validity criteria. The stress-response has been argued to meet the diagnostic criteria for panic disorder [49, 77]. In particular, the paradigm appears to model situationally-bound panic disorder "as the symptom onset is rapid, intense and brief with clear etiological origins" (pg. 582, [77]). Accompanying the measured behavioral response, the physiological response of corticosterone release that takes place in panic disorder also occurs in the model [17]. Additionally, the model contains no obvious behavioral or physiological dissimilarities with panic disorder. Thus, the model achieves the criteria of sign and symptom homology. Also, as subjects show signs and symptoms in the presence of a potentially threatening stimulus, the model achieves the criteria of etiological homology.

The chick SSP has also effectively demonstrated its possession of treatment similarity. As would be expected if the model simulates situationally-bound panic disorder, the chick SSP successfully screens high-potency benzodiazepines [77, 78], benzodiazepines [19], tricyclic antidepressants (imipramine: [18, 19, 77]; maprotiline: [18]) and an MAO-I (phenelzine: [18, 77]) at doses similar to human treatment [18, 19, 77, 78], while failing to screen scopalamine, caffeine, chlorpromazine, haloperidol [19], buspirone [19, 77], and trazodone [77]. Additionally, the chick SSP fails to screen fluoxetine [77]; a drug used clinically for the treatment of panic disorder [26, 43, 44] that is ineffective in the treatment of situationally-bound panic disorder [72]. Based on the chick SSP successfully screening the appropriate drugs at the appropriate doses, this model achieves the criteria of pharmacological sensitivity.

The chick SSP is a high utility model as it possesses attributes that reduce the economic and labor cost of conducting an experiment. Specifically, the model allows the use of a low-cost animal ($0.50 a chick, [56]), requires one short (i.e., 3 minute) test session, uses an apparatus that allows high-throughput screening (i.e., up to 6 chicks per test session), and employs a simple observational measurements. Additionally, the model also fully achieves the 3 R's criterion. The model *reduces* and *replaces* the number of purpose-bred research animals with a non-purpose bred phylogenetically lower species. Additionally, the model has been *reduced* the number of chicks necessary for a study because it uses mirrors instead of untreated companion chicks in the low-stress condition. The model has a *refined* methodology as it uses a single, short test session.

General Discussion

The purpose of this chapter was to develop a method of evaluating animal models, use it to assess five models of anxiety, and to compare the evaluation outcomes. The developed evaluation methodology assessed animal models in terms of their validity, utility and ethical justification. The validity assessment consisted of animal models being evaluated on their similarity to a human anxiety disorder in terms of etiology, signs and symptoms and treatment profile. The utility assessment evaluated the economic and labor costs that are associated with using the animal model. And the ethical justification assessment evaluated how effectively the model addressed the 3 R's.

Each of the animal models was able to achieve the criterion of etiological homology. It should be noted that etiology homology was evaluated in a very narrow manner. The current theoretical understanding of anxiety disorders supports an etiology that is biopsychosocial in nature [30]. That is, anxiety disorders are believed to be the product of the interaction of individual genetic vulnerabilities and resiliencies with environmental stressors. In the future, it is essential that this view of anxiety disorder etiology be incorporated into anxiety simulations through the use of varied strains, genetic manipulation, and the administration of different amounts of a stressor (e.g., social interaction model and chick SSP).

Also, each of the animal models was able to achieve the criterion of sign and symptom homology. The assessment of the simulations revealed differences in the specific anxiety disorders the paradigms simulated. Three of the models simulated phobic disorder (i.e., social interaction, elevated-plus maze, and Vogel conflict), while the startle response simulated PTSD, and the chick SSP simulated situationally-bound panic disorder.

The criterion of pharmacological sensitivity was difficult to attain as it was only fully achieved by the elevated-plus maze and the chick SSP and partially achieved by the social interaction model. That is, the elevated-plus maze and the chick SSP were able to detect the pharmacological agents within the treatment profile of the modeled anxiety disorder without displaying false positives or false negatives. The social interaction model only detected one of the two pharmacological agents within the treatment profile of the simulated anxiety disorder. The startle response model failed to achieve the pharmacological sensitivity criterion due to the model failing to screen one of the drugs within the treatment profile of the simulated anxiety disorder. Additionally, this model detected drugs that were not within the treatment profile of the simulated syndrome. Similarly, the Vogel conflict model failed to achieve the pharmacological sensitivity criterion due to the model detecting drugs that were not within the treatment profile of the simulated syndrome.

It should be noted that failing to achieve the criterion of pharmacological sensitivity may harm the validity of a model as a simulation but not the function of the model as a pharmacological screening assay. For example, the startle response model and Vogel conflict model do not achieve the pharmacological sensitivity criteria as simulations because they fail to match an anxiety disorder's treatment profile. However, both of these models have the ability to detect at least three different classes of anxiolytics. For a researcher considering the use of a screening assay, these models' pharmacological detection profile could likely be of interest. Among the reviewed the models, the chick SSP had the ability to detect the broadest array of pharmacological classes (i.e., benzodiazepines, high-potency benzodiazepines, tricyclics, and MAO-Is) and the social interaction model had the narrowest detection pattern (i.e., benzodiazepines).

Diversity was also seen in the utility the models possessed. Three models were able to demonstrate that they were high utility (i.e., social interaction, elevated-plus maze, and chick SSP). That is, these models used experimental techniques that were cost-effective (e.g., allowed the use of mice or non-mammalian subjects instead rats), were not labor intensive (e.g., only required one short testing session) and used simple behavioral measures. The Vogel conflict model also possessed each of these traits, but it requires a expensive and technologically driven apparatus which lowers the model's utility.

Only one model, the chick SSP, was able to satisfy the criteria of the 3 R's. This model was unique in that it employs a non-purpose bred non-mammalian species which allows it to meet the *reduce* and *replace* criteria. Three of the models (i.e., social interaction, elevated-plus maze and chick SSP) were able to meet the *refine* criterion by using short, non-invasive experimental techniques. This suggests that improvements have been made in the ethical implementation of biomedical research.

It can not be emphasized enough that the evaluations conducted on these anxiety models do not create a "best buy" list. The evaluations simply inform researchers on the validity, utility and ethical justification that a model provides. Thus, each model can be effective if used within the epistemic boundaries established by the assessment. For example, the startle response is a model of PTSD that possesses the validity criteria of sign and symptom homology and etiological homology. However, due to its difficulty with the pharmacological sensitivity criteria, it is unable to provide much information on the pharmacological treatment of PTSD. Hopefully, such uses of animal model evaluations will aid in the development of new animal models and the proper employment of those already developed.

ACKNOWLEDGMENTS

The authors would like to express their gratitude to Kyla D. Warnick for her insightful reviews of several early versions of this manuscript.

REFERENCES

[1] Abramson LY, Seligman MEP. Modeling psychopathology in the laboratory: History and rationale. In Maser JD, Seligman MEP, editors. *Psychopathology: Experimental models.* San Francisco, W. H. Freeman and Company, 1977, 1-26.

[2] American Psychiatric Association. *Diagnostic and statistical manual of mental disorders IV-TR.* Washington DC, American Psychiatric Association, 2000.

[3] Bond NW. Animal models in psychopathology: An introduction. In Bond NW, editor, *Animal models in psychopathology*. Sydney, Academic Press, 1984, 1-21.

[4] Borsini F, Podhorna J, Marazziti D. Do animals models of anxiety predict anxiolytic-like effects of antidepressants? *Psychopharmacology* (Berl), 2002, 163, 121-141.

[5] Bowlby J. The making and breaking of affectional bonds. I. Aetiology and psychopathology in the light of attachment theory. An expanded version of the 50th Maudsley Lecture, delivered before the Royal College of Psychiatrists, 19 November 1976. *Br J Psychiatry*, 1977, 130, 201-210.

[6] Brain PF, Marrow L. Rodent models of human neuroses and psychoses. In Haug M, Whalen RE, editors, *Animal models of human emotion and cognition.* Washington DC, American Psychological Association, 1999, 59-73.

[7] Brown JS, Kalish HI, Farber IE. Conditioned fear as revealed by magnitude of startle response to an auditory stimulus. *J Exp Psychol*, 1951, 41, 317-328.

[8] Campeau S, Falls WA, Cullinan WE, Helmreich DL, Davis M, Watson SJ. Elicitation and reduction of fear: Behavioural and neuroendocrine indices and brain induction of the immediate-early gene c-fos. *Neuroscience*, 1997, 78, 1087-1104.

[9] Carden SE, Hofer MA. Socially mediated reduction of isolation distress in rats pups is blocked by naltrexone but not by RO 15-1788. *Behav Neurosci*, 1990, 104, 457-463.

[10] Cassella J, Davis M. Fear-enhanced acoustic startle is not attenuated by acute or chronic imipramine treatment in rats. Psychopharmacology (Berl), 1985, 87, 278-282.

[11] Commissaris RL, Hill TJ. High-dose subchronic imipramine treatment: Effects on anxiety-like (conflict) behavior in rats. Anxiety, 1994, 1, 109-113.

[12] de Angelis L, File SE. Acute and chronic effects of three benzodiazepines in the social interaction anxiety test in mice. Psychopharmacology (Berl), 1979, 64, 127-129.

[13] Dulcan MK, Bregman JD, Weller EB, Weller RA. Treatment of childhood and adolescent disorders. In Schatzberg AF, Nemeroff CB, editors, *The American Psychiatric Press textbook of psychopharmacology*. Washington, DC, American Psychiatric Press, 1995, 669-706.

[14] Ellenbroek BA, Cools AR. Animal models with construct validity for schizophrenia. *Behav Pharmacol*, 1990, 1, 469-490.

[15] Feltenstein MW, Ford NG, Freeman KB, Sufka KJ. Dissociation of stress behaviors in the chick social-separation-stress procedure. *Physiol Behav*, 2002, 75, 675-679.

[16] Feltenstein MW, Lambdin LC, Ganzera M, Dharmaratne HRW, Nanayakkara NPD, Khan IA, Sufka KJ. Anxiolytic properties of piper methysticum extract samples and fractions in the chick social-separation-stress procedure. *Phytother Res*, 2003, 17, 210-216.

[17] Feltenstein MW, Lambdin LC, Webb HE, Warnick JE, Khan SI, Khan IA, Acevedo EO, Sufka, KJ. Corticosterone response in the chick separation-stress paradigm. *Physiol Behav*, 2003, 78, 489-493.

[18] Feltenstein MW, Sufka KJ. Screening antidepressants in the chick separation-stress paradigm. *Psychopharmacology* (Berl), 2005, 181, 153-159.

[19] Feltenstein MW, Warnick JE, Guth AN, Sufka KJ. The chick separation stress paradigm: A validation study. *Pharmacol Biochem Behav*, 2004, 77, 221-226.

[20] File SE. The use of social interaction as a method for detecting anxiolytic activity of chlordiazepoxide-like drugs. *J Neurosci Methods*, 1980, 2, 219-238.

[21] File SE. The validation of animal tests of anxiety: Pharmacological implications. *Pol J Pharmacol Pharm*, 1984, 36, 505-512.

[22] File SE, Hyde JRG. Can social interaction be used to measure anxiety? *Br J Pharmacol,* 1978, 62, 19-24.

[23] File SE, Hyde JRG. A test of anxiety that distinguishes between the actions of benzodiazepines and those of other minor tranquilisers and of stimulants. *Pharmacol Biochem Behav*, 1979, 11, 65-69.

[24] File SE, Peet LA. The sensitivity of the rat corticosterone response to environmental manipulations and to chronic chlordiazepoxide treatment. *Physiol Behav*, 1980, 25, 753-758.

[25] File SE, Seth P. A review of 25 years of the social interaction test. *Eur J Pharmacol*, 2003, 463, 35-53.

[26] Gorman JM, Liebowtiz MR, Fyer AJ, Goetz D, Campeas RB, Fyer MR, Davies SO, Klein DF.An open trial of fluoxetine in the treatment of panic attacks. *J Clin Psychopharmacol*, 1987, 7: 329-332.

[27] Green S, Hodges H. Animal models of anxiety. In Willner P, editor, *Behavioural models in psychopharmacology: Theoretical, industrial and clinical perspectives.* Cambridge, Cambridge University Press, 1991, 21-49.

[28] Greenberg PE, Sisitsky T, Kessler RC, Finkelstein SN, Berndt ER, Davidson JR, Ballenger JC, Fyer AJ. The economic burden of anxiety disorders in the 1990s. *J Clin Psychiatry*, 1999, 60, 427-435.

[29] Griez E. Experimental models of anxiety: Problems and perspectives. *Acta Neurol Psychiatr Belg*, 1984, 84, 511-532.

[30] Gross C, Hen R. Genetic and environmental factors interact to influence anxiety. *Neurotox Res*, 2004, 6, 493-501.

[31] Gurski JC, Davis K, Scott JP. Interaction of separation discomfort with contact comfort and discomfort in the dog. *Dev Psychobiol*, 1980, 13, 463-467.

[32] Guy AP, Gardner CR. Pharmacological characterisation of a modified social interaction model of anxiety in the rat. *Neuropsychobiology*, 1985, 13, 194-200.

[33] Handley SL, Mithani S. Effects of alpha-adrenoceptor agonists and antagonists in a maze-exploration model of 'fear'-motivated behaviour. *Naunyn Schmiedebergs Arch Pharmacol*, 1984, 327, 1-5.

[34] Harris JC. Experimental animal modeling of depression and anxiety. *Psychiatr Clin North Am*, 1989, 12, 815-836.

[35] Hijzen TH, Houtzager SW, Joordens RJ, Olivier B, Slangen JL. Predictive validity of the potentiated startle response as a behavioral model for anxiolytic drugs. *Psychopharmacology* (Berl), 1995, 118, 150-154.

[36] Kalueff AV, Tuohimaa, P. Experimental modeling of anxiety and depression. *Acta Neuobiol Exp*, 2004, 64, 439-448.

[37] Kehne JH, Cassella JV, Davis M. Anxiolytic effects of buspirone and gepirone in the fear-potentiated startle paradigm. *Psychopharmacology* (Berl), 1988, 94, 8-13.

[38] Koga Y, Takahashi H, Oikawa D, Tachibana T, Denbow DM, Furuse M. Brain creatine functions to attenuate acute stress responses through GABAnergic system in chicks. *Neuroscience*, 2005, 132, 65-71.

[39] Lader M. Animal models of anxiety: A clinical perspective. In Willner P, editor, *Behavioural models in psychopharmacology: Theoretical, industrial and clinical perspectives*. Cambridge: Cambridge University Press, 1991, 76-88.

[40] Levine S, Wiener SG. Psychoendocrine aspects of mother-infant relationships in nonhuman primates. *Psychoneuroendocrinology*, 1988, 13, 143-154.

[41] Liebowitz MR, Fyer AJ, Gorman JM, Campeas RB, Sandberg DP, Hollander E, Papp LA, Klein DF. Tricyclic therapy of the DSM-III anxiety disorders: a review with implications for further research. *J Psychiatr Res*, 1988, 22 Suppl, 7-31.

[42] McKinney WT, Bunney WE. Animal model of depression: Review of evidence and implications for research. *Arch Gen Psychiatry*, 1969, 21, 240-248.

[43] Michelson D, Lydiard RB, Pollack MH, Tamura RN, Hoog SL, Tepner R, Demitrack MA, Tollefson GD. Outcome assessment and clinical improvement in panic disorder: Evidence from a randomized controlled trial of fluoxetine and placebo. *Am J Psychiatry*, 1998, 155, 1570-1577.

[44] Michelson D, Pollack M, Lydiard RB, Tamura R, Tepner R, Tollefson G. Continuing treatment of panic disorder after acute response: Randomised, placebo-controlled trial with fluoxetine. *Br J Psychiatry*, 1999, 174, 213-218.

[45] Millan MJ, Brocco M. The Vogel conflict test: Procedural aspects, gamma-aminobutyric acid, glutamate and monoamines. *Eur J Pharmacol*, 2003, 463, 67-96.

[46] Montgomery KC. The relation between fear induced by novelty stimulation and exploratory behaviour. *J Comp Physiol Psychol*, 1955, 48, 254-260.

[47] Murphy DL. *Animal models for human psychopathology: Observations from the vantage point of clinical psychopharmacology*. In Serban G, Kling A, editors, Animal models in human psychobiology. New York: Plenum Press, 1976, 265-271.

[48] Office of Laboratory Animal Welfare. *Institutional animal care and use committee guidebook*, 2nd edition. Bethesda MD, Office of Laboratory Animal Welfare, 2002.

[49] Panksepp J. *Affective neuroscience: The foundations of human and animal emotions.* New York: Oxford University Press, 1998.

[50] Panksepp J, Meeker R, Bean NJ. The neurochemical control of crying. *Pharmacol Biochem Behav*, 1980, 12, 437-443.

[51] Panksepp J, Vilberg T, Bean NJ, Coy DH, Kastin AJ. Reduction of distress vocalization in chicks by opiate-like peptides. *Brain Res Bull*, 1978, 3, 663-667.

[52] Pellow S, Chopin P, File SE, Briley M. Validation of open:closed arm entries in an elevated plus-maze as a measure of anxiety in the rat. *J Neurosci Methods*, 1985, 14, 149-167.

[53] Pericic D, Pivac N. Sex differences in conflict behaviour and in plasma corticosterone levels. *J Neural Transm Gen Sect*, 1995, 101, 213-221.

[54] Regier DA, Boyd JH, Burke JD, Rae DS, Myers JK, Kramer M, Robins LN, George LK, Karno M, Locke BZ. One-month prevalence of mental disorders in the United States. Based on five epidemiologic catchment area sites. *Arch Gen Psychiatry*, 1988, 45, 977-986.

[55] Reiman EM. Anxiety. In Gelenberg AJ, Bassuk EL, editors, *The practitioner's guide to psychoactive drugs*, 4th edition. New York, Plenum Medical Book Company, 1998, 213-264.

[56] Roach JT, Sufka KJ. Characterization of the chick carrageenan response. Brain Res, 2003, 994, 216-225.

[57] Rodgers RJ. Animal models of 'anxiety': Where next? *Behav Pharmacol*, 1997, 8, 477-496.

[58] Rodgers RJ, Cao BJ, Dalvi A, Holmes A. Animal models of anxiety: An ethological perspective. *Braz J Med Biol Res*, 1997, 30, 289-304.

[59] Rodgers RJ, Cole JC. The elevated plus-maze: Pharmacology, methodology and ethology. In Cooper SJ, Hendrie CA, editors, *Ethology and psychopharmacology*. Chichester, New York, John Wiley and Sons, 1994, 9-44.

[60] Sanger DJ. Animal models of anxiety and the screening and development of novel anxiolytic drugs. In Boulton A, Baker G, Martin-Iverson M, editors, *Neuromethods* Volume 19: Animal models in psychiatry II. Clifton NJ, Humana Press, 1991, 147-198.

[61] Sarter M, Bruno JP. Animal models in biological psychiatry. In D'Haenen H, den Boer JA, Willner P, editors, *Biological psychiatry*. Boston, John Wiley and Sons, 2002, 1-8.

[62] Shekhar A, Sims LS, Bowsher RR. GABA receptors in the region of the dorsomedial hypothalamus of rats regulate anxiety in the elevated plus-maze test. II. Physiological measures. *Brain Res*, 1993, 627, 17-24.

[63] Smith KK, Dharmaratne HRW, Feltenstein MW, Broom SL, Roach JT, Nanayakkara NPD, Khan IA, Sufka KJ. Anxiolytic effects of kava extract and kavalactones in the chick social separation-stress paradigm. *Psychopharmacology* (Berl), 2001, 155, 86-90.

[64] Söderpalm B, Eriksson E, Engel JA. Anticonflict and rotarod impairing effects of alprazolam and diazepam in rat after acute and subchronic administration. *Prog Neuropsychopharmacol Biol Psychiatry*, 1989, 13, 269-283.

[65] Stephen DN, Andrews JS. Screening for anxiolytic drugs. In Willner P, editor, *Behavioural models in psychopharmacology: Theoretical, industrial and clinical perspectives*. Cambridge, Cambridge University Press, 1991, 50-75.

[66] Sufka KJ, Hughes RA. Differential effects of handling on isolation-induced vocalizations, hypoalgesia, and hyperthermia in domestic fowl. *Physiol Behav*, 1991, 50, 129-133.

[67] Sufka KJ, Hughes RA, McCormick TM, Borland JL. Opiate effects on isolation stress in domestic fowl. *Pharmacol Biochem Behav*, 1994, 49, 1011-1015.

[68] Sufka KJ, Roach JT, Chambliss WG, Broom SL, Feltenstein MW, Wyandt CM, Zeng L. Anxiolytic properties of botanical extracts in the chick social separation-stress procedure. *Psychopharmacology* (Berl), 2001, 153, 219-224.

[69] Sufka KJ, Weed NC. Construct validation of behavioral indices of isolation stress and inflammatory nociception in young domestic fowl. *Physiol Behav*, 1994, 55, 741-746.

[70] Taylor CB. Treatment of anxiety disorders. In Schatzberg AF, Nemeroff CB, editors, *The American Psychiatric Press textbook of psychopharmacology*. Washington DC, American Psychiatric Press, 1995, 641-655.

[71] Treit D. Animal models for the study of anti-anxiety agents: A review. *Neurosci Biobehav Rev*, 1985, 9, 203-222.

[72] Uhlenhuth EH, Matuzas W, Warner TD, Paine S, Lydiard RB, Pollack MH. Do antidepressants selectivity suppress spontaneous (unexpected) panic attacks? A replication. *J Clin Psychopharmacol*, 2000, 20, 622-627.

[73] van der Staay FJ. Animal models of behavioral dysfunctions: Basic concepts and classifications, and an evaluation strategy. *Brain Res Rev*, 2006, 52, 131-159.

[74] Vogel JR, Beer B, Clody DE. A simple and reliable conflict procedure for testing anti-anxiety agents. *Psychopharmacologia*, 1971, 21, 1-7.

[75] Vollmayer B, Henn FA. Learned helplessness in the rat: Improvements in validity and reliability. *Brain Res Protoc*, 2001, 8, 1-7.

[76] Warnick JE, McCurdy CR, Sufka KJ. Opioid receptor function in social attachment in young domestic fowl. *Behav Brain Res*, 2005, 160, 277-285.

[77] Warnick JE, Wicks RT, Sufka KJ. Modeling anxiety-like states: Pharmacological characterization of the chick separation stress paradigm. *Behav Pharmacol*, 2006, 17, 581-587.

[78] Watson GS, Roach JT, Sufka KJ. Benzodiazepine receptor function in the chick social-separation-stress procedure. *Exp Clin Psychopharmacol*, 1999, 7, 83-89.

[79] Willner P. Validation criteria for animal models of human mental disorders: Learned helplessness as a paradigm case. *Prog Neuropsychopharmacol Biol Psychiatry*, 1986, 10, 677-690.

[80] Willner P. *Behavioural models in psychopharmacology. In Willner P, editor, Behavioural models in psychopharmacology: Theoretical, industrial and clinical perspectives.* Cambridge, Cambridge University Press, 1991, 3-18.

[81] Willner P. Methods for assessing the validity of animal models of human psychopathology. In Boulton A, Baker G, Martin-Iverson M, editors, *Neuromethods Vol. 18: Animal models in psychiatry I.* Clifton NJ, Humana Press, 1991, 1-23.

In: Behavioral Models in Stress Research
Editors: A.V. Kalueff and J.L. LaPorte, pp. 73-96
ISBN: 978-1-60456-361-0

Chapter 4

THE OPEN-FIELD MODEL OF EMOTIONAL REACTIVITY AND PROGRESS IN THE MOLECULAR GENETICS OF FEAR

Saffron A.G. Willis-Owen*
Imperial College, London, UK

INTRODUCTION

Despite high levels of heritability, the genetic origins of inter-individual variation in emotional reactivity; both in terms of normal-range variation and extreme (clinical) manifestations, remain largely uncharacterised. This failure to identify the genetic contributions towards complex (in this case behavioral) traits is likely to result from a combination of experimental and aetiological factors; including small individual effect sizes [49], gene-gene interactions (also known as epistasis), gene-environment interactions (including epigenetic modes of transmission), and difficulties in establishing appropriate phenotype definitions, modes of measurement and accurate phenotype translation across populations (and even species).

In order to circumvent some of these limitations a number of investigators have chosen to focus on behavioral models in genetically more tractable species such as the mouse (*mus musculus*). The use of these model systems enables the control of known confounds that originate from the genetic background and rearing environment, and the reduction of behavioral phenotypes into a finite collection of directly quantifiable elements.

A vast array of these models, founded on principals of active and passive avoidance, autonomic activation, and behavioral inhibition (the halting of species-typical behaviors such as grooming, exploration and consumption in anxiogenic / novel environments), have been developed to assay murine emotional reactivity. One such model, the 'open-field test' (OFT), has emerged as one of the most widely utilized and highly evidenced. The open-field is a circular white, brightly lit, fully-enclosed arena [37] in which behavior; specifically

* Email address: s.willis-owen@imperial.ac.uk. NHLI, Guy Scadding Building, Dovehouse Street, London. SW3 6LY, UK

locomotion, defecation, rearing and thigmotaxis (wall-seeking behavior) are monitored over a brief (5 or 10 minute) interval. Ambulation and defecation within this apparatus are typically inversely correlated [19, 15], and mouse lines that have been bi-directionally selected over multiple generations for extremities of ambulation in this apparatus exhibit concomitant and inverse extremities in defecation [19].

Since there is no known overlap in the peripheral nervous system control of ambulation and defecation, their co-coordinated activity within the OFT apparatus is thought to reflect the orchestrating action of a central mechanism or psychological trait [25]. Based on principals of anthropomorphism, given that intense fear can result in defecation, urination [90] and immobility in humans [10] it has historically been assumed that a combination of behavioral inactivity and defecation may provide a basic index of heightened fear or emotional reactivity in mice, partially related to, although substantially less complex than it's human counterparts. Consistent with this interpretation, repeated exposures to the OFT environment yield reductions in both defection and avoidance of the central (anxiogenic) areas [77]; that is a habituation towards the anxiogenic environment. Furthermore, comparable neuroanatomical structures appear to be involved in rodent OFT performance and human emotionality [5], and pharmacological agents that are clinically effective in the treatment of human anxiety disorders also appear to influence rodent performance in this apparatus [88].

Over the last decade, the open-field behavioral model (specifically the co-variation between defecation and ambulation within this system) has formed the basis of an intensive research program aimed to identify the genetic determinants of emotionality. In 2004 this program culminated in the first successful cloning of a small-effect behavioral locus; a gene which encodes the regulator of G-protein signaling, *RGS2* located on mouse chromosome 1. In this chapter we will outline research which led to the identification of this locus, discuss the informative value of the underlying behavioral model (the OFT), and the down-stream relevance of this information towards the characterization of related phenotypes in humans.

The Building Blocks of Emotionality: A Reductionist Approach

Numerous models of murine emotionality are currently in usage, assessing various, in some cases overlapping, aspects of the emotionality phenotype. These models include fear conditioning (measuring the response to cues or contexts previously associated with a negative stimulus) [108]; the light-dark box [9] and elevated plus maze (examining preference for anxiogenic (light and/or open) versus safe (dark and/or enclosed) environments) [12]; hyponeophagia (latency to consume a novel food substance when food deprived) [98]; forced swim and tail suspension tests (two hopelessness paradigms) [113]; and the acoustic startle response (quantifying magnitude of response towards an acoustic stimulus) [54]. Whilst the data derived from these paradigms often align, with for example behavior in the elevated plus maze exhibiting a predictive relationship with behavior in at least two other models of anxiety (hyponeophagia and the acoustic startle) [98], there is increasing evidence that behavior both across and within these tests may be more informatively divided into a small number of biologically distinct, but correlated dimensions.

This hypothesis is supported by a recent factorial analysis of >100 behavioral phenotypes derived from a battery of 5 etiological tests (including the open-field) using more than 1,600 mice [40]. This analysis yielded 5 primary dimensions of emotionality which transcend test type and demonstrate genetic separation through QTL (quantitative trait locus) mapping. These dimensions include: (i) inactivity in 'safe areas', (ii) avoidance of anxiogenic areas, (iii) suppression of rearing, (iv) latency to enter novel areas, and (v) autonomic activity in novel environments. On average between 4 and 6 genetic loci were found to influence each measure, with each locus conferring an individually diminutive effect (less than a quarter of loci contributed >2% of phenotypic variance to any given trait). These data indicate that anxiety, per se cannot be considered a single phenotype, but rather an amalgam of individual (in some cases correlated) behavioral elements. Consequently research that selectively focuses on a small number of these elements may, through reduced heterogeneity, accrue sufficient statistical power to identify individual (i.e., low magnitude) genetic effects.

Control of Extraneous Factors

Inbred mouse strains, a diverse collection of lines which have been inbred over many generations through repeated brother-sister matings, represent a unique resource for the investigation of genetic contributions towards complex traits. More than 450 inbred strains currently exist, many of which have been maintained for more than 150 generations. These strains can be considered essentially isogenic because the mouse genome reaches homozygosity at more than 98.6% of loci after around 20 generations [7]. A high quality draft sequence of the mouse euchromatic genome is already publicly available for one strain (C57BL/6J) [107], with a further 15 common strains subject to on-going re-sequencing efforts. These resources provide an excellent foundation for well characterized transgenics and heterogeneous (outbred) stocks in which genetic background and ancestry can be fully defined. However some degree of caution must be exercised given the potential for genetic drift and accumulation of mutations in inappropriately managed stock resources [91].

Likewise, the use of model organisms also enables control of confounding environmental factors, where those factors are known to the experimenter. Laboratory by strain interactions were established in the late 1990s [16, 104], and more recently data have also highlighted significant within-laboratory genotype by environment interactions; although fewer for behavioral as opposed to physiological traits [102]. Specifically with regard to the OFT apparatus, investigators have identified significant interactions with season [102] and environmental enrichment conditions [99]. These data are consistent with observations that the effect of stress on behavior in an open-field paradigm is subject to the moderatory effect of season [66]. Clearly in order to generate reliable, reproducible behavior in this paradigm, these (and other potentially influential) factors need to be considered in experimental design.

Examining the Heritability of Open-Field Behaviour

Prior to undertaking genetic mapping studies it is necessary to confirm trait heritability. If genetic variation underlies a trait or disease, then it should be subject to selection (be it natural or experimental). Consequently, even where no clear phenotypic difference exists between inbred strains (e.g. due to factors which operate in opposing directions) a program of intercrossing combined with artificial selection based on extreme trait values should result in an increased frequency of trait-associated alleles, and a consequent shift in the trait mean away progenitor phenotype values. Ultimately after sufficient generations selected (and linked) alleles should become fixed in the population. Artificial selection therefore provides an ideal method for determining the nature and extent of genetic influences on a phenotype.

In the 1970s DeFries reported a bidirectional selection experiment based on ambulation in the OFT arena. Two inbred mouse strains were crossed (BALB/cJ and C57BL/6J) and the F_3 (filial 3, or third generation) progeny selected for trait extremes over 30 generations. The four resultant lines (two high-active and two low-active) exhibited divergent (non-overlapping) phenotype distributions, with a greater than 30-fold difference in mean activity, and a 7-fold difference in defecation (in the opposite direction) [17-19]. This evidence demonstrates the heritability of open-field activity, and confirms the presence of an inverse biological relationship between activity and defecation in the open-field. Similar investigations have been conducted on other features of open-field behavior, such as thigmotaxis. Whilst these experiments also demonstrate a significant response to selection (i.e. trait heritability), they also confirm the presence of phenotypic sub-structure within the open-field; since lines selected for high and low levels of thigmotaxis do not also show consistent variation in ambulation [58, 59].

Mapping a Murine Emotionality Locus

Having established that behavioral measures of murine emotionality are heritable, it now remains to identify those genes and variants that underlie the genetic effect. To date there have been at least 15 attempts to map murine emotionality using the OFT apparatus, 14 of which utilize measures of ambulation, defecation and / or the co-variation between these factors (termed EMO) (see reference [109] for a full description of the methods used and data derived from them). These approaches have led to the identification of candidate emotionality loci (i.e. attaining a $\leq$5% significance level) on one sex chromosome (X) and all but one of the mouse autosomes (chromosome 13). However, as figure 1 demonstrates, by far the greatest proportion of these loci map to mouse chromosome 1 (between 50 and 110cM).

Evidence of an OFT emotionality locus located on mouse chromosome 1 was originally identified in an F_2 mapping experiment involving an intercross between two phenotypically opposing open-field ambulation selection lines (derived from the DeFries experiment briefly described earlier). This locus was mapped to a 65.6Mb region between markers D1MIT218 and D1MIT293, and was found to moderate both ambulation and defecation in the open-field [25] with less pronounced effects on other species-typical behaviors such as latency to enter the centre of the OFT apparatus and rearing in the periphery [40]. Subsequent experiments which have utilised replicates of this F_2 population have confirmed the location of this linkage signal; consistently identifying evidence of linkage between 72 and 82cM [100, 101]. However, the DeFries selection lines utilised for all the mapping studies described thus far, were constructed from a two strain intercross between BALB/cJ and C57BL/6J. Consequently these data cannot be considered fully informative of the genetic complexity underlying OFT behavior (since only a small proportion of contributory loci are likely to segregate between these two closely related strains). More recent mapping approaches have aimed to resolve this issue by maximizing initial progenitor diversity using as many as 8 different progenitor strains to create heterogeneous (pseudo-outbred) or true outbred stocks. Two such

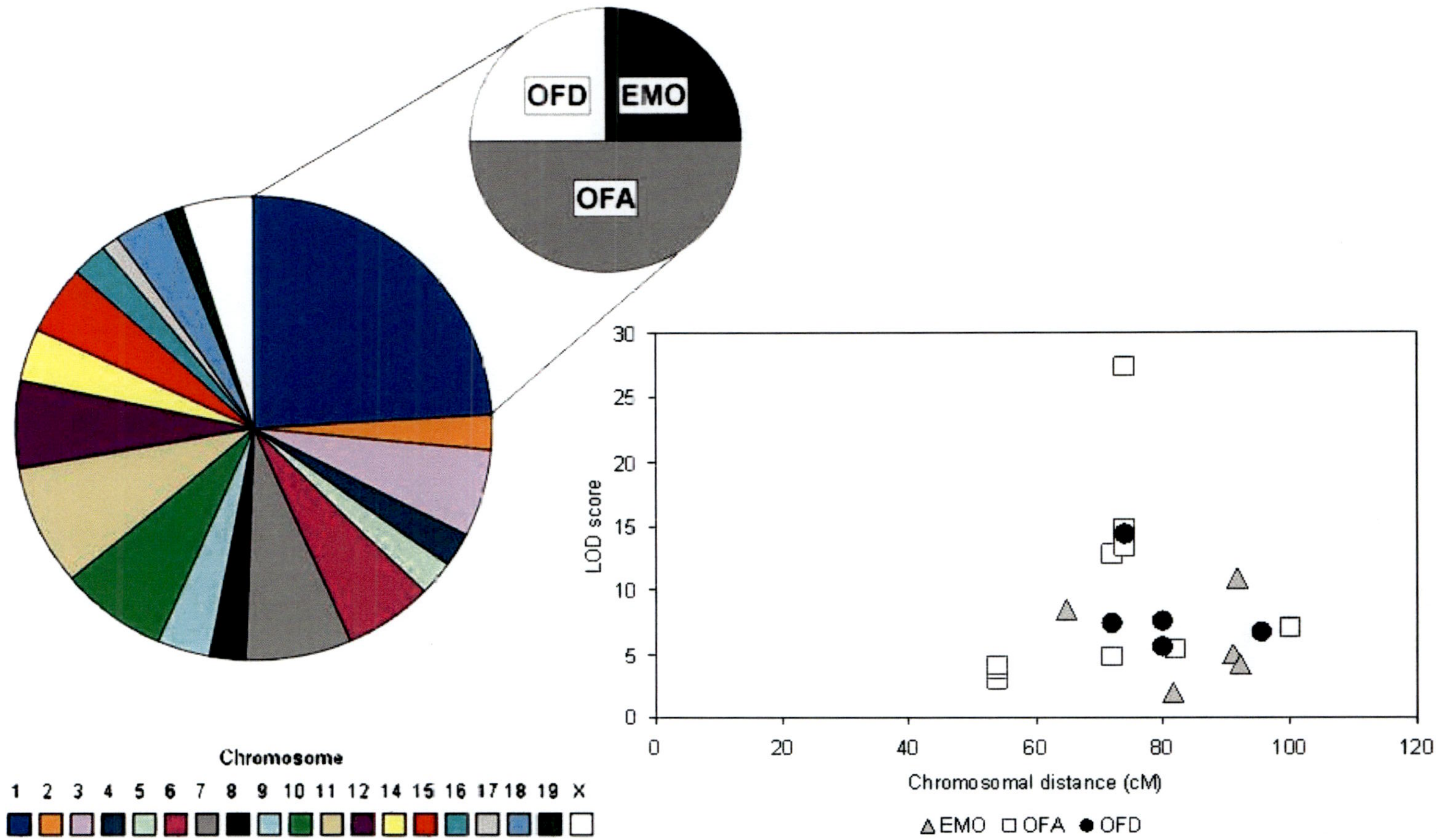

Figure 1. Open-field ambulation and defecation quantitative trait loci by chromosome. Where LOD scores were not available, -Log10(P) values have been presented. OFT (open-field test), EMO (emotionality as inferred from the co-variation between ambulation and defecation in the open-field arena), OFA (open-field ambulation), OFD (open-field defecation), LOD (logarithm of the odds), -Log10(P) (the base 10 logarithm of the p-values).

experiments have been published; one mapping EMO to a 0.77cM region surrounding the marker D1MIT264 located at 71.5cM on chromosome 1 [92] (a novel locus which does not segregate between BALB/cJ and C57BL/6J), and the other resolving the original QTL interval into three independent loci collectively spanning 0.82cM [111]. These QTL are shown in figure 2.

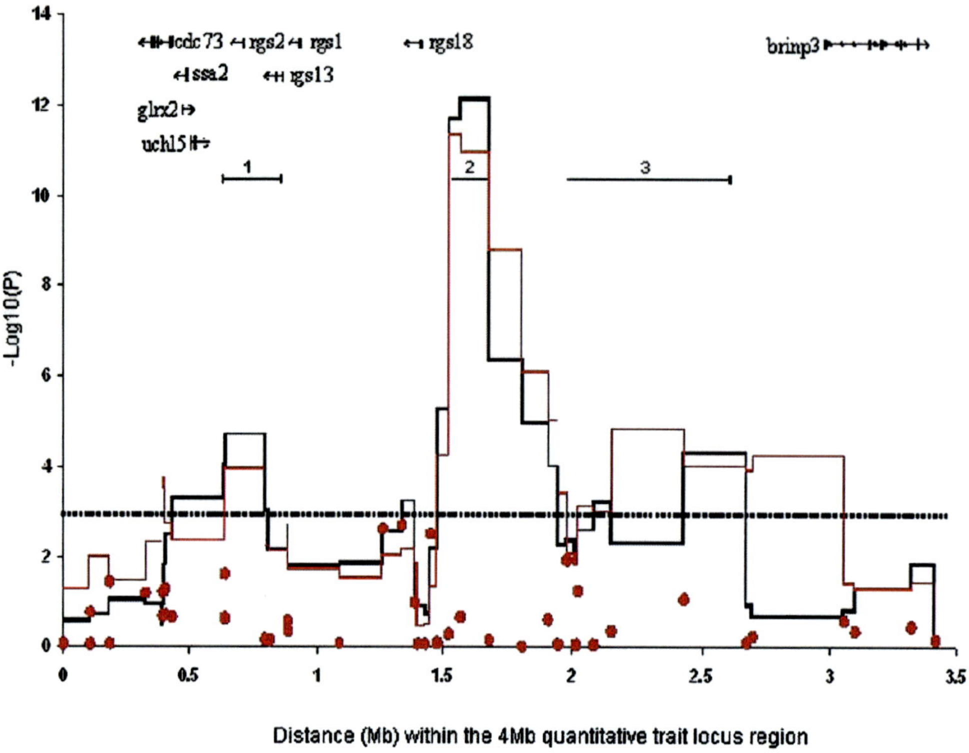

Figure 2. High resolution mapping of the chromosome 1 emotionality quantitative trait locus. The 95% confidence intervals of the QTL are shown as horizontal bars labeled 1-3. Single-point results are shown as red circles and multi-point results are shown as solid lines based on either 4 (black) or 8 (red) progenitors. The Bonferroni corrected 5% significance threshold is shown as a broken horizontal line. Mb (megabases), -Log10(P) (the base 10 logarithm of the P values).

Together these three loci are estimated to account for 12% of variance in the EMO trait, with no individual locus contributing more than 5% of the total phenotypic variance. Since all three peaks remain significant when mapped using either simultaneous or conditional algorithms, and no two peaks can be mapped to the same progenitor haplotype, all three loci have been considered to represent independent genetic effects.

Over 90% of the mouse and human genomes can be portioned into regions of mouse-human synteny [1]. Of particular note, the region encapsulating the three mouse EMO loci outlined above demonstrates conserved synteny with a large replicated [26, 71, 72] human QTL also on chromosome 1, mapped using the personality trait Neuroticism; a baseline index of emotionality reflecting propensity towards neurotic breakdown under stress. Because of this overlap, the reduction of these mouse QTL intervals to their constituent genes (or even variants) represents a challenge of considerable biomedical relevance. The high level of mapping resolution afforded by the use of outbred mouse stocks has already enabled the reduction of this broad region to just a handful of candidate genes. Specifically only two genes have coding sequence within the 95% confidence intervals of the QTL boundaries, with one further gene located in close proximity (*RGS2*, *RGS13* and *RGS18* respectively). Remarkably, all three of these genes are members of the regulator of G-protein signaling, (RGS) family, first identified in yeast cells in the mid 1990s [22].

The Regulators of G-Protein Signalling (RGS) Genes

Heterotrimeric G-protein coupled receptors, or GPCR (where G stands for GTP, the downstream signaling molecule) represent a large family of receptors characterized by seven transmembrane helices [20, 84]. The intracellular loops which connect these helices constitute a G-protein binding domain, with an N-terminal fragment on the exoplasmic aspect, and a C-terminal fragment on the cytosolic aspect of the plasma membrane. The G-proteins which associate with the cytoplasmic aspect of the receptor play a fundamental role in the transduction of ligand-binding to downstream intracellular effectors. The interaction of an agonist with the cell-surface receptor induces a conformational change which promotes the attachment of intracellular G-proteins. This coupling favors the rapid exchange of Gα-bound GDP for GTP, thereby facilitating the dissociation of the G-protein α subunit from its βγ complex. Both these activated units are then able to communicate with downstream intracellular effector molecules [30, 62] including both enzymes and ion channels.

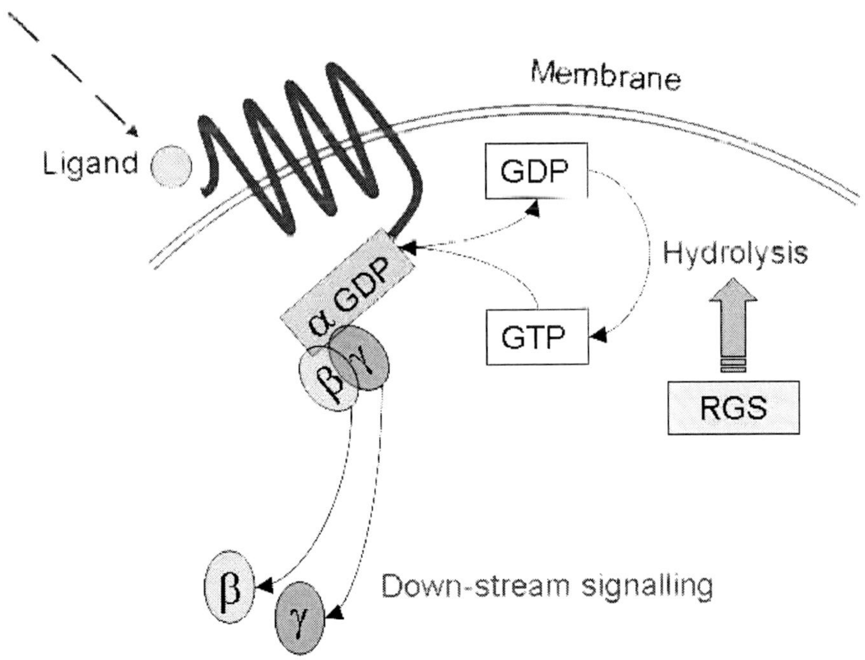

Figure 3. G-protein receptor activity and regulators of G-protein signaling. RGS (regulator of G-protein signaling), GDP (guanosine diphosphate), GTP (guanosine triphosphate).

Signal amplitude and longevity are determined by the intrinsic GTPase activity of the G-protein α subunit (Gα). The signal is terminated when the γphosphate is removed from the Gα-bound GTP, leaving GDP occupying the guanine nucleotide exchange pocket, and culminating in the re-association of Gα and Gβγ subunits (i.e. a return to the basal conditions). This process of deactivation has been observed to occur substantially faster *in vivo* than predicted from rates of Gα GTP hydrolysis observed when using purified components [2, 115]. The unidentified factors now known to modify the kinetics of G-protein mediated signaling events are Regulator of G-protein Signaling (RGS) proteins. This family of proteins have been demonstrated to accelerate the Gα-catalyzed GTP hydrolytic turnover by up to 1000-fold [68], without compromising the strength of steady-state signaling.

High concentrations of GPCRs, such as may be found at synaptic terminals, can lead to a local saturation of the capacity for GDP-GTP exchange; making the speed of GTP hydrolysis a rate limiting factor in signal transduction. Because the GTPase action of RGS proteins serves to simultaneously deactivate the G-protein and increase local availability of Gα GDP, these proteins enable repeated rapid receptor activation without degradation of signal intensity [117]. Consistent with these features RGS proteins have been found to play important roles in physiological systems characterized rapid response kinetics [97, 73].

RGS proteins share a common domain consisting of approximately 125 amino-acids [21, 55]. This core domain binds to the switch regions on Gα stabilizing them towards hydrolysis of GTP [96]. Currently in excess of 30 mammalian genes have been identified which contain the 23 conserved hydrophobic residues found at the core of the RGS domain which are necessary, but not always sufficient for RGS action *in vivo* [14]. These have been divided into 6 families on the basis of phylogeny [116], and structural similarity [81]. *RGS2*, *13* and *18*, all three of those genes that either underlie or localize in close proximity to murine EMO QTL, are members of the B/R4 family which predominantly exhibit short NH_2 (N) and COOH (C) terminal regions. These sites include a variety of motifs which govern cellular localization, and Gα selectivity.

RGS2

The function of the various RGS proteins is governed at least in part by their specificity towards particular Gα species. RGS2 has demonstrable GAP properties towards both Gαi and Gαq, although potency of this action is substantially greater for Gαq. This preference is mediated by intrinsic structural properties of the protein's G-protein binding interface, and to a lesser extent by variations in expression levels. Minimal expression of *RGS2* is sufficient to attenuate Gαq-mediated signaling, whereas higher levels are required to impact upon Gαi. The relative preference of RGS2 towards these 2 Gα species can be modified by point mutation of 3 residues of the N and C termini which differ between *RGS2* and another RGS protein with opposite Gα preferences (RGS4) [41]. These residues are thought to affect the geometry of the threonine-binding pocket of the RGS protein; a shallower pocket in RGS2 conferring preferential docking of the Gαq switch1 over Gαi.

The Gαq signaling pathway revolves around the activation of phospholipaseC-β which hydrolyzes phosphatidylinositol 4,5-bisphosphate (PIP_2) to form Inositol 1,4,5-trisphosphate (IP_3). IP_3 mobilizes calcium ions from intracellular stores, enabling activation of store-operated Ca^{2+} channels located in the plasma membrane. The resulting calcium influx activates ATPase pumps situated both in the sarco/endoplasmic reticulum and plasma membrane thereby terminating the signaling event. These intracellular oscillations in calcium concentration, which control a wide range of cell functions [8], can be agonist-induced [44, 70], or subject to control by RGS proteins, specifically RGS2 [105].

At a subcellular level, RGS2 has been shown to localize predominantly in the nucleus [13] by passive diffusion, unless a receptor or activated G-protein (Gαq) are co-expressed, in which case RGS2 is observed to undergo recruitment to the plasma membrane [42, 82]. A conserved N-terminal domain is both necessary and sufficient for this signal-induced redistribution, and is required for attenuation of Gαq signaling [42]. This domain has been characterized as an amphipathic α-helix which binds vesicles containing acidic phospholipids. Because recruitment of RGS2 to the plasma membrane is not essential for the termination of a Gαq mediated signal, it is likely that other important functions of the N-terminal domain have not yet been elucidated.

At a structural and cellular level the expression of *RGS2* is spatially, and temporally specific. In situ hybridization analysis using a subtype-specific S-labeled cRNA probe in the developing postnatal rat brain, has demonstrated a concomitant reduction and shift from diffuse to specific expression patterns with increasing age [48]. The highest signal intensities are transiently localized in the superficial layers of the neocortex between postnatal days 2 and 10, a region corresponding to the cortical plate. An early abundance of *RGS2* in both the amygdala and caudate putamen, although maintained in the caudate putamen, becomes reduced and differentiated in the amygdala from P10 onwards. Likewise, hippocampal expression although diffuse at P2, is redistributed to the pyramidal layer of the hippocampus and granule cell layer of the dentate gyrus from P18 through to adulthood. Thalamic expression of *RGS2* is consistently low, and the initially high levels of *RGS2* expression in the hypothalamus are no longer apparent by P10. Throughout development cerebellar *RGS2* expression is confined to granule cell layers.

Uniquely amongst the RGS protein family, *RGS2* appears to play a role in synaptic plasticity in the hippocampus. The transcription of *RGS2* is dynamically up-regulated in response to neuronal stimulation, both in terms of single fibre pulse application [47], the induction of an electroconvulsive seizure [31], experimental cortical impact injury [53], and administration of MECS, a pharmacological agent known to induce expression of immediate early genes and cause long term enhancement of synaptic contacts in the hippocampus [47]. Importantly this up-regulation is not seen when protein synthesis is blocked by co-administration of CH with MECS, or when single fibre stimulation falls below a frequency sufficient for the induction of long term potentiation. In each case *RGS2* expression is maximal approximately 2 hours post-insult, and has returned to baseline by 8-24 hours [47, 31, 53]. At 2 hours post-event the greatest effects on expression are seen in the pyramidal cell layer of the hippocampus and the granule cell layer of the dentate gyrus [47]. Recent data suggest that *RGS2* may exert its effect on synaptic plasticity in the hippocampus through a regulation of Gαi/o mediated pre-synaptic calcium channel inhibition [38].

Although *RGS2* has often been associated with Gαq-mediated signaling, it remains unclear which specific receptor populations / signaling pathways are subject to its moderation. Currently there is evidence for *RGS2* involvement in a range of systems including angiotensin II [33, 61], serotonin [29], corticotrophin-releasing factor receptor 1 [79], dopamine [28, 94, 95], glutamate [52], adrenaline [36] and P2Y (and therefore indirectly GIRK) [24]. Consistent with known Gα preferences, *RGS2* has been shown to attenuate 5-

HT2A receptor signaling (Gαq-mediated) without any quantifiable effect on 5-HT1A (a Gαi –mediated signaling pathway) [29]. *RGS2* also co-localizes with both dopamine D1 and D2 receptors, with its expression level directly and inversely related to the agonist/antagonist occupancy of the two receptor populations [94, 95]. In confirmation of *RGS2's* role in angiotensin signaling (another Gαq-mediated pathway), *RGS2* has been found to be present at increased levels in the mononuclear cells of human patients with abnormalities in vascular control (Bartter's/Gitelman's syndrome) [11], and to exhibit a number of missence mutations in hypertensive patients [112].

A further interaction has recently been identified between *RGS2* and the dendritic spine protein Spinophilin (SPL) [106]. *SPL* binds to the N-terminal domain of RGS2 and the third intracellular loop of several GPCRs, actively recruiting *RGS2* to the GPCR complex. This interaction is of interest for several reasons: first *SPL* is known to moderate fast excitatory synaptic transmission through both NMDA and AMPA glutamate receptors [35] (both of which are known to participate in learning and memory [4, 6]). Second *SPL* null mice exhibit both neuronal pathologies comparable to those seen in *RGS2* null mice (abnormal hippocampal dendritic spine densities [23]) and a failure to learn simple taste aversions [87], and third *SPL* transcription is reduced in the hippocampus of human mood disorder patients [57].

RGS2 null mice are both viable and fertile, and exhibit a range of both physiological and behavioral deficits including reduced T cell proliferation and IL-2 production [76], renovascular abnormalities including a persistent constriction of the resistance vasculature, and prolonged response of the vasculature to vasoconstrictors reflecting a slowed pace of signal transduction in the smooth musculature [43]. These effects culminate *in vivo* to produce an impaired antiviral immunity [76], and intensely hypertensive phenotype [43].

RGS2 null mice are reported to display normal circadian rhythmicity, motor coordination, exploratory behavior, spatial learning and memory. However a greater preference for the dark in the light/dark test, and increased response to the acoustic startle has been taken as evidence of abnormal emotional reactivity in these animals [76]. Structurally an effect of the *RGS2* mutation has only been observed in the dendritic morphology and branching of CA1 hippocampal neurons. Specifically relative to *RGS2*$^{+/-}$ mice, homozygous null mice showed a reduced apical and basilar spine density in this population of neuronal cells. This translates physiologically to a significant reduction in the electrical input/output of these cells.

RGS13

RGS13 displays GTPase activating properties for several Gα species, including both $G\alpha_q$ and $G\alpha_i$. The capacity for *RGS13* to also inhibit receptor-stimulated cAMP generation [50] indicates a further interaction with $G\alpha_s$, although the lack of direct $G\alpha_s$ GAP activity suggests that *RGS13* may block this signal at the level of adenylyl cyclase. Sub-cellular *RGS13* localization is dependent on the cell type assayed, and the presence/absence of G-protein species. Co-expression of active $G\alpha_i$ causes *RGS13* to accumulate at the intracellular junctions, whereas in the presence of $G\alpha_s$ *RGS13* shifts to the nucleus. $G\alpha_q$ does not appear to have a substantial effect [86].

RGS13 is most abundantly expressed in immune tissues including the spleen, lymph node, thymus medulla and tonsil [86], and until recently its highly localized central expression pattern had resulted in variable reports of prevalence in the brain [32, 74, 50, 56].

More detailed characterization, as achieved by *in situ hybridization* in the rat, has revealed an expression pattern predominantly correlated with that of *RGS2*, although of lower amplitude. There is an enrichment of *RGS13* mRNA in the hippocampal formation (specifically in the granule cell layers of CA1-CA3, and the dentate gyrus), amygdala, mammillary nuclei, and both the pontine and interpeduncular nuclei [32].

Although the physiological role of *RGS13* is as yet largely unknown, variation in expression has been associated with several immune functions such as chemokine responsiveness and desensitization. However, as yet the neurological role of *RGS13* remains to be elucidated.

RGS18

RGS18 consists of 235 amino acids, with an RGS domain located at residues 86 through 202. Further putative motifs include a consensus site for phosphorylation by both cAMP- and cGMP-dependent protein kinases (residues 213-216), 5 additional sites for casein kinase II phosphorylation, and a C-terminus CAAX motif (residues 221-224) which may be important for association with the nucleoplasmic surface of the inner nuclear membrane.

RGS18 is expressed from a single gene, and shows tissue-specific distribution patterns which are comparable between human and mouse orthologs. Experimental over-expression of *RGS18* in both mammalian and yeast cells has confirmed the protein's capacity for moderation of Gα signaling [114]. Specifically *RGS18* has an *in vitro* guanosine triphosphatase activity preference for Gα_q [78] (which activates Phospholipase β) and Gα_{i1}, Gα_{i2}, and Gα_{i3} [27] (which bind and inhibit adenylate cyclase) but not for Gα_s, Gα_{12} or Gα_z [27].

Distribution of expression indicates a principal physiological role in haematogenesis. *RGS18* is expressed in bone marrow, specifically in megakaryocytes, and significantly in haematopoietic stem cells [69], with expression also observed later in platelets, monocytes and granulocytes. During embryonic stages of development, abundant expression is additionally reported in the liver; the primary organ for haematopoiesis [78, 114]. *RGS18* is not observed in cells of lymphoid lineage. Whilst there is little published evidence in favor of *RGS18* expression in the Central Nervous System (CNS), recent in situ issue hybridization data derived from several murine inbred strain (C3H) brain sections and a probe spanning exons 4 and 5 indicated low levels of expression limited to the hippocampus and basal cerebral peduncles. No expression has been identified in the amygdala (Fullerton, J personal communication).

Candidate Gene Selection

The predominant expression of *RGS2*, and to a lesser extent *RGS13* in a structure central to anxiety (the hippocampus [34]) is consistent with a role in emotionality, and therefore the open-field EMO trait. Rodent hippocampal lesions yield a characteristic spectrum of anxiolytic effects which can be dissociated from and therefore do not appear to be mediated by abnormal amygdaloid function [65]. Ventral hippocampal lesions lead to reductions in emotionality as quantified by both the plus-maze and hyponeophagia measures, and whilst dorsal hippocampal lesions influence emotional behavior in the plus-maze they have no discernable impact on assessments of hyponeophagia [3]. Since cortical plasticity events within the hippocampus are likely to be of fundamental importance to this structure's function, and *RGS2* exhibits a dynamic up-regulation in response to such events [31], *RGS2* emerges as the primary candidate for investigation in relation to the emotionality locus on murine chromosome 1. This hypothesis is further supported by the fact that *RGS2* is the only gene which has its entire coding sequence located within the boundaries of a chromosome 1 QTL.

Gene Candidacy Testing: Quantitative Complementation

If *RGS2* can account for at least a proportion of the phenotypic variation associated with the murine chromosome 1 locus (either through coding sequence variation or polymorphisms in more distal control elements) then this contribution should be detectable using an experimental design known as quantitative complementation [64] (also known as the knockout-interaction test). This experimental design has previously been used to advance from QTL to gene in drosophila [63, 39] and yeast species [89], but until the pioneering research of Yalcin and colleagues published in 2004 [111], had not previously been applied to mammalian genetics.

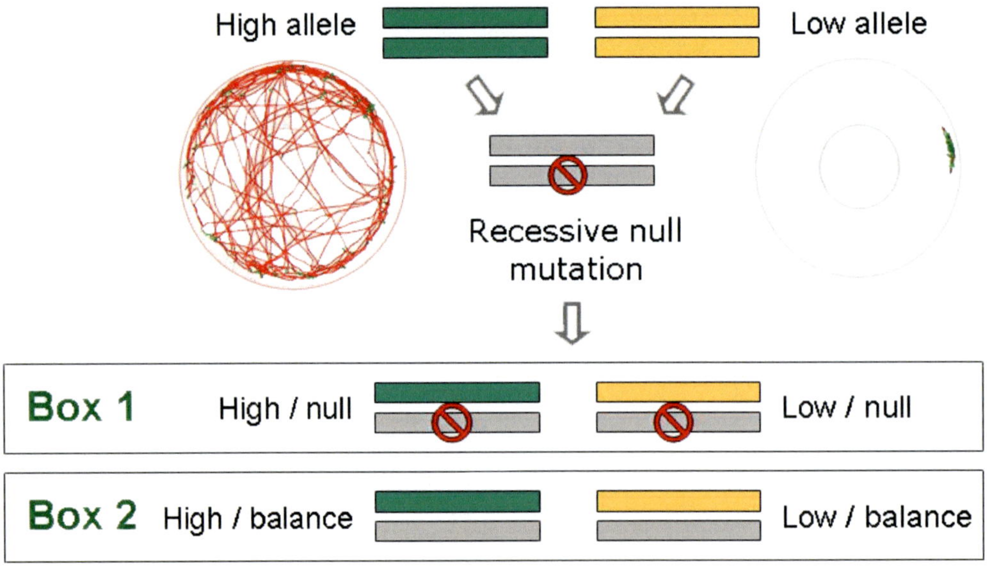

Figure 4. The quantitative complementation design.

Quantitative complementation looks for an interaction between a mutated positional candidate gene and a segregating QTL through analysis of variance. This design requires an animal that is heterozygous for a recessive mutation targeted to the gene of interest, and two inbred strains between which the QTL of interest is known to segregate (referred to as high and low). By crossing the heterozygous animal with these two inbred strains an F_1 generation can be created which consists of 4 genotypes at the candidate locus: high/null, low/null, high/balance, and low/balance (where balance represents the wild-type allele). The difference between high and low alleles can then be compared in the presence (box 1) or absence (box 2) of the targeted mutation. If these differences are unequal then the gene holding the mutation must contribute towards the QTL (either through allelism or epistasis); an observation otherwise known as a quantitative failure to complement.

A viable null mutant exists for the *RGS2* gene [76]. This mutant was generated by Oliveira-Dos-Santos and colleagues by replacing the 4th and 5th exons of *RGS2* (a region corresponding to the RGS domain) with PGK-Neo^{R} in antisense orientation. This process was performed in embryonic stem cells derived from the 129/P2OLA strain, with successfully recombined cells subsequently injected into C57BL/6 blastocysts and backcrossed to C57BL/6 for a further 15 generations. For ease of generating embryonic stem cells and efficiency of colonizing developing embryos, null mutations are often generated on a 129 background [85] and backcrossed onto the reproductively more successful, and genetically better characterized C57BL/6 prior to phenotyping. This approach however, adds an extra layer of complexity to the concept of complementation since as a consequence of linkage disequilibrium the mutated allele is likely to co-transmit with a surrounding region of unknown proportions derived from the 129 progenitor strain as opposed to the C57BL/6 backcross strain. Consequently, in the absence of a litter-mate control, genetic effects located within the co-transmitted region could potentially provide an alternative explanation for the mutation's failure to complement (provided that the QTL of interest segregates between the two strains).

Because of its low reproductive performance 129P2-OLA has not been routinely used in behavioural phenotyping experiments, and as such its relationship with the chromosome 1 QTL is currently unknown. However, the genomes of mouse inbred strains are closely related [103, 83] and can be thought of as an intricate collage of alternating sequence homology and difference. Consequently the problem of a mixed strain background can, in theory be surmounted by demonstrating that the sequence of the mutant strain, outside of the target mutation itself (*RGS2*), is identical to that of the control (C57BL/6J) strain; either in terms of actual base-pair homology, or comparability of genetic effect. In this way even though sequence differences might exist between these two strains, if the genetic effect remains the same then the two strains could be said to carry the same QTL. Sequence similarity can be explored by genotyping, focusing on those markers which distinguish between the two strains of interest (in this case C57BL/6J and 129P2-OLA). Once a region of dissimilarity has been broadly delimited, sequence analysis can provide an indication of genetic effect through comparison of the mutant strain with other inbred strains of known relationship to the QTL. Fortunately, due to the use of heterogeneous stocks in the mapping of the chromosome 1 locus, a wealth of information exists regarding inbred strain sequence homology across this region. Current mapping data indicates that the chromosome 1 QTL segregates between A/J and C3H on the one hand, and C57BL/6, AKR, DBA/2, I, RIII and BALB/c on the other. Equivalence of genetic effect could therefore be inferred so long as the *RGS2* mutant strain was not found to show any substantial divergence from C57BL/6, AKR, DBA/2, I, RIII or BALB/c across the QTL region. To this end Yalcin and colleagues conducted a detailed sequence analysis of the chromosome 1 QTL region in the *RGS2* null mutant (using dispersed 1.2Kb amplimers), in combination with a more wide-scale assessment of sequence homology between the mutant animal and C57BL/6 using microsatellite markers located randomly across the genome. Using this approach the authors found that that the *RGS2*-null strain differs intermittently from C57BL/6 between 100.3 and 171.9Mb on chromosome 1 (a region encapsulated by markers D1MIT386 - D1MIT113). Within this region the *RGS2*-null strain was found to alternate between C57BL/6-homologous and DBA/2-homologous sequence blocks. Specifically, with reference to the three fine-scale loci identified in the outbred mapping experiments, the sequence of the *RGS2*-null strain was found to be identical to C57BL/6 over both the first two loci, and varied between C57BL/6 and DBA/2 homology over the third locus. Since mapping studies have not previously identified any QTL in this region which segregate between DBA/2 and C57BL/6 [80, 67, 45, 93] the authors showed that the *RGS2* mutant could be considered functionally equivalent to C57BL/6. Thus in the absence of a littermate control, inbred C57BL/6 could be used as the wild-type balance allele in a quantitative complementation experiment.

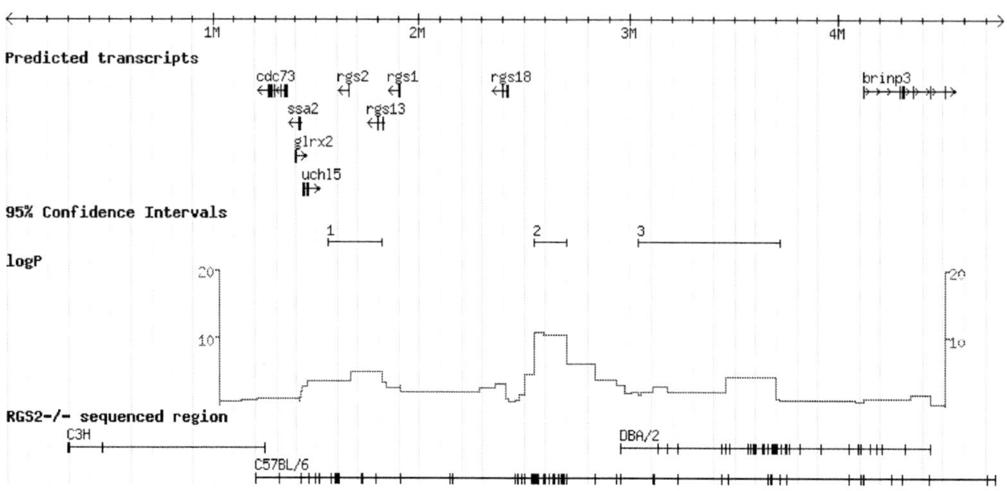

Figure 5. Murine quantitative trait loci on chromosome 1. This figure details the location of known landmarks across the 4.7Mb region of interest on murine chromosome 1 including genes and emotionality QTL estimates based on the MF1 data [111]. The positions of those segments sequenced in the RGS2-null mutant are provided at the base of the figure with an indication as to the sequence homology (DBA/2, C57BL/6 or C3H). Abbreviations: LogP (the base 10 logarithm of the p-values), RGS2-/- (RGS2 null).

The quantitative complementation experiment utilised C57BL/6 as the wild-type, or balance allele, and the *RGS2* null variant as the mutant allele. An interaction effect was then sought between the null variant and two inbred strain alleles between which the QTL was known to differ [67]; C3H (a high emotionality strain) and C57BL/6 (a low emotionality strain). Using this approach, and the combined open-field ambulation and defecation phenotype (EMO) a significant quantitative failure to complement was observed (f =8.2 [1], p=0.005). Importantly, ruling out indirect effects on physical activity, no such failure to complement was seen for ambulation in a home-cage (i.e. non-anxiogenic) environment. By way of negative control, a quantitative failure to complement was not also seen when the behavioural effect of a null allele was compared in two strains which are known to share same QTL (C57BL/6 and DBA). This data confirms that *RGS2* does contribute towards the EMO QTL in this region on chromosome 1, and that this effect is specific to emotionality as opposed to general activity phenotypes.

Relevance to Human Research

As mentioned previously, a region of mouse-human homology extends from 186.8 Mb to 191.8 Mb on human chromosome 1 based on the NCBI build 34 assembly of the human genome (July 2004). Although there is a general inversion of orientation across this region (telomeric in mouse as opposed to centromeric in human), both gene order and gene orientation are preserved. There are 9 known genes in the 4.78Mb region of QTL synteny - including the three regulators of G-protein signaling that are located proximal to the mouse EMO QTL; *RGS2*, *RGS13* and *RGS18*. This region has been implicated in a variety of human emotionality-related phenotypes through linkage analysis approaches, including both normal range variation (as assayed by the personality trait neuroticism [26, 71, 72]) and clinical manifestations such as alcoholism or depression [75]. Of particular interest, genetic variants located in two murine candidate genes; *RGS2* and *RGS13* have been successfully associated with panic disorder [60], with the greatest levels of association identified in subsets of patients with panic disorder and co-morbid agoraphobia [46]. These data clearly suggest that an evolutionarily conserved mechanism for the moderation of emotional reactivity may reside with this chromosome 1 region (see [110] for a full discussion of this argument). However these studies are small are therefore require further replication.

Conclusions

The co-variation between ambulation and defecation in the open-field test arena represents a robust and heritable measure of murine emotional reactivity. Attempts to map this trait have proven successful, with multiple potential contributory regions identified across the genome. By far the most compelling of these loci is a region located on mouse chromosome 1. Not only has this region been replicated in several independent populations and crosses, but the region also exhibits preserved synteny with a human emotionality locus mapped using the personality trait neuroticism.

Enhanced levels of mapping resolution that are available in the mouse have recently enabled the reduction of the murine chromosome 1 emotionality locus to just a handful of candidate genes; all of which are regulators of G-protein signaling (*RGS2*, *RGS13* and *RGS18*). The candidacy of one of these genes to the QTL has been confirmed by a process called quantitative complementation, allowing the authors to conclude that this gene, *RGS2*, is a quantitative trait gene for murine emotionality (at least as inferred from ambulation and defecation in the open-field) [111].

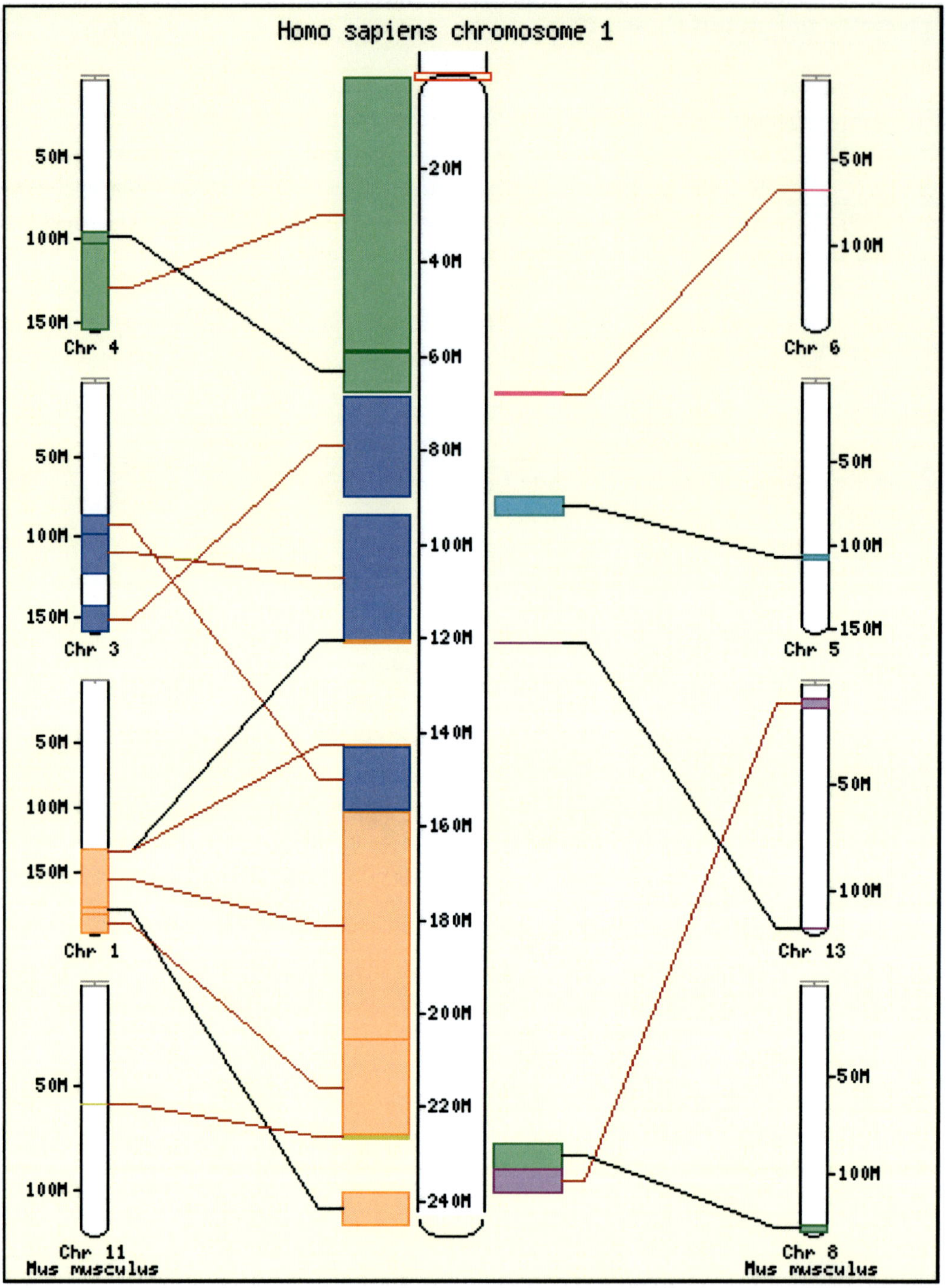

Figure 6. Mouse homology to Human chromosome 1. (Ensembl Human *SyntenyView,* Ensembl release 44, April 2007). http://www.ensembl.org/Homo_sapiens/syntenyview Abbreviations: M (Megabases), Chr (chromosome).

Because of the synteny which exists with the human genome across this chromosome 1 region, investigators have begun considering the potential contribution of genetic variants in *RGS2* (and *RGS13*) towards various indices of human emotionality. These research efforts have resulted in a growing body of evidence which suggests that variants in these regulators of G-protein signaling might contribute towards the genetic etiology of Panic Disorder, in particular Panic Disorder with co-morbid agoraphobia [60, 46, 51]. These data demonstrate that murine models of human behavior, in particular OFT ambulation and defecation, are likely to offer vital insights into the genetic origins of basic evolutionarily conserved behavioural mechanisms; not least emotional reactivity and its associated clinical phenotypes.

ACKNOWLEDGEMENTS

The author is supported by the Wellcome Trust, and would like to thank Dr. Jonathan Flint, MD, for his comments on an earlier version of this manuscript.

REFERENCES

[1] Abril JF, Agarwal P, Alexandersson M, Antonarakis SE, Baertsch R, Berry E, Birney E, Bork P, Bray N, Brent MR, et al: Initial sequencing and comparative analysis of the mouse genome. *Nature* 2002; 420:520-562.

[2] Arshavsky VY, Pugh EN, Jr.: Lifetime regulation of G protein-effector complex: emerging importance of RGS proteins. *Neuron* 1998; 20:11-14.

[3] Bannerman DM, Deacon RM, Offen S, Friswell J, Grubb M, Rawlins JN: Double dissociation of function within the hippocampus: spatial memory and hyponeophagia. *Behav Neurosci* 2002; 116:884-901.

[4] Bannerman DM, Deacon RM, Seeburg PH, Rawlins JN: GluR-A-Deficient mice display normal acquisition of a hippocampus-dependent spatial reference memory task but are impaired during spatial reversal. *Behav Neurosci* 2003; 117:866-870.

[5] Bannerman DM, Grubb M, Deacon RM, Yee BK, Feldon J, Rawlins JN: Ventral hippocampal lesions affect anxiety but not spatial learning. *Behav Brain Res* 2003; 139:197-213.

[6] Bannerman DM, Deacon RM, Brady S, Bruce A, Sprengel R, Seeburg PH, Rawlins JN: A comparison of GluR-A-deficient and wild-type mice on a test battery assessing sensorimotor, affective, and cognitive behaviors. *Behav Neurosci* 2004; 118:643-647.

[7] Beck JA, Lloyd S, Hafezparast M, Lennon-Pierce M, Eppig JT, Festing MF, Fisher EM: Genealogies of mouse inbred strains. *Nat Genet* 2000; 24:23-25.

[8] Berridge MJ, Lipp P, Bootman MD: The versatility and universality of calcium signalling. *Nat Rev Mol Cell Biol* 2000; 1:11-21.

[9] Bourin M, Hascoet M: The mouse light/dark box test. *Eur J Pharmacol* 2003; 463: 55-65.

[10] Buss KA, Davidson RJ, Kalin NH, Goldsmith HH: Context-specific freezing and associated physiological reactivity as a dysregulated fear response. *Dev Psychol* 2004; 40:583-594.

[11] Calo LA, Pagnin E, Davis PA, Sartori M, Ceolotto G, Pessina AC, Semplicini A: Increased expression of regulator of G protein signaling-2 (RGS-2) in Bartter's/Gitelman's syndrome. A role in the control of vascular tone and implication for hypertension. *J Clin Endocrinol Metab* 2004; 89:4153-4157.

[12] Carobrez AP, Bertoglio LJ: Ethological and temporal analyses of anxiety-like behavior: the elevated plus-maze model 20 years on. *Neurosci Biobehav Rev* 2005; 29:1193-1205.

[13] Chatterjee TK, Fisher RA: Cytoplasmic, nuclear, and golgi localization of RGS proteins. Evidence for N-terminal and RGS domain sequences as intracellular targeting motifs. *J Biol Chem* 2000; 275:24013-24021.

[14] Chen C, Lin SC: The core domain of RGS16 retains G-protein binding and GAP activity in vitro, but is not functional in vivo. *FEBS Lett* 1998; 422:359-362.

[15] Clement Y, Martin B, Venault P, Chapouthier G: Involvement of regions of the 4th and 7th chromosomes in the open-field activity of mice. *Behav Brain Res* 1995; 70:51-57.

[16] Crabbe JC, Wahlsten D, Dudek BC: Genetics of Mouse Behavior: Interactions with Laboratory Environment. *Science* 1999; 284:1670-1672.

[17] DeFries JC, Wilson JR, McClearn GE: Open-field behavior in mice: selection response and situational generality. *Behav Genet* 1970; 1:195-211.

[18] DeFries JC, Hegmann JP, Halcomb RA: Response to 20 generations of selection for open-field activity in mice. *Behav Biol* 1974; 11:481-495.

[19] DeFries JC, Gervais MC, Thomas EA: Response to 30 generations of selection for open-field activity in laboratory mice. *Behav Genet* 1978; 8:3-13.

[20] Dohlman HG, Bouvier M, Benovic JL, Caron MG, Lefkowitz RJ: The multiple membrane spanning topography of the beta 2-adrenergic receptor. Localization of the sites of binding, glycosylation, and regulatory phosphorylation by limited proteolysis. *J Biol Chem* 1987; 262:14282-14288.

[21] Dohlman HG, Song J, Ma D, Courchesne WE, Thorner J: Sst2, a negative regulator of pheromone signaling in the yeast Saccharomyces cerevisiae: expression, localization, and genetic interaction and physical association with Gpa1 (the G-protein alpha subunit). *Mol Cell Biol* 1996; 16:5194-5209.

[22] Druey KM, Blumer KJ, Kang VH, Kehrl JH: Inhibition of G-protein-mediated MAP kinase activation by a new mammalian gene family. *Nature* 1996; 379:742-746.

[23] Feng J, Yan Z, Ferreira A, Tomizawa K, Liauw JA, Zhuo M, Allen PB, Ouimet CC, Greengard P: Spinophilin regulates the formation and function of dendritic spines. *Proc Natl Acad Sci* U S A 2000; 97:9287-9292.

[24] Filippov AK, Fernandez-Fernandez JM, Marsh SJ, Simon J, Barnard EA, Brown DA: Activation and inhibition of neuronal G protein-gated inwardly rectifying K(+) channels by P2Y nucleotide receptors. *Mol Pharmacol* 2004; 66:468-477.

[25] Flint J, Corley R, DeFries JC, Fulker DW, Gray JA, Miller S, Collins AC: A simple genetic basis for a complex psychological trait in laboratory mice. *Science* 1995; 269:1432-1435.

[26] Fullerton J, Cubin M, Tiwari H, Wang C, Bomhra A, Davidson S, Miller S, Fairburn C, Goodwin G, Neale MC, et al: Linkage analysis of extremely discordant and concordant sibling pairs identifies quantitative-trait loci that influence variation in the human personality trait neuroticism. *Am J Hum Genet* 2003; 72:879-890.

[27] Gagnon AW, Murray DL, Leadley RJ: Cloning and characterization of a novel regulator of G protein signalling in human platelets. *Cell Signal* 2002; 14:595-606.

[28] Geurts M, Hermans E, Maloteaux JM: Opposite modulation of regulators of G protein signalling-2 RGS2 and RGS4 expression by dopamine receptors in the rat striatum. *Neurosci Lett* 2002; 333:146-150.

[29] Ghavami A, Hunt RA, Olsen MA, Zhang J, Smith DL, Kalgaonkar S, Rahman Z, Young KH: Differential effects of regulator of G protein signaling (RGS) proteins on serotonin 5-HT1A, 5-HT2A, and dopamine D2 receptor-mediated signaling and adenylyl cyclase activity. *Cell Signal* 2004; 16:711-721.

[30] Gilman AG: G proteins: transducers of receptor-generated signals. *Annu Rev Biochem* 1987; 56:615-649.

[31] Gold SJ, Heifets BD, Pudiak CM, Potts BW, Nestler EJ: Regulation of regulators of G protein signaling mRNA expression in rat brain by acute and chronic electroconvulsive seizures. *J Neurochem* 2002; 82:828-838.

[32] Grafstein-Dunn E, Young KH, Cockett MI, Khawaja XZ: Regional distribution of regulators of G-protein signaling (RGS) 1, 2, 13, 14, 16, and GAIP messenger ribonucleic acids by in situ hybridization in rat brain. *Brain Res Mol Brain Res* 2001; 88:113-123.

[33] Grant SL, Lassegue B, Griendling KK, Ushio-Fukai M, Lyons PR, Alexander RW: Specific regulation of RGS2 messenger RNA by angiotensin II in cultured vascular smooth muscle cells. *Mol Pharmacol* 2000; 57:460-467.

[34] Gray JA, McNaughton N: *The neuropsychology of anxiety : an enquiry into the functions of the septo-hippocampal system.* 2nd edn. Oxford: Oxford University Press; 2000.

[35] Greengard P: The neurobiology of slow synaptic transmission. *Science* 2001; 294:1024-1030.

[36] Hague C, Bernstein LS, Ramineni S, Chen Z, Minneman KP, Hepler JR: Selective Inhibition of {alpha}1A-Adrenergic Receptor Signaling by RGS2 Association with the Receptor Third Intracellular Loop. *J Biol Chem* 2005; 280:27289-27295.

[37] Hall CS: Emotional behavior in the rat, I.,Defecation and urination as measures of individual differences in emotionality. *Journal of comparative psychology* 1934; 18:385-403.

[38] Han J, Mark MD, Li X, Xie M, Waka S, Rettig J, Herlitze S: RGS2 determines short-term synaptic plasticity in hippocampal neurons by regulating Gi/o-mediated inhibition of presynaptic Ca2+ channels. *Neuron* 2006; 51:575-586.

[39] Harbison ST, Yamamoto AH, Fanara JJ, Norga KK, Mackay TF: Quantitative trait loci affecting starvation resistance in Drosophila melanogaster. *Genetics* 2004; 166:1807-1823.

[40] Henderson ND, Turri MG, DeFries JC, Flint J: QTL analysis of multiple behavioral measures of anxiety in mice. *Behav Genet* 2004; 34:267-293.

[41] Heximer SP, Srinivasa SP, Bernstein LS, Bernard JL, Linder ME, Hepler JR, Blumer KJ: G protein selectivity is a determinant of RGS2 function. *J Biol Chem* 1999; 274:34253-34259.

[42] Heximer SP, Lim H, Bernard JL, Blumer KJ: Mechanisms governing subcellular localization and function of human RGS2. *J Biol Chem* 2001; 276:14195-14203.

[43] Heximer SP, Knutsen RH, Sun X, Kaltenbronn KM, Rhee MH, Peng N, Oliveira-dos-Santos A, Penninger JM, Muslin AJ, Steinberg TH, et al: Hypertension and prolonged vasoconstrictor signaling in RGS2-deficient mice. *J Clin Invest* 2003; 111:445-452.

[44] Hirose K, Kadowaki S, Tanabe M, Takeshima H, Iino M: Spatiotemporal dynamics of inositol 1,4,5-trisphosphate that underlies complex Ca2+ mobilization patterns. *Science* 1999; 284:1527-1530.

[45] Hitzemann R, Malmanger B, Cooper S, Coulombe S, Reed C, Demarest K, Koyner J, Cipp L, Flint J, Talbot C, et al: Multiple cross mapping (MCM) markedly improves the localization of a QTL for ethanol-induced activation. *Genes Brain Behav* 2002; 1:214-222.

[46] Hohoff C, Freitag C, Leygraf A, Willis-Owen SA, Fuellen G, Bennour P, Keck ME, Erhardt A, Binder EB, Krakowitzky P, et al: *Relevance of the Mouse Anxiety QTL Genes Rgs2 and Rgs13 for Human Panic Disorder.* AJHG In Submission.

[47] Ingi T, Krumins AM, Chidiac P, Brothers GM, Chung S, Snow BE, Barnes CA, Lanahan AA, Siderovski DP, Ross EM, et al: Dynamic regulation of RGS2 suggests a novel mechanism in G-protein signaling and neuronal plasticity. *J Neurosci* 1998; 18:7178-7188.

[48] Ingi T, Aoki Y: Expression of RGS2, RGS4 and RGS7 in the developing postnatal brain. *Eur J Neurosci* 2002; 15:929-936.

[49] Ioannidis JP, Trikalinos TA, Khoury MJ: Implications of small effect sizes of individual genetic variants on the design and interpretation of genetic association studies of complex diseases. *Am J Epidemiol* 2006; 164:609-614.

[50] Johnson EN, Druey KM: Functional characterization of the G protein regulator RGS13. *J Biol Chem* 2002; 277:16768-16774.

[51] Jung A, Hohoff C, Freitag C, Willis-Owen SAG, Krakowitzky P, Fritze J, Franke P, Bandelow B, Fimmers R, Nöthen MM, et al: Are Rgs2 gene polymorphisms modulators of anxiety in humans? *Molecular Psychiatry* In submission.

[52] Kammermeier PJ, Ikeda SR: Expression of RGS2 alters the coupling of metabotropic glutamate receptor 1a to M-type K+ and N-type Ca2+ channels. *Neuron* 1999; 22:819-829.

[53] Kobori N, Clifton GL, Dash P: Altered expression of novel genes in the cerebral cortex following experimental brain injury. *Brain Res Mol Brain Res* 2002; 104:148-158.

[54] Koch M: The neurobiology of startle. *Prog Neurobiol* 1999; 59:107-128.

[55] Koelle MR, Horvitz HR: EGL-10 regulates G protein signaling in the C. elegans nervous system and shares a conserved domain with many mammalian proteins. *Cell* 1996; 84:115-125.

[56] Larminie C, Murdock P, Walhin JP, Duckworth M, Blumer KJ, Scheideler MA, Garnier M: Selective expression of regulators of G-protein signaling (RGS) in the human central nervous system. *Brain Res Mol Brain Res* 2004; 122:24-34.

[57] Law AJ, Weickert CS, Hyde TM, Kleinman JE, Harrison PJ: Reduced spinophilin but not microtubule-associated protein 2 expression in the hippocampal formation in schizophrenia and mood disorders: molecular evidence for a pathology of dendritic spines. *Am J Psychiatry* 2004; 161:1848-1855.

[58] Leppanen PK, Ewalds-Kvist SB, Selander RK: Mice selectively bred for open-field thigmotaxis: life span and stability of the selection trait. *J Gen Psychol* 2005; 132:187-204.

[59] Leppanen PK, Ravaja N, Ewalds-Kvist SB: Twenty-three generations of mice bidirectionally selected for open-field thigmotaxis: selection response and repeated exposure to the open field. *Behav Processes* 2006; 72:23-31.

[60] Leygraf A, Hohoff C, Freitag C, Willis-Owen SA, Krakowitzky P, Fritze J, Franke P, Bandelow B, Fimmers R, Flint J, Deckert J: Rgs 2 gene polymorphisms as modulators of anxiety in humans? *J Neural Transm* 2006; 113:1921-1925.

[61] Li Y, Hashim S, Anand-Srivastava MB: Angiotensin II-evoked enhanced expression of RGS2 attenuates Gi-mediated adenylyl cyclase signaling in A10 cells. *Cardiovasc Res* 2005; 66:503-511.

[62] Logothetis DE, Kurachi Y, Galper J, Neer EJ, Clapham DE: The beta gamma subunits of GTP-binding proteins activate the muscarinic K+ channel in heart. *Nature* 1987; 325:321-326.

[63] Lyman RF, Mackay TF: Candidate quantitative trait loci and naturally occurring phenotypic variation for bristle number in Drosophila melanogaster: the Delta-Hairless gene region. *Genetics* 1998; 149:983-998.

[64] Mackay TF: The genetic architecture of quantitative traits. Annu Rev Genet 2001; 35:303-339.

[65] McHugh SB, Deacon RM, Rawlins JN, Bannerman DM: Amygdala and ventral hippocampus contribute differentially to mechanisms of fear and anxiety. *Behav Neurosci* 2004; 118:63-78.

[66] Meyer L, Caston J, Mensah-Nyagan AG: Seasonal variation of the impact of a stressful procedure on open field behaviour and blood corticosterone in laboratory mice. *Behav Brain Res* 2006; 167:342-348.

[67] Mott R, Talbot CJ, Turri MG, Collins AC, Flint J: A method for fine mapping quantitative trait loci in outbred animal stocks. *Proc Natl Acad Sci U S A* 2000; 97:12649-12654.

[68] Mukhopadhyay S, Ross EM: Rapid GTP binding and hydrolysis by G(q) promoted by receptor and GTPase-activating proteins. *Proc Natl Acad Sci U S A* 1999; 96:9539-9544.

[69] Nagata Y, Oda M, Nakata H, Shozaki Y, Kozasa T, Todokoro K: A novel regulator of G-protein signaling bearing GAP activity for Galphai and Galphaq in megakaryocytes. *Blood* 2001; 97:3051-3060.

[70] Nash MS, Young KW, Challiss RA, Nahorski SR: Intracellular signalling. Receptor-specific messenger oscillations. *Nature* 2001; 413:381-382.

[71] Nash MW, Huezo-Diaz P, Williamson RJ, Sterne A, Purcell S, Hoda F, Cherny SS, Abecasis GR, Prince M, Gray JA, et al: Genome-wide linkage analysis of a composite index of neuroticism and mood-related scales in extreme selected sibships. *Hum Mol Genet* 2004.

[72] Neale BM, Sullivan PF, Kendler KS: A genome scan of neuroticism in nicotine dependent smokers. *Am J Med Genet B Neuropsychiatr Genet* 2005; 132:65-69.

[73] Nishiguchi KM, Sandberg MA, Kooijman AC, Martemyanov KA, Pott JW, Hagstrom SA, Arshavsky VY, Berson EL, Dryja TP: Defects in RGS9 or its anchor protein R9AP in patients with slow photoreceptor deactivation. *Nature* 2004; 427:75-78.

[74] Norlin EM, Berghard A: Spatially restricted expression of regulators of G-protein signaling in primary olfactory neurons. *Mol Cell Neurosci* 2001; 17:872-882.

[75] Nurnberger JI, Jr., Foroud T, Flury L, Su J, Meyer ET, Hu K, Crowe R, Edenberg H, Goate A, Bierut L, et al: Evidence for a locus on chromosome 1 that influences vulnerability to alcoholism and affective disorder. *Am J Psychiatry* 2001; 158:718-724.

[76] Oliveira-Dos-Santos AJ, Matsumoto G, Snow BE, Bai D, Houston FP, Whishaw IQ, Mariathasan S, Sasaki T, Wakeham A, Ohashi PS, et al: Regulation of T cell activation, anxiety, and male aggression by RGS2. *Proc Natl Acad Sci U S A* 2000; 97:12272-12277.

[77] Ossenkopp K-P, Sorenson L, Mazmanian DS: Factor analysis of open-field behavior in the rat (Rattus norvegicus): application of the three-way PARAFAC model to a longitudinal data set. *Behavioural Processes* 1994; 31:129-144.

[78] Park IK, Klug CA, Li K, Jerabek L, Li L, Nanamori M, Neubig RR, Hood L, Weissman IL, Clarke MF: Molecular cloning and characterization of a novel regulator of G-protein signaling from mouse hematopoietic stem cells. *J Biol Chem* 2001; 276:915-923.

[79] Peeters PJ, Gohlmann HW, Van Den Wyngaert I, Swagemakers SM, Bijnens L, Kass SU, Steckler T: Transcriptional Response to Corticotropin-releasing Factor in AtT-20 Cells. *Mol Pharmacol* 2004.

[80] Plomin R, McClearn GE, Gora-Maslak G, Neiderhiser JM: Use of recombinant inbred strains to detect quantitative trait loci associated with behavior. *Behav Genet* 1991; 21:99-116.

[81] Ross EM, Wilkie TM: GTPase-activating proteins for heterotrimeric G proteins: regulators of G protein signaling (RGS) and RGS-like proteins. *Annu Rev Biochem* 2000; 69:795-827.

[82] Roy AA, Lemberg KE, Chidiac P: Recruitment of RGS2 and RGS4 to the plasma membrane by G proteins and receptors reflects functional interactions. *Mol Pharmacol* 2003; 64:587-593.

[83] Sakai T, Kikkawa Y, Miura I, Inoue T, Moriwaki K, Shiroishi T, Satta Y, Takahata N, Yonekawa H: Origins of mouse inbred strains deduced from whole-genome scanning by polymorphic microsatellite loci. *Mamm Genome* 2005; 16:11-19.

[84] Schertler GF, Villa C, Henderson R: Projection structure of rhodopsin. *Nature* 1993; 362:770-772.

[85] Seong E, Saunders TL, Stewart CL, Burmeister M: To knockout in 129 or in C57BL/6: that is the question. *Trends Genet* 2004; 20:59-62.

[86] Shi GX, Harrison K, Wilson GL, Moratz C, Kehrl JH: RGS13 regulates germinal center B lymphocytes responsiveness to CXC chemokine ligand (CXCL)12 and CXCL13. *J Immunol* 2002; 169:2507-2515.

[87] Stafstrom-Davis CA, Ouimet CC, Feng J, Allen PB, Greengard P, Houpt TA: Impaired conditioned taste aversion learning in spinophilin knockout mice. *Learn Mem* 2001; 8:272-278.

[88] Stefanski R, Palejko W, Kostowski W, Plaznik A: The comparison of benzodiazepine derivatives and serotonergic agonists and antagonists in two animal models of anxiety. *Neuropharmacology* 1992; 31:1251-1258.

[89] Steinmetz LM, Sinha H, Richards DR, Spiegelman JI, Oefner PJ, McCusker JH, Davis RW: Dissecting the architecture of a quantitative trait locus in yeast. *Nature* 2002; 416:326-330.

[90] Stouffer SA: *The American soldier*. Princeton: Princeton University Press; 1949.

[91] Taft RA, Davisson M, Wiles MV: Know thy mouse. *Trends Genet* 2006; 22:649-653.

[92] Talbot CJ, Nicod A, Cherny SS, Fulker DW, Collins AC, Flint J: High-resolution mapping of quantitative trait loci in outbred mice. *Nat Genet* 1999; 21:305-308.

[93] Talbot CJ, Radcliffe RA, Fullerton J, Hitzemann R, Wehner JM, Flint J: Fine scale mapping of a genetic locus for conditioned fear. Mamm Genome 2003; 14:223-230.

[94] Taymans JM, Leysen JE, Langlois X: Striatal gene expression of RGS2 and RGS4 is specifically mediated by dopamine D1 and D2 receptors: clues for RGS2 and RGS4 functions. *J Neurochem* 2003; 84:1118-1127.

[95] Taymans JM, Kia HK, Claes R, Cruz C, Leysen J, Langlois X: Dopamine receptor-mediated regulation of RGS2 and RGS4 mRNA differentially depends on ascending dopamine projections and time. *Eur J Neurosci* 2004; 19:2249-2260.

[96] Tesmer JJ, Berman DM, Gilman AG, Sprang SR: Structure of RGS4 bound to AlF4--activated G(i alpha1): stabilization of the transition state for GTP hydrolysis. *Cell* 1997; 89:251-261.

[97] Thirunavukkarasu K, Halladay DL, Miles RR, Geringer CD, Onyia JE: Analysis of regulator of G-protein signaling-2 (RGS-2) expression and function in osteoblastic cells. *J Cell Biochem* 2002; 85:837-850.

[98] Trullas R, Skolnick P: Differences in fear motivated behaviors among inbred mouse strains. *Psychopharmacology* (Berl) 1993; 111:323-331.

[99] Tucci V, Lad HV, Parker A, Polley S, Brown SD, Nolan PM: Gene-environment interactions differentially affect mouse strain behavioral parameters. *Mamm Genome* 2006; 17:1113-1120.

[100] Turri MG, Datta SR, DeFries J, Henderson ND, Flint J: QTL analysis identifies multiple behavioral dimensions in ethological tests of anxiety in laboratory mice. *Curr Biol* 2001; 11:725-734.

[101] Turri MG, Henderson ND, DeFries JC, Flint J: Quantitative trait locus mapping in laboratory mice derived from a replicated selection experiment for open-field activity. *Genetics* 2001; 158:1217-1226.

[102] Valdar W, Solberg LC, Gauguier D, Cookson WO, Rawlins JN, Mott R, Flint J: Genetic and environmental effects on complex traits in mice. *Genetics* 2006; 174:959-984.

[103] Wade CM, Kulbokas EJ, 3rd, Kirby AW, Zody MC, Mullikin JC, Lander ES, Lindblad-Toh K, Daly MJ: The mosaic structure of variation in the laboratory mouse genome. *Nature* 2002; 420:574-578.

[104] Wahlsten D, Metten P, Phillips TJ, Boehm SL, 2nd, Burkhart-Kasch S, Dorow J, Doerksen S, Downing C, Fogarty J, Rodd-Henricks K, et al: Different data from different labs: lessons from studies of gene-environment interaction. *J Neurobiol* 2003; 54:283-311.

[105] Wang X, Huang G, Luo X, Penninger JM, Muallem S: Role of RGS2 in Ca2+ oscillations and adaptation of Ca2+ signaling to reduce excitability of RGS2-/- cells. *J Biol Chem* 2004.

[106] Wang X, Zeng W, Soyombo AA, Tang W, Ross EM, Barnes AP, Milgram SL, Penninger JM, Allen PB, Greengard P, Muallem S: Spinophilin regulates Ca2+ signalling by binding the N-terminal domain of RGS2 and the third intracellular loop of G-protein-coupled receptors. *Nat Cell Biol* 2005; 7:405-411.

[107] Waterston RH, Lindblad-Toh K, Birney E, Rogers J, Abril JF, Agarwal P, Agarwala R, Ainscough R, Alexandersson M, An P, et al: Initial sequencing and comparative analysis of the mouse genome. *Nature* 2002; 420:520-562.

[108] Wehner JM, Radcliffe RA, Rosmann ST, Christensen SC, Rasmussen DL, Fulker DW, Wiles M: Quantitative trait locus analysis of contextual fear conditioning in mice. *Nat Genet* 1997; 17:331-334.

[109] Willis-Owen SA, Flint J: The genetic basis of emotional behaviour in mice. *Eur J Hum Genet* 2006; 14:721-728.

[110] Willis-Owen SA, Flint J: Identifying the genetic determinants of emotionality in humans; insights from rodents. *Neurosci Biobehav Rev* 2007; 31:115-124.

[111] Yalcin B, Willis-Owen SA, Fullerton J, Meesaq A, Deacon RM, Rawlins JN, Copley RR, Morris AP, Flint J, Mott R: Genetic dissection of a behavioral quantitative trait locus shows that Rgs2 modulates anxiety in mice. *Nat Genet* 2004; 36:1197-1202.

[112] Yang J, Kamide K, Kokubo Y, Takiuchi S, Tanaka C, Banno M, Miwa Y, Yoshii M, Horio T, Okayama A, et al: Genetic variations of regulator of G-protein signaling 2 in hypertensive patients and in the general population. *J Hypertens* 2005; 23:1497-1505.

[113] Yoshikawa T, Watanabe A, Ishitsuka Y, Nakaya A, Nakatani N: Identification of multiple genetic loci linked to the propensity for "behavioral despair" in mice. *Genome Res* 2002; 12:357-366.

[114] Yowe D, Weich N, Prabhudas M, Poisson L, Errada P, Kapeller R, Yu K, Faron L, Shen M, Cleary J, et al: RGS18 is a myeloerythroid lineage-specific regulator of G-protein-signalling molecule highly expressed in megakaryocytes. *Biochem J* 2001; 359:109-118.

[115] Zerangue N, Jan LY: G-protein signaling: fine-tuning signaling kinetics. *Curr Biol* 1998; 8:R313-316.

[116] Zheng B, De Vries L, Gist Farquhar M: Divergence of RGS proteins: evidence for the existence of six mammalian RGS subfamilies. *Trends Biochem Sci* 1999; 24:411-414.

[117] Zhong H, Wade SM, Woolf PJ, Linderman JJ, Traynor JR, Neubig RR: A spatial focusing model for G protein signals. Regulator of G protein signaling (RGS) protien-mediated kinetic scaffolding. *J Biol Chem* 2003; 278:7278-7284.

In: Behavioral Models in Stress Research
Editors: A.V. Kalueff and J.L. LaPorte, pp. 97-109
ISBN: 978-1-60456-361-0

Chapter 5

Mouse Lines Selected for Difference in Sensitivity to ß-CCM also Differ in Spatial Memory, Corticosterone Activation and Fear Reactivity

Daniel Beracochea[1*], Patrice Venault[2], Christophe Pierard[3] and Georges Chapouthier[2]

[1]CNIC, Laboratoire de Neurosciences Intégratives et Cognitives,
Universités de Bordeaux 1et de Bordeaux 2, UMR CNRS,
Batiment Biologie Animale, Avenue des Facultes, Talence Cedex, France.
[2]CNRS UMR 7593 "Vulnérabilité, Adaptation et Psychopathologie",
CHU Pitié-Salpêtrière, Paris, France.
[3] IMASSA, Physiologie Intégrée, BP73, Bretigny sur Orge-cédex, France.

Introduction

Methyl ß-carboline-3-carboxylate (or ß-CCM) is a benzodiazepine receptor ligand with inverse agonist properties. Unlike benzodiazepines which are anticonvulsant [31, 44], anxiolytic [3, 45] and amnesic [30], ß-CCM in mice is convulsant at high doses (10 mg/kg i.p.) [42; 37], anxiogenic at moderate (1-2 mg/kg i.p.) doses [29;43;48], and promnestic at low doses (0.2-0.5 mg/kg i.p.) [31;56].

Experiments were conducted on two strains of mice, one selected for sensitivity and one for resistance to the convulsive action of ß-CCM (4mg/kg i.p.): the ß-CCM-sensitive strain (BS) and the ß-CCM-resistant strain (BR) [8]. Breeding started with a heterogeneous but genetically controlled pool first obtained by crossing four ß-CCM-sensitive strains (BALB/cBy, CBA/H, C3H/HeJ, DBA/2J) and four relatively resistant strains (C57BL/6J, C57BL/10J, XLII, NZB/B1NJ). Two lines were selected based on the shortest convulsion

* E-mail address: d.beracochea@cnic.u-bordeaux1.fr. Phone: 33-0540002439. CNIC, Laboratoire de Neurosciences Intégratives et Cognitives, UMR CNRS 5228 ; Universités de Bordeaux 1 et de Bordeaux 2 , Bâtiment Biologie Animale, Avenue des Facultés, 33405 Talence Cedex, France

latencies (the β-CCM sensitive group, BS) and failure to convulse within a 360-s time limit (the β-CCM resistant group, BR). Mating was designed so as to minimise inbreeding.

Differences between the two strains were observed for several biochemical and pharmacological characteristics: BR mice were less aggressive [25] and also displayed greater resistance to the convulsive effect of other convulsant compounds affecting the gamma-aminobutyric acid-benzodiazepine (GABA-BDZ) receptor complex [45], as well as to the anxiolytic and sedative effects of benzodiazepines [45]. Comparisons of BR and BS mice in standard tests used to test antidepressant compounds showed the profile of BR mice to be less "depressed-like" [18]. Differences in behavior associated with anxiety showed a dose-dependent link between the effects of ß-CCM, i.e. for convulsive, anxiogenic and learning-enhancing actions. Suaudeau et al. [52] reported on standard tests used to assess anxiety in rodents (elevated plus-maze, light-dark discrimination test, staircase and holeboard), and observed that BR mice were more "anxious-like" than BS mice. BR mice spent more time in the closed arms of the elevated plus-maze, less time in the lit box in the light-dark discrimination task, and registered more rearings in the staircase test. In the holeboard test, BR mice explored less than BS mice, which may be interpreted as a strategy designed to avoid potential situations of anxiety.

The more anxious BR strain was also found to be less aggressive with a profile that could be described as less "depressed-like", which is in contrast with standard behavior patterns; while most mouse strains [25;26] display aggressive behavior in connection with anxiety rather than the opposite. Although Suaudeau observed clear differences in anxiety-related behavior, Rinaldi et al. [45] did not replicate these findings. This highlights the need for further investigation of these behavioral phenomena, studying the two strains.

One hypothesis based on biochemical studies in these mice is that resistance to seizures may be related to binding, where low levels of B_{max} in the BR strain could be an adaptation response, modifying the number of benzodiazepine receptors after ß-CCM is administered, thereby making the mice ß-CCM-resistant [9]. However, binding cannot explain other behavioral phenomena observed in these strains [11]; e.g. no clear evidence can be found of any relationship between the more "anxious" behavior displayed by BR mice and the lower levels of B_{max}. Further studies targeting other biological markers are therefore needed.

The present study explored the behavioral–induced activation of the hypothalamus-pituitary-adrenal (HPA) axis in the two strains. Indeed, the level of plasma corticosterone has been already demonstrated to be critically involved in retrieval memory processes [8;15;16;41] and to be a physiological index of the emotional state of the subjects. No direct hypothesis was formulated for any possible trend expected in the results (i.e. whether BR or BS would perform better or would differ in corticosterone activity).

Material and Methods

Animals

A total of 80 mice were used for the study. The animals were housed and reared under standard conditions: 22°+1°C, a 12:12 h photoperiod with lights on at 8:00 am, tap water and Souriffarat® (IM UAR) food available ad libitum, and dust-free soft wood sawdust bedding. Litters were culled to 7 subjects at birth. From birth to weaning, the animals were kept with their mothers only. The sires were removed from the mating cages one or two days before parturition. Male and female offspring were separated when weaned at 30 ± 2 days. The animals were 3 to 5 months old (30-35 g) at the time of the study. Only males were used. All tests were performed on animals from generations G17 and G18. All the experiments were carried out between 8.30 am and 12.30 pm and respected the ethical guidelines laid down by the French Ministry of Agriculture.

Before behavioral testing, animals were housed individually with free access to food and water for 15 days. Then, they were submitted to the food deprivation schedule as described below.

Spatial Delayed Discrimination in a 4-Hole Board Apparatus

All tests were performed in a four-hole board apparatus (45 cm x 45 cm x 30 cm high) enclosed by grey Plexiglass [21]. The four-hole board apparatus was placed on the floor of the room (3.0m X 3.0m X 2.40 m high). On the floor, 4 holes opening on a food cup (3 cm diameter X 2.5 cm in depth) were located 6 cm away from the sidewalls. The apparatus was placed in a room exposed to a 40 dB background noise and a light centered over the apparatus provided a 20 lux intensity at the position of the apparatus. Spatial cues were provided with 2 reverse patterns (black vertical stripes on a white background and white horizontal stripes on a black background) symmetrically opposite one another on the surrounding walls. The apparatus was cleaned with 95% ethanol, then with water before each mouse behavioral testing. Photocells placed in each hole measured the duration and number of head dips and recorded the data on a computer. Each session had two phases: the acquisition phase and the test phase. Throughout the study, all subjects were maintained at 82-84% of their ad libitum body weight. During the partial food deprivation period (over 3 consecutive days) each subject was handled by the experimenter for 10 min. each day. The following day, the acquisition phase began: the subject was placed in a black cylinder (10 cm diameter x 20 cm high) in the center of the hole board and left there for 30 sec. This procedure meant that the subject's initial orientation was at random. The cylinder was then removed, and the mouse was free to explore the hole board for 6 min. During the acquisition phase, ten 20 mg food pellets were available, but only from one hole. The rewarded hole changed from one subject to another. The subject was removed either at the end of the 6 min period, or immediately after the last food pellets had been taken if done within the 6 min period. At the end of the acquisition phase, the animals were returned to their home cage during the retention interval.

The test phase was carried out either 5 minutes or 24 hours later, using independent groups of BS and BR mice (N = 8 for each retrieval delay and groups). After initial observations, an additional group of BS mice was used to assess retrieval after an even longer delay (48 hrs; N = 8). Retention was assessed by measuring exploration frequency (number of head dips per hole) and the exploration time for each hole; this phase lasted 3 min and no food pellets were in the apparatus.

Corticosterone Measurement

This method has been already described in full [40]. Blood sampling was performed between 08:00 and 12:30 a.m. Experimental subjects were beheaded 1 min after the end of the 5 min or 24 hours memory testing sessions (memory condition) and trunk blood was collected for corticosterone analysis. They were compared to subjects maintained in the colony room ("Quiet" condition; N=6 per groups) and to subjects having explored the hole-board for 3 minutes, without any food in the apparatus ("active" condition; N=6 per groups) before sacrifice. Plasma corticosterone was quantified on plasma samples of 50 µl, using an original high-performance liquid chromatography (HPLC) method with fluorometric detection (λ_{ex} = 375 nm ; λ_{em} = 485 nm), preceded by 2 liquid-liquid extractions with ethyl acetate. This method was validated according to AFNOR guidelines XP T 90-210.

Elevated Plus Maze

The plus maze, which was constructed of grey Plexiglass, consisted of four arms arranged in the shape of a plus sign. Each arm was 30 cm long, 7cm wide, and was elevated 40 cm above the ground. The four arms were joined at the center by a 7-cm square platform. Two opposite arms of the plus maze were enclosed by sidewalls 17 cm high, but open on the top. The remaining arms did not have sidewalls. These walls did not extend from the center of the maze.

The experiment was performed between 08:30 and 12:00 a.m. The elevated plus maze test was performed using independent groups of mice (N=8 per groups). At the beginning of the session, mice were placed at the center of the plus-maze in a cylinder (8 cm diameter, 17 cm high) for 30 s. Then, the cylinder was removed and mice were allowed to freely explore all arms of the maze for 6 min. An entry was counted only when a mouse entered an arm with all four paws. Two measures of "anxiety-like" behavior were taken. The first was the ratio of the time spent in the open arms divided by the total time spent in all arms of the maze (Latency ratio). The second was the ratio of entries into the open arms divided by the total number of entries in all arms (entry ratio). Results were expressed in percentages (ratio x 100).

Statistics

Statistical analyses were performed using the Statview ® 5.0.1 software. In the 4-hole board experiment, the correct response ratio (number of head dips in the rewarded hole/total number of head dips x 100) was used for statistical analysis. The homogeneity of variance has been assessed by the Bonferroni-Dunn test. All statistical analyses were factorial ANOVA followed by post-hoc paired comparisons using the Scheffe F-test.

RESULTS

3.1. Spatial Delayed Discrimination in the 4-Hole Board

Acquisition phase. Analysis of pooled data showed that BR mice spent more time on the rewarded hole, compared to BS mice ($F(1,30) = 9.5$; $P < 0.02$). No other significant behavioral differences in exploratory patterns were observed between the two strains. More specifically, the total number of head-dips in the four holes was similar (31±5.6 versus 35±4.9 for BR and BS mice respectively as well as the number of head-dips in the rewarded hole (14 ±3.5 versus 12 ±3.5 for BR and BS mice respectively; $F < 1.0$ in all comparisons).

Test phase. The results are presented in Figure 1. A non-significant statistical difference was found between the two strains of mice ($F(1,28) = 0.24$) with correct choice ratios being affected by the retention intervals ($F(1,28) = 15.5$; $p = 0.0005$). The performance of BR mice declined faster, however, as a function of the length of the retention intervals, as compared to BS mice (strains x delays: ($F(1,28) = 8.9$; $p = 0.005$). While no between-strain difference was observed after the 5 min retention interval ($F(1,14) = 3.2$; $p=0.09$), significant differences were observed at the 24 hrs retention interval: $F(1,14) = 5.8$; $p<0.029$). An additional analysis showed that the deficit of BR mice as compared to BS mice was observed at the first ($p<0.01$), the second ($p<0.05$) and the third minute ($p<0.01$). Within-strain analyses showed that the slump in the performance of the BR strain was significant over the two time intervals ($F(1,14)=43.0$; $p=0.001$); a non-significant decline in the performance of the BS strain was observed for the same time intervals ($F(1,14)=0.32$). For the additional third time interval (48 hrs), BS mice showed a significant drop in performance after 48 hrs as compared to the 5 min interval ($F(1,14)=4.2$; $p=0.01$). Performance by BS mice after 48 hrs was however not statistically different as compared to their performance at 24 hrs ($F < 1.0$).

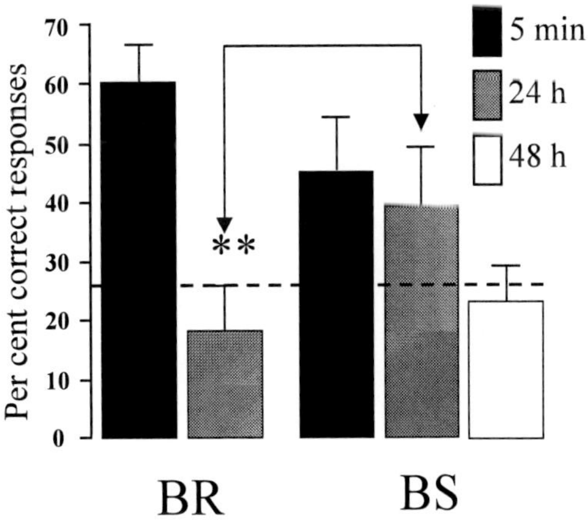

Figure 1. Retention performance (mean ±S.E.) in a delayed spatial discrimination task in the 4-hole board. Percentage of correct responses per group of subjects and for the two strains BR and BS (n = 8 per group), after different time intervals: 5 min, 24 or 48 hrs. With four possible holes, the chance level is 25%. ** p < 0.03 as compared to the respective BS group.

Corticosterone

Results are represented in Figure 2. An overall analysis showed that BR mice exhibited higher levels of plasma corticosterone as compared to BS mice (strain difference; ($F(1,50)=9.6$; $p=0.003$). Corticosterone levels also depended on the conditions (Quiet, Active and Memory testing; $F(2, 50) = 16.8$; $p = 0.0001$); more specifically, separate factorial analysis showed that the two strains significantly differed in the activity and memory testing conditions ($p<0.01$ in all analyses) but not in the quiet condition ($F(1,10)=3.6$; $p=0.08$). No significant interaction between strains and conditions was found ($F(2,50) < 1.0$) (see Figure 2A).

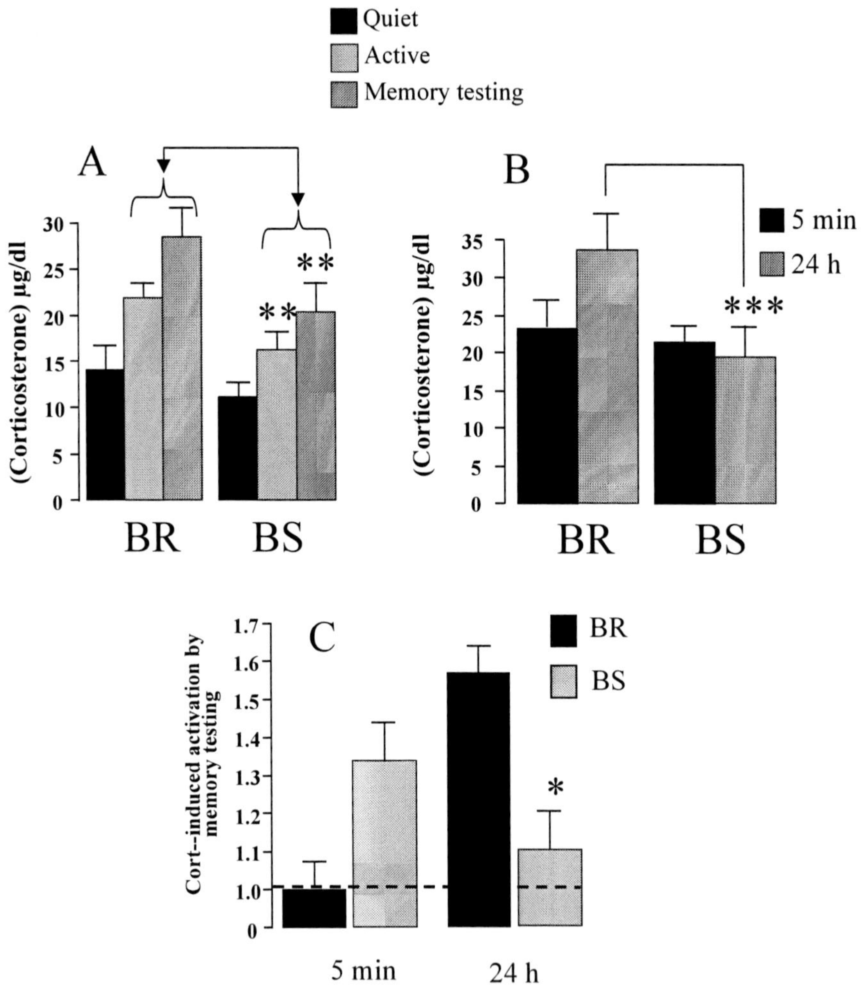

Figure 2. A: Comparisons of plasma corticosterone levels in BS and BR mice in three conditions: Quiet, Active, or Memory. **:p<0.01 to respective BR groups; B: Comparisons of plasma corticosterone levels in BS and BR mice after memory testing (5min or 24 hrs intervals) ; ***:p<0.001 as compared to respective BR groups; C: Magnitude of the corticosterone-induced activation by memory testing in BS and BR mice as compared to the mean corticosterone level of each strain at the active condition used as baseline; *p<0.05 as compared to the BR group

Among subjects submitted to memory testing, a significant interaction between strains and delay intervals was found (F(1,28)=7.1; p=0.01) (Figure 2B). More precisely, BR mice exhibited a plasma corticosterone level at the 5 min retention interval similar to the one observed in BS mice (F(1,14)=0.32) ; in contrast, BR mice exhibited a higher level of plasma corticosterone at the 24 hrs interval (F(1,14) = 16.5; p=0.001).

An additional analysis was carried out to determine the relative magnitude of the corticosterone activation induced by memory testing at either 5 min or 24 hrs, relative to the mean score exhibited by each strain in the active condition, used as a reference baseline. Thus, the following ratio was calculated for each strain: ("Individual Corticosterone scores at memory condition / mean corticosterone score at the active condition"). We showed a significant interaction between strains and delay intervals (F(1,28)=6.5; p=0.01) (Figure 2C). More precisely, BR mice exhibited a significant corticosterone activation of greater magnitude at the 24-h delay as compared to BS mice (p<0.05) whereas an opposite pattern, though *not statistically significant,* was observed at the 5 min delay.

Elevated Plus Maze

Results are presented in Fig.3. BR mice were found to be more "anxious" as operationally defined by the decrease of entry and latency ratios. Thus BR mice exhibited a significantly lower entry ratio (F(1,14)=32.3; p = 0.001) and only a trend toward a lower latency ratio (F(1,14)=3,5; p=0.08), as compared to BS mice respectively.

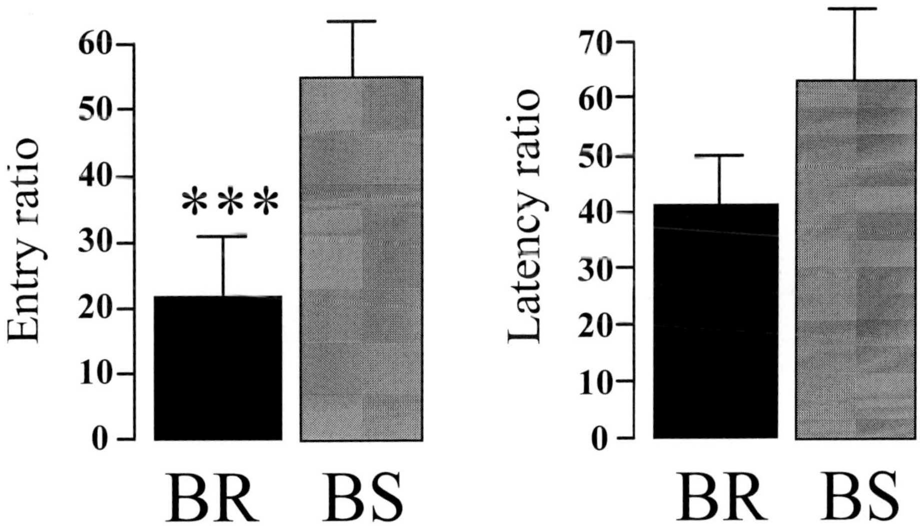

Figure 3. Entry and latency ratios in the elevated plus maze task in BS and BR mice; ***:p<0.001 as compared to the BR group.

Discussion

Previous results [11, 12] have provided evidence of several behavioral differences between the BR and BS strains. In the present study, we extended these earlier observations to cover three situations, namely: (1) spatial delayed discrimination on the 4-hole board, (2) behavioral-induced corticosterone activation and (3) emotional reactivity in a plus-maze.

In the first test, BR and BS strains exhibit a similar high percentage of recognition after a 5 min interval. In contrast, at the 24-h interval, BR mice responded at chance while BS mice performed better for the same time interval; the difference was significant. After 48 hours, BS mice resorted to chance responses. Thus, BR mice exhibited an accelerated time-dependent rate of forgetting as compared to BS mice.

In the other two experiments, BR mice exhibited substantial increased levels of plasma corticosterone after memory testing or after exploration of the hole-board, as compared to BS mice, which were also found to be significantly less "anxious" in the elevated-plus maze.

Many studies have been conducted comparing learning abilities in different strains of mice [2, 7, 57]. Many comparisons of mouse strains in anxiety models tested in the open field, in free exploratory behavior tests [6, 53, 19], the elevated plus maze [19,31] and light dark preference [25, 32] have been conducted, but to our knowledge, there has not been any direct systematic study of the relationship between learning, corticosterone level and anxious-like behavior. Beuzen and Belzung [5] reported that emotional memory in mice is linked to "state" but not "trait" anxiety. Fernandes et al. [20] tested in an open-field situation and a Morris water maze, comparing PWD/Ph (an inbred mouse strain derived from wild mice) and C57BL/6J mice. While the PWD/Ph were more anxious than the C57BL/6J mice in the open field at the beginning, they subsequently displayed much higher levels of exploration and lower anxiety than the C57BL/6J mice after a novel object was introduced. In the Morris water maze, no differences were observed in overall cognitive ability (spatial learning) between PWD/Ph and C57BL/6J mice. In Tg13592 transgenic mice, Lalonde et al. [32] reported a loss in the motivation to explore novel environmental stimuli, plus impaired acquisition of place learning in a Morris water maze. They linked both deficits to changes in regional brain metabolism in selective brain regions as a result of overexpression of the C99 fragment of APP.

In the introduction, we maintained that no direct hypothesis could be formulated for the possible direction expected with the results, i.e. whether BR or BS would perform better. Our data provide evidence for better performance by the BS strain. Several explanations can be presented.

One is that the two strains may differ in the emotional processes involved, with BR being more anxious than BS. Our results fit well with this hypothesis, since BR mice exhibited greater corticosterone levels as compared to BS mice, mainly in a behavioral situation (memory or exploration of the hole-board). In addition, they were also found to be more "anxious" in the elevated plus maze. Certain findings suggest that memory processes may be related to anxiety [4, 31]. Interestingly, in the present study, the greater increase of plasma corticosterone in BR mice was observed at the long-term delay interval, when memory performance was impaired. As the anxious BR strain is less successful, the assumption could be that BR emotionality is too high and, instead of improving, impairs memory processes. We do not have any clear explanations for the relatively weak level of plasma corticosterone at the 5 min interval as compared to the one observed at the 24 hrs interval in BR mice; it is possible that the reduced time interval between the acquisition and test sessions in the 5 min protocol attenuates the HPA axis response to handling or to the re-exposure to the apparatus and its spatial environmental cues. On the contrary, the re-exposure to the behavioral situation after the long-term interval (24-h) could maximize its relative "novelty" and as a result, increased the HPA axis response. Whatever the explanations, the high levels of plasma corticosterone in BR mice at the long-delay interval (see figure 2C) suggests that the level of anxiety might be responsible, at least in part, of the long-term memory impairments observed in the BR strain. This hypothesis is strengthened by numerous data showing that modifications of the levels of corticosterone may interfere with memory performance [8; 15;16] as well as with the benzodiazepine receptors. Indeed, it has been shown that the pre-stress administration of compounds acting at the GABA-A receptors such as diazepam interact with corticosterone levels [55] and also reduced the stress-induced c-Fos expression in various brain areas [17]. Such an interaction between steroids and GABA-A receptors has been evidenced both in human [51] and rat hippocampus [55], a key structure involved in memory consolidation processes.

These latter data suggest a more direct explanation, not involving anxiety as a primary cause of the memory impairments. The relationships between corticosterone levels and compounds acting at the GABA-A receptors have already been demonstrated and may contribute to the difference in memory performance observed between BS and BR mice. Thus, an increase of GABA-A inhibition can be associated with learning deficits [13;34;35;38]. Benzodiazepines are known to induce learning deficits in rodents [47;49;50;54] and the anxiolytic effect is opposite to the effect produced by several compounds in the β carboline family that increase memory performance [10]. It could be hypothesized that a strain with higher sensitivity to ß-CCM, such as BS, may also have brain mechanisms yet to be identified and which contribute to better memory performance. Endogenous ligands have memory-enhancing properties analogous to ß-CCM. For example, diazepam binding inhibitor (DBI) or similar ligands [1,21,24,30,36,46] could play a role in these hypothetical mechanisms, and would be expected to have a more pronounced action in a sensitive strain such as BS. These endogenous ligands are still controversial, and there is no quantitative assessment of them in the brains of BR and BS mice, but this hypothesis cannot be excluded as an interpretation of the data. In this case, BS mice would be expected to have more ß-CCM-like endogenous ligands, acting on the inverse agonist side of the benzodiazepine receptors.

Further work will be needed to choose between the possible explanations. No clear conclusion can be drawn from the data presented here as to whether an "emotional" explanation of the differences observed is a more convincing hypothesis than a purely "cognitive" one. Therefore, our future studies will be aimed at studying the effects of a pre-test reduction of plasma corticosterone on memory, whether pharmacologically or behaviorally–induced [23;28], in BR mice to test more directly the emotional hypothesis.

Conclusion

The present results show that genetic selection for sensitivity to the convulsing properties of an inverse BZD agonist, ß-CCM, also produces differences involving memory processing in the delayed discrimination task. Data also show that the two strains differ in behavioral-induced corticosterone activation as well as in behavioral emotional reactivity.

Further experiments are needed to identify the genetic basis underlying the behavioral differences in memory performance as observed between BR and BS strains, studying a number of behavioral phenomena including aggressiveness, anxiety and seizures. GABA-A receptors appear to be involved in anxiety [13, 27] and also in memory processes [39,40], and an analysis of the coding sequence of hetero-oligomeric protein complex receptor GABA-A subunits is currently being conducted in a bid to detect any mutation that might explain the phenotype described above.

Acknowledgements

The authors thank Frances Ash (ashberac@free.fr) for her help with improving the English text, and Bruno Huart (IMASSA) for his help with corticosterone analysis. The study was supported by a grant from DGA (Paris, France) and the CNRS.

References

[1] Ahboucha S, Pomier-Layrargues G, Butterworth RF. Increased brain concentrations of endogenous (non-benzodiazepine) GABA-A receptor ligands in human hepatic encephalopathy. *Metab Brain Dis* 2004;19:241-251.

[2] Ammassari-Teule M, Castellano C. Strains of rodents and the pharmacology of learning and memory. *Neural Plast* 2004;11:205-216.

[3] Argyropoulos SV, Nutt DJ. The use of benzodiazepines in anxiety and other disorders. *Eur Neuropsychopharmacol* 1999;9 suppl.6:S407-S412.

[4] Beracochea D, Celerier A, Pierard C. Beta-CCM but not physostigmine enhancement of memory retrieval depends on emotional processes in mice. *Psychopharmacology* (Berl) 2004;176:66-73.

[5] Beuzen A, Belzung C. Link between emotional memory and anxiety states: A study by principal component analysis. *Physiol Behav* 1995;58:111-118.

[6] Bothe GW, Bolivar VJ, Vedder MJ, Geistfeld JG. Behavioral differences among fourteen inbred mouse strains commonly used as disease models. *Comp Med* 2005;55:326-334.

[7] Bovet D, Bovet-Nitti F, Oliverio A. Genetic aspect of learning and memory in mice. *Science* 1969;163:139-149.

[8] Celerier A, Pierard C, Rachbauer D, Sarrieau A, Beracochea, D. *Learning and memory* 2004; 11: 196-204.

[9] Chapouthier G, Launay JM, Venault P, Breton C, Roubertoux PL, Crusio WE. Genetic selection of mouse lines differing in sensitivity to a benzodiazepine receptor inverse agonist. *Brain Res* 1998;787:85-90.

[10] Chapouthier G, Negroni J, Venault P. The pharmacology of anxiety and its consequences: epilepsy, balance control and memory processes. *Curr Top Pharmacol* 2004;8: 219-229.

[11] Chapouthier G, Venault P. A pharmacological link between epilepsy and anxiety ? *Trends Pharmacol Sci* 2001;22:491-493.

[12] Chapouthier G, Venault P. Mouse strains selected for differences in sensitivity to a benzodiazepine receptor inverse agonist: pharmacological and behavioural responses. *Biogenic Amines* 2003;17:185-197.

[13] Clément Y, Chapouthier G. Biological bases of anxiety. *Neurosci Biobehav Rev* 1998;22: 623-633.

[14] Costa RM, Federov NB, Kogan JH, Murphy GG, Stern J, Ohno M, Kucherlapati R, Jacks T, Silva AJ. Mechanism for the learning deficits in a mouse model of neurofibromatosis type 1. *Nature* 2002;415:526-530.

[15] De Quervain DJF, Roozendaal B, Nitsch RM, Mc Gaugh JL, Hock C. Acute corticone administration impairs retrieval of long-term declarative memory in humans. *Nat. Neurosci* 2000; 3: 313-314.

[16] De Quervain DJF, Henke DJF, Aerni A, Treyer V, McGaugh JL, Berthold T, Nitsch RM, Buck A, Rozendaal B, Hock C. Glucocorticoid-induced impairment of declarative memory retrieval is associated with reduced blood flow in the medial temporal lobe. *Eur. J. Neurosci* 2003; 17: 1296-1302.

[17] De Medeiros MA, Reis C, Eugenio-Mello L. Stress-induced c-Fos expression is differentially modulated by dexamethasone, diazepam and imipramine. *Neuropsychopharmacology*, 2005; 30: 1246-1256.

[18] Do-Rego JC, Suaudeau C, Chapouthier G, Costentin J. Mouse lines differing in sensitivity to ß-CCM differ in tasks used for testing antidepressants. *Pharmacol Biochem Behav* 2002;72: 411-416.

[19] Ducottet C, Belzung C. Correlations between behaviours in the elevated plus-maze and sensitivity to unpredictable subchronic mild stress: evidence from inbred strains of mice. *Behav Brain Res* 2005;156:153-162.

[20] Fernandes C, Liu L, Paya-Cano JL, Gregorova S, Forejt J, Schalkwyk LC. Behavioral characterization of wild derived male mice (Mus musculus musculus) of the PWD/Ph inbred strain: high exploration compared to C57BL/6J. *Behav Genet* 2004;34:621-630.

[21] Ferrero P, Costa E, Conti-Tronconi B, Guidotti A. A diazepam binding inhibitor (DBI)-like neuropeptide is detected in human brain. *Brain Res* 1986;399:136-142.

[22] Galey D, Jaffard R. Post-training medial septal stimulation improves spatial information processing in BALB/c mice. *Neurosci Lett* 1992;143:87-90.

[23] Gellermann LW. Chances orders of alternating stimuli in visual discrimination experiments. *J Genet Psychol* 1933;42:206-208.

[24] Guidotti A, Toffano G, Costa E. An endogenous protein modulates the affinity of GABA and benzodiazepine receptors in rat brain. *Nature* 1978;275:553-535.

[25] Guillot PV, Chapouthier G. Intermale aggression and dark/light preference in ten inbred mouse strains. *Behav Brain Res* 1996;77:211-213.

[26] Guillot PV, Sluyter F, Crusio WE, Chapouthier G. Mice selected for differences in sensitivity to a benzodiazepine receptor inverse agonist vary in intermale aggression. *Neurogenetics* 1999;2:171-175.

[27] Holmes A. Targeted gene mutation approaches to the study of anxiety-like behavior in mice. *Neurosci Biobehav Rev* 2001;25:261-273.

[28] Izidio GS, Lopes DM, Spricigo L Jr, Ramos A. Common variations in the pretest environment influence genotypic comparisons in models of anxiety. *Genes Brain Behav* 2005;4:412-419.

[29] Kalueff AV, Nutt DJ. Role of GABA in anxiety and depression. *Depression and Anxiety* 2006; 0: 1-23.

[30] Kavaliers M, Hirst M. An octadecaneuropeptide (ODN) derived from diazepam binding inhibitor increases aggressive interactions in mice. *Brain Res* 1986;383:343-349.

[31] Krazem A, Borde N, Beracochea D. Effects of diazepam and ß-CCM on working memory in mice: relationships with emotional reactivity. *Pharmacol Biochem Behav* 2001;68:235-244.

[32] Lalonde R, Dumont M, Fukuchi K. Strazielle C. Transgenic mice expressing the human C99 terminal fragment of betaAPP: effects on spatial learning, exploration, anxiety, and motor coordination. *Exp Gerontol* 2002;37:1401-1412.

[33] Lepicard E, Joubert C, Hagneau I, Perez-Diaz F, Chapouthier G. Differences in anxiety-related behavior and response to diazepam in BALB/cByJ and C57BL/6J strains of mice. *Pharmacol Biochem Behav* 2000;67:739-748.

[34] Lister RG. The amnesic action of benzodiazepines in man. *Neurosci Biobehav Rev* 1985;9: 87-94.

[35] Macdonald RL, McLean MJ. Anticonvulsant drugs: mechanisms of action. *Adv Neurol* 1986;44:713-736.

[36] Marquardt H, Todaro GJ, Shoyab M. Complete amino acid sequences of bovine and human endozepines. Homology with rat diazepam binding inhibitor. *J Biol Chem* 1986;261: 9727-9731.

[37] Martinez JL Jr, Jensen RA, McGaugh JL. Facilitation of memory consolidation. In: Deutsch, JA, editor. *The physiological basis of memory* (2nd ed). New York: Academic Press, 1983, pp. 49-70.

[38] Marvel CL, Paradiso S. Cognitive and neurological impairment in mood disorders. *Psychiatr Clin North Am* 2004;27:19-36.

[39] Maubach K. GABA(A) receptor subtype selective cognition enhancers. Curr Drug Targets *CNS Neurol Disord* 2003;2:233-239.

[40] Muir WJ. Genetics advances and learning disability. Br J Psychiatry 2000;176:12-19.

[41]Pierard C, Liscia P, Valleau M, Drouet I, Chauveau F, et al. Modafinil-induced modulation of working memory and plasma corticosterone in chronically-stressed mice. *Pharmacol.Biochem.Behavior* 2006; 83:1-8.

[42] Prado de Carvalho L, Grecksch G, Cavalheiro EA, Dodd RH, Chapouthier G, Rossier J. Characterization of convulsions induced by methyl-ß-carboline-3-carboxylate in mice. *Eur J Pharmacol* 1984;103:287-293.

[43] Prado de Carvalho L, Grecksch G, Chapouthier G, Rossier J. Anxiogenic and non-anxiogenic benzodiazepine antagonists. *Nature* 1983;301:64-66.

[44] Raffalli-Sébille MJ, Chapouthier G, Venault P, Dodd RH. Methyl ß-carboline-3-carboxylate enhances performance in a multiple-trial learning task in mice. *Pharmacol Biochem Behav* 1990;35:281-284.

[45] Rinaldi D, Boutrel B, Adrien J, Venault P, Chapouthier G. Two mouse lines selected for differential sensitivities to ß-carboline-induced seizures are also differentially sensitive to various pharmacological effects of other GABA(A) receptor ligands. *Behav Genet* 2000;30: 277-284.

[46] Rommelspacher H, Nanz C, Borbe HO, Fehske KJ, Müller WE, Wollert U. 1-methyl-ß-carboline (Harmane), a potent endogenous inhibitor of benzodiazepine receptor binding. *Naunyn Schmiedebergs Arch Pharmacol* 1980;314:97-100.

[47] Savic MM, Obradovic DI, Ugresic ND, Cook JM, Yin W, Bokonjic DR. Bidirectional effects of benzodiazepine binding site ligands in the passive avoidance task: differential antagonism by flumazenil and beta-CCt. *Behav Brain Res* 2005 158;293-300.

[48] Sarter M, Bruno JP, Berntson GG. Psychotogenic properties of benzodiazepine receptor inverse agonists. *Psychopharmacology* (Berl) 2001;156:1-13.

[49] Schmidt D. Benzodiazepines - diazepam. In: Woodbury DM, Penry JK, Pippenger CE, editors. *Antiepileptic drugs*. New-York, Raven Press, 1982, pp. 711-773.

[50] Sternbach LH. The benzodiazepine story. *J Med Chem* 1979;22:1-7.

[51] Stromberg J, Backstron T, Lundgren P. Rapid non genomic effect of glucocorticoid metabolites and neurosteroids on the gamma-amynobutyric acid -A-receptor. *Eur J Neurosci* 2005; 21: 2083-2088.

[52] Suaudeau C, Rinaldi D, Lepicard E, Venault P, Crusio WE, Costentin J, Chapouthier G. Divergent levels of anxiety in mice selected for differences in sensitivity to a convulsant agent. *Physiol Behav* 2000;71:517-523.

[53] Tang X, Xiao J, Liu X, Sanford LD. Strain differences in the influence of open field exposure on sleep in mice. *Behav Brain Res* 2004;154:137-147.

[54] Thiébot MH. Some evidence for amnesic-like effects of benzodiazepines in animals. *Neurosci Biobehav Rev* 1985;9:95-100.

[55] Tritos N, Kitraki E, Philipidis H, Stylianopoulou F Neurotransmitter modulation of glucocorticoid receptor mRNA levels in the rat hippocampus. *Neuroendocrinology* 1999; 69: 324-330.

[56] Venault P, Chapouthier G, Prado de Carvalho L, Simiand J, Morre M, Dodd RH, Rossier J. Benzodiazepine impairs and ß-carboline enhances performance in learning and memory tasks. *Nature* 1986;321:864-866.

[57] Wahlsten D, Cooper SF, Crabbe J. Different rankings of inbred mouse strains on the Morris maze and a refined 4-arm water escape task. *Behav Brain Res* 2005;165:36-51.

In: Behavioral Models in Stress Research
Editors: A.V. Kalueff and J.L. LaPorte, pp. 111-157
ISBN: 978-1-60456-361-0

Chapter 6

OPTIMIZATION OF THE CHRONIC STRESS DEPRESSION MODEL IN C57 BL/6 MICE: EVIDENCES FOR IMPROVED VALIDITY

Tatyana Strekalova *
Department of Psychiatry and Psychotherapy
Animal Behavior Division, RWTH Aachen University, Aachen, Germany

INTRODUCTION

Depression is a devastating disorder, the prevalence of which in the economically developed world is above 12%; more than 15% of the population experiences depression at least once during their lifetime. A large representation of depressive disorders in a population results in major medical, social and economic consequences [60, 22, 61, 62]. By the year 2020, depression is projected to become the second major cause of disability in the world [80]. This suggests a high relevance for the further investigation of the biological basis of depression, which would allow for improved prevention and treatment of depressive disorders.

According to the Diagnostic and Statistical Manual (DSM) IV-TR, depression is defined by the presence of at least one core symptom lasting minimally 2 weeks that is typically accompanied by a number of subsidiary symptoms. Anhedonia, a decreased ability to experience pleasures, and depressive mood are generally considered as the key symptoms of clinical depression [43, 63, 96].

Most of the symptoms of human depression, such as anhedonia, coping deficits, loss of interest to novelty, cognitive impairment, disturbances in circadian rhythmicity, specific changes in the sleep pattern and some vegetative features could be simulated in animals [for a review, see 148, 27, 28, 74, 36]. Because anhedonia is a core phenomenon of depressive disorders and can be evoked in rodents, the hedonic deficit can be regarded as a primary feature to be addressed while modeling depression in animals.

* E-mail address: t_strekalova@yahoo.com. Phone: 49 (241) 8080 968. Pauwelsstrasse 30, Aachen D-52074, Germany

Models of Stress-Induced Anhedonia: Advantages and Drawbacks

Among a number of different experimental approaches, chronic stress is regarded as one of the most elaborate methods for inducing anhedonia in rodents [for a review, see 7, 27, 149]. The etiological relevance is another important feature of chronic stress models, since stress is a major environmental factor in the development of depressive disorders [84, 61].

Originally, Katz and co-workers employed a 21-day long stress protocol in rats, comprised of rather intense stressors: application of electric foot shock, food and water deprivation, exposure to a heat, swimming in cold water and others. These procedures caused a decrease in intake of sucrose solution that was interpreted as a sign of the hedonic deficit. Anhedonia in rats was reversible by antidepressants, but not by neuroleptics and anxiolytics [58, 59].

The unpredictable chronic mild stress model of Willner is probably the most established and broadly used paradigm of stress-induced anhedonia in rodents [145-149]. To obtain a more realistic model of depression, Willner modified the protocol of Katz, mimicking more closely the analogues of human stressors. Therefore, he substantially reduced the severity of stressors using only mild ones, such as: soiled cage, presence of a foreign object, restricted access to food, constant lighting and others, and extended the duration of the stress procedure up to 3 months. The new protocol provided a longer lasting decrease in the sensitivity to rewards, as shown by a decrease in sucrose intake and preference and an increase in the thresholds of intracranial self-stimulation. Hedonic deficit was again shown to be specifically reversible by antidepressants, but not by compounds with other activities. However, this sophisticated model later was described to be highly sensitive to the minor environmental influences in the animals' surroundings and criticized for insufficient reproducibility [141, 18, 85, 39].

Based on the originally proposed theoretical and methodological principles of simulating depression with a chronic stress paradigm, i.e., employment of a hedonic deficit as a criterion of depressive-like state and the use of a prolonged application of uncontrollable and unpredictable stressors of mild intensity for anhedonia induction, a number of chronic stress procedures were elaborated and validated in rats and mice [78, 96, 89, 46, 64, 138, 44, 82, 33, 42, 55, 9, 133]. In addition to anhedonia, various chronic stress procedures were described to induce subsidiary features of depressive status, such as reduced latency and increased duration of REM sleep [79, 23], decreased sexual and increased submissive behavior [149, 39], locomotor inhibition in stressful conditions [59], impaired circadian rhythmicity [113, 111], increased cortisol levels [58].

Meanwhile, a number of reports evidenced inconsistencies between the induction of the hedonic deficit and the behavioral, physiological and molecular effects of the chronic stress procedures. Overall, during the last years, concerns have been raised over the validity of the chronic stress depression model [85, 83, 76] and sucrose test as a paradigm reflecting an anhedonic state in rodents [73, 142, 46, 50].

Another apparent difficulty of stress-induced anhedonia paradigms is regarded to be the insufficient accuracy of the sucrose test. Typically, in mice, available paradigms of the sucrose test can reveal the differences in the reward sensitivity between the experimental groups, but not between the individual animals. The low resolution of the commonly used protocols of this test is believed to be due to the overlooked physiological and physical artifacts in evaluating drinking behavior in rodents [93, 131, 57, 129].

Importantly, many behavioral studies, using chronic stress depression models, resulted in conflicting findings and failed to define a consistent phenotype seen in chronically stressed rats and mice with hedonic deficit. Data on their locomotion, anxiety, exploration, and other behaviors often demonstrated paradoxical and inconsistent behavioral changes [30, 20, 101, 64, 47-49, 99]. This greatly complicated the characterization of behavioral correlates for stress-induced anhedonia, which is induced with the chronic stress method in rodents.

Besides methodological problems, application of the chronic stress approach encountered some conceptual drawbacks. Perhaps the most obvious one consists in the fact that in previously proposed models, all effects observed in groups of chronically stressed animals with signs of a decreased sensitivity to reward are attributed to the anhedonic state. Meanwhile, stress *per se* can evoke a number of physiological alterations, which are not associated with a depressive-like state. Since available chronic stress models did not provide a control for the effects of chronic stress alone, it was not possible to relay findings obtained in chronically stressed animals selectively to anhedonia. Thus, strictly speaking, specific biological correlates of hedonic deficit could not be addressed with earlier developed chronic stress depression models.

Here, we established a mouse model of stress-induced anhedonia with an internal control for the effects of stress alone and attempted to resolve some important methodological drawbacks of the chronic stress depression paradigm in mice. Elaboration of this approach was encouraged by numerous evidences of remarkable inter-individual variability in animals' response to stress observed in our own and others' studies [145, 146, 5, 114, 4, 53]. We aimed to establish an anhedonia-evoking stress regimen of such intensity and duration, upon which not all individuals would develop a hedonic deficit. Employment of the stressed non-anhedonic group as an internal control for the effects of chronic stress *per se* was expected to provide a possibility to re-evaluate the validity of the chronic stress depression paradigm in a more refined way.

General Conditions of Experiments

In all studies, 3.5 months old male C57 BL/6 mice were used. Animals were housed individually for two weeks before the start of experiments (since their transportation from a supplier) under a reverse 12 h: 12 h light-dark cycle (lights on: 21:00 h) in standard laboratory conditions. Behavioral experiments were carried out during the dark phase of the animals's light cycle. Independent data sets were analyzed by the Kruskal-Wallis test, followed by the Mann-Whitney U test; repeated measurements were evaluated by the Wilcoxon test. Qualitative data were analyzed by the Fischer's exact test. The level of confidence was set at 95% ($p < 0.05$). Unless specified, data on graphs are expressed as mean ± standard error of measurement (SEM). More details on experimental conditions, number of animals per group and statistical analysis are presented elsewhere [121, 125, 128].

New Chronic Stress Procedure Induces Anhedonia in a Subgroup of Mice

In order to induce anhedonia in mice, we applied a chronic stress procedure. Application of a 4-week chronic stress paradigm, consisting of uncontrollable and unavoidable stressors (such as exposure to a rat, tail suspension and restraint stress) in different procedural variations (*see below*), resulted in a profound decrease in sucrose preference, a putative parameter of a hedonic state in rodents. The proposed stress procedure, repeated now over twelve times in four different laboratory environments, led to anhedonia defined as a decrease of preference to a 1% sucrose solution over tap water ≤ 65% by the 4th week of continuous stress application in 50-70% of mice [116-130].

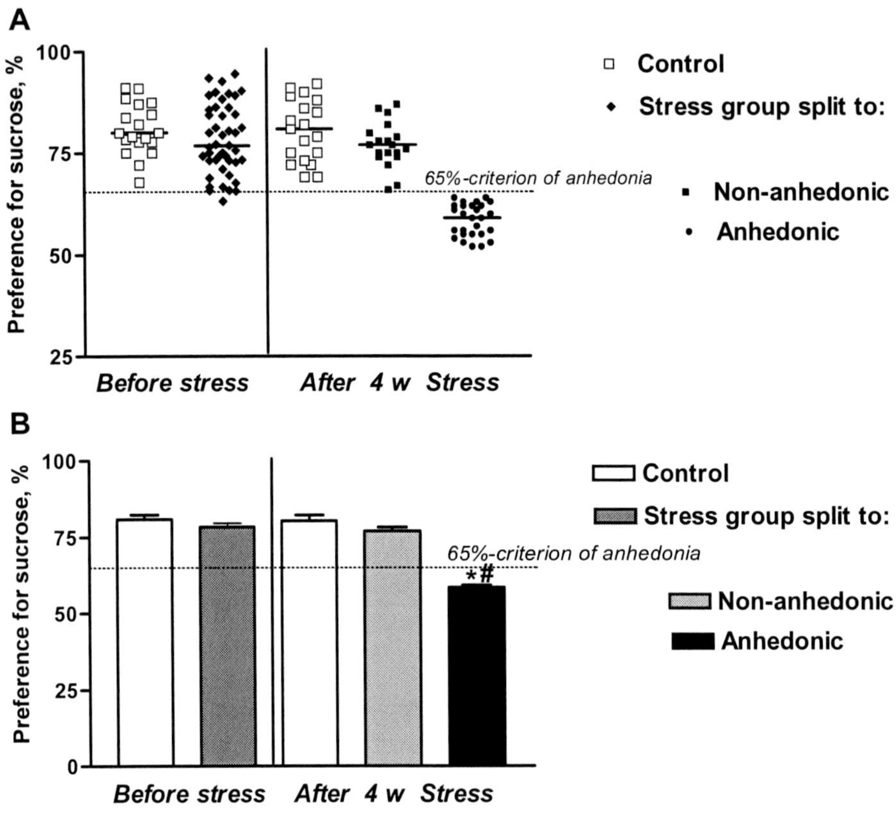

Figure 1. Chronic stress leads to a decrease in sucrose preference in a subgroup of mice. (A) Individual data show that chronic stress causes a drop of sucrose preference, measured after the termination of a 4-week stress, in a subgroup of mice. Stressed mice split into anhedonic and non-anhedonic subgroups, according to the criterion of 65% preference for sucrose solution (see the text). Bars indicate medians of the groups. (B) After stress, the mean sucrose preference of mice, which were defined as anhedonic is significantly lower compared to control and non-anhedonic groups (*p < 0.05 vs. control group and # p < 0.05 vs. non-anhedonic group; Mann-Whitney). There were no significant differences between non-anhedonic and control groups (p > 0.05). Data on graphs are expressed as mean ± standard error of measurement (SEM).

Figure 1 represents changes of sucrose preference in one of the typical chronic stress experiments (adapted from [128]). Based on the chosen criterion of 65% sucrose preference, mice are assigned to the anhedonic and non-anhedonic groups. A criterion for anhedonia is based on our results, which indicated that mice with a sucrose preference ≤ 65% demonstrate a depressive-like syndrome shown by elevated floating in the forced swim test, increased immobilization in the tail suspension test, decreased novelty exploration and other depressive-like changes, while stressed mice with a sucrose preference above this value did not display this behavioral phenotype [116-130].

Before the onset of stress, all three groups (control, non-anhedonic and anhedonic mice) had a similar preference for sucrose over water and absolute intake of water and sucrose solution. After stress, mice from the anhedonic group are characterized by a decrease in consumption of sucrose solution, whereas in non-anhedonic animals these parameters are not different from the control values [128]. Multiple regression analysis showed that a drop of sucrose preference is accompanied by a decrease in sucrose consumption and increase of water intake. In contrast, for many stress protocols, reduction in sucrose preference is solely due to an increase in water intake that is thought to be a sign of stress-induced polydipsia. However, the link between polydypsia and a state of anhedonia is questionable [50, 138, 9]. Correlation analysis revealed no relationship between stress-induced loss of body weight and sucrose intake. Together, these results led us to suggest that a decrease in sucrose preference after chronic stress, particularly in our model, is not determined by metabolic disturbances and alterations of consummatory behavior in animals, as it has been discussed in a literature [37, 73, 97], but particularly in our chronic stress paradigm, may reflect diminished sensitivity to reward.

Behavioral Correlates of Stress and Stress-Induced Anhedonia

As the proposed chronic stress procedure induces anhedonia only in a subgroup of stressed animals, the stressed non-anhedonic group is used as an internal control for the stress effects, which are not associated with hedonic deficit. This enabled us to assess behavioral correlates of anhedonia separately from the effects of chronic stress. To our knowledge, this approach is the first attempt to segregate behavioral features of stress and stress-induced anhedonia. Behavioral analysis of control, non-anhedonic and anhedonic mice in our study was aimed at the investigation of parameters of "behavioral despair" / coping, exploration, anxiety and locomotion.

For these experiments, we modified "standard" protocols of test procedures that were earlier calibrated for experiments with non-stressed C57 BL6 mice [115]. As was demonstrated in chronically stressed mice, subtle stressors like light trigger hyperlocomotion that confounds all types of behavioral testing [122, 125]. Therefore, for a behavioral characterization of the mouse model of stress-induced anhedonia we developed "mild" protocols of standard tests, in which stress of testing procedure is essentially diminished (*see below*). Employed test battery included open field, activity boxes, forced swim and tail suspension tests, new object exploration paradigm, novel cage, zero-maze and a dark/light box.

In the "mild" version of the forced swim test, mice are introduced to a large transparent pool filled with warm water. The depth of water is reduced in comparison to the standard procedure [91]; testing is performed under the red lighting [121]. Floating behavior scored in this test is regarded as a sign of "behavioral despair"/impaired coping in rodents. Preliminary experiments showed that testing under the conditions described above during the first 2 min of a single exposure of animals to a swimming session is most informative for the behavioral analysis of the employed mouse strain.

Anhedonic mice exhibit a pronounced decrease of the latency to begin floating in the forced swim test, as compared to control and non-anhedonic animals; there are no differences in latency of floating between the non-anhedonic and control groups [121]. The total duration of floating is significantly elevated in anhedonic mice versus control and non-anhedonic groups; no differences are observed in total duration of floating behavior between the non-anhedonic and control mice (Fig.2; adapted from [128]).

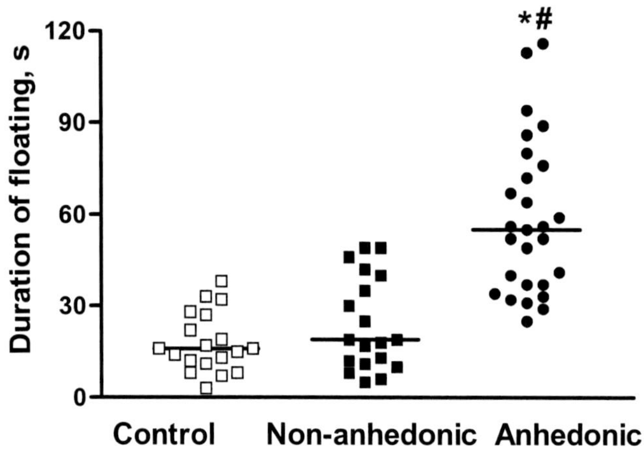

Figure 2. Forced swim test: floating is elevated in the anhedonic, but not in the non-anhedonic mice. Individual data show a significant increase of total duration of floating in anhedonic mice, while behavior of non-anhedonic mice is similar to control. Bars indicate medians of the groups (*$p < 0.05$ vs. control group and # $p < 0.05$ vs. non-anhedonic group; Mann-Whitney).

As non-anhedonic animals and non-stressed control mice exhibit similar behavior in this test, one can conclude that an increase of floating behavior in stressed anhedonic animals is selectively related to the hedonic deficit. These results show that in our paradigm, stressed non-anhedonic group of mice can generally serve as a proper control for the effects of stress, which are not associated with a depressive-like state. This finding also suggests a selectivity of a proposed protocol of the forced swim test in a differentiation of depressive-like behavior in chronically stressed mice.

The data obtained in the forced swim test are in line with the results of testing in the tail suspension test [118]. In this test, anhedonic mice demonstrate reduced latency of the first episode of immobilization and increased duration of immobilization behavior, measured over a time period of 2 min, as compared to control and non-anhedonic mice.

As decreased interest in novelty is an important feature of human depression, we assessed behavior of control, non-anhedonic and anhedonic mice in two paradigms of novelty exploration, novel cage and new object exploration tests. Experiments with chronically stressed animals were performed under lighting of weak intensity (5 Lux) in quiet rooms, in order to provide "mild" testing conditions. Mice were allowed to explore a new object (an artificial flower) in a restricted area. Anhedonic mice display a decreased total duration of new object exploration measured over a 15-min period of testing, as compared to control and non-anhedonic mice (Fig. 3A; adapted from [121]).

In the novel cage, animals were assessed for their activity to explore a new environment. In this test, mouse exploratory rearings are scored during a 5-min trial in a novel cage (the same size of a home cage) lit with a red light. Anhedonic animals reveal a decreased number of exploratory rearings versus non-stressed controls and non-anhedonic animals (Fig.3B; adapted from [121]). Exploratory behavior of non-anhedonic mice is not different from controls. These data provide evidence that decreased exploratory activity in stressed mice is selectively associated with a presence of anhedonia.

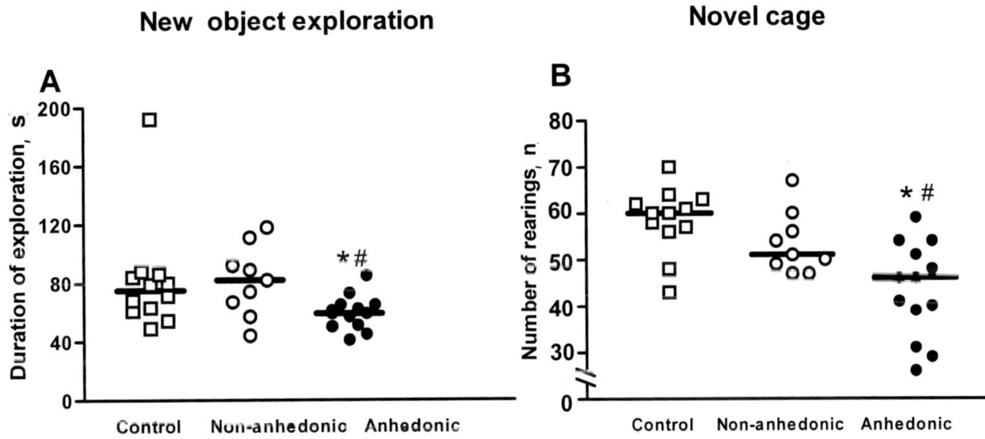

Figure 3. Reduced novelty exploration is associated with anhedonia, but not with chronic stress alone. (A) Duration of the new object exploration is decreased in anhedonic mice (*$p < 0.05$ vs. control group and # $p < 0.05$ vs. non-anhedonic group; Mann-Whitney), but not in the non-anhedonic group ($p > 0.05$ vs. control group). (B) The mean number of exploratory rearings in the novel cage test is significantly reduced in anhedonic animals (*$p < 0.05$ vs. control group and # $p < 0.05$ vs. non-anhedonic group), but not in non-anhedonic mice ($p > 0.05$ vs. control group). Bars indicate medians of the groups. Reproduction of this material is permitted by Macmillan Publishers Ltd.

In order to evaluate anxiety-like behavior of stressed mice with and without hedonic deficit, "mild" protocols of the elevated Zero-maze and dark/light box were applied (in both tests, illumination strength was 5 Lux). Both anhedonic and non-anhedonic mice show similar behavioral alterations, consisting in a decrease of time spent in anxiety-related areas, the open arms of the Zero-maze and in the lit box of the dark/light box (Fig. 4, adapted from [121]). Anhedonic and non-anhedonic animals have fewer exits to the anxiety-related compartments in comparison to a non-stressed control group. Thus, our data suggest that in a proposed model of anhedonia, elevated anxiety of stressed mice is a consequence of chronic stress, which is not related to the anhedonic status.

This finding was replicated under the same and slightly different testing conditions in several experiments [119, 124, 127]. Our results support the clinical view on comorbidity of anxiety and depressive disorders, where core pathogenetic mechanisms are thought to be distinct [38, 86]. To our knowledge, this is the first experimental evidence in rodents that separates anxiety from behaviors related to anhedonia.

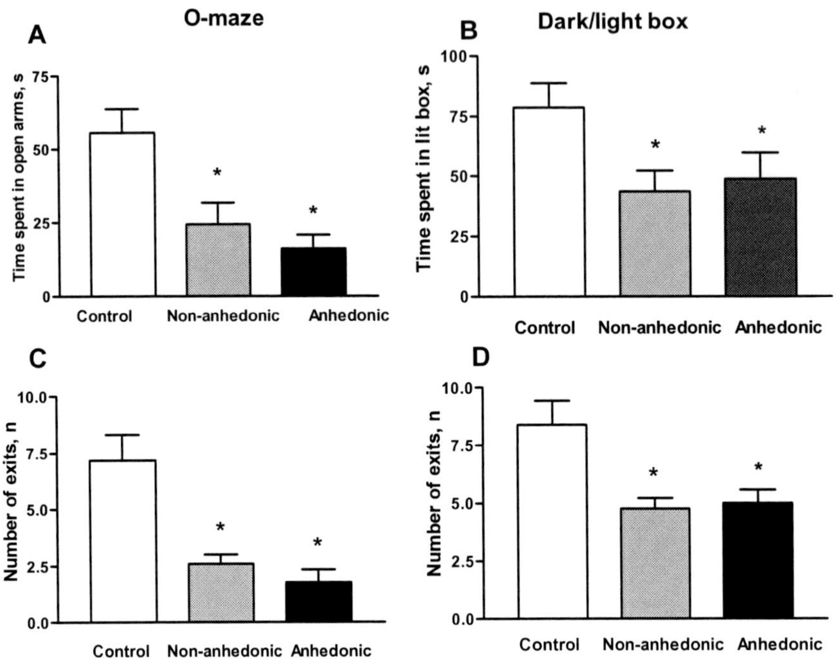

Figure 4. Increased anxiety is a consequence of chronic stress that occurs independently from anhedonia. (A, B) Time spent in open arms of the Zero-maze and in the lit box of the dark/light box is significantly decreased in both the anhedonic and non-anhedonic mice in comparison to control group (*$p < 0.05$ vs. control group; Mann-Whitney). (C, D) Numbers of exits to the open arms in Zero-maze and to the lit box of the dark/light box are significantly diminished in anhedonic and non-anhedonic groups as compared to controls (*$p < 0.05$ vs. control group). Data on graphs are expressed as mean ± standard error of measurement (SEM). Reproduction of this material is permitted by Macmillan Publishers Ltd.

Locomotor disturbances are another feature of chronic stress and depression. In the open field box lit with light of modest intensity (25 Lux), anhedonic and non-anhedonic mice show an increased total distance moved in comparison to control animals (Fig.5A, adapted from [121]). Thus, chronic stress leads to hyperactivity in mice, irrespective to the presence of hedonic deficit. Under weak illumination (5 Lux), stressed mice demonstrate no hyperactivity (Fig.5B). Anhedonic and non-anhedonic animals do not differ in either protocol of the open field test in mean total distance moved. In complete darkness, anhedonic and non-anhedonic mice have similarly decreased activity scores as compared to the non-stressed control group (Fig.5C).

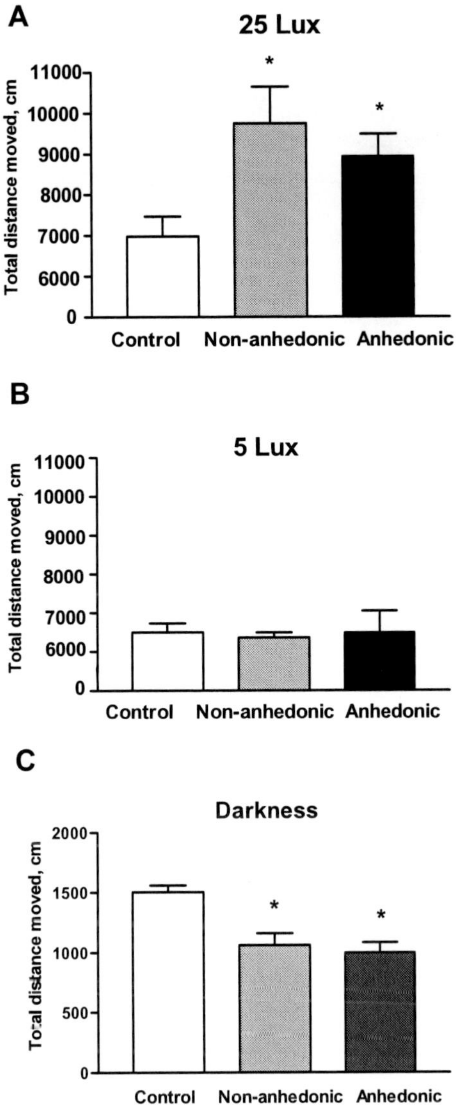

Figure 5. Stress-induced locomotor disturbances are not related to anhedonia. (A) Total distance moved is significantly increased both in anhedonic and non-anhedonic mice in the open field illuminated with light of 25 Lux as compared to the controls (*$p < 0.05$ vs. control group; Mann-Whitney). (B) Total distance moved in the open field lit with light of 5 Lux intensity does not differ between non-anhedonic, anhedonic and control mice ($p > 0.05$). (C) Total distance moved measured in dark activity boxes is significantly reduced both in anhedonic and non-anhedonic mice as compared to a control group (*$p < 0.05$ vs. control group). Data on graphs are expressed as mean ± standard error of measurement (SEM). Reproduction of this material is permitted by Macmillan Publishers Ltd.

The open field test, repeated in several chronic stress experiments at different time points after the termination of the stress regimen confirmed the phenomenon of hyperlocomotion induced by the light. There were no differences in locomotor activity observed between the anhedonic and non-anhedonic groups [116, 124] (unpublished results). Thus, altered locomotion is not related to anhedonia in a proposed mouse model of depression.

In summary, behavioral analysis of chronically stressed mice with and without anhedonia revealed specific correlates of a hedonic deficit. Anhedonia was found to be selectively associated with depressive-like behaviors, such as signs of "behavioral despair" / coping deficits in the forced swim and tail suspension tests and decreased novelty exploration. Importantly, these behavioral deficits were not observed in stressed non-anhedonic animals. Increased anxiety and changes in locomotion were found to be the consequences of chronic stress, which are detected both in stressed mice with and without hedonic deficit.

Obviously, distinct behavioral characteristics of the anhedonic and non-anhedonic groups cannot be attributed merely to the differences in sensitivity to stress, as the mean body weight of anhedonic and non-anhedonic groups is reduced comparably throughout the stress experiment (Fig.6; adapted from [128]); every chronic stress experiment replicated this finding [116-130].

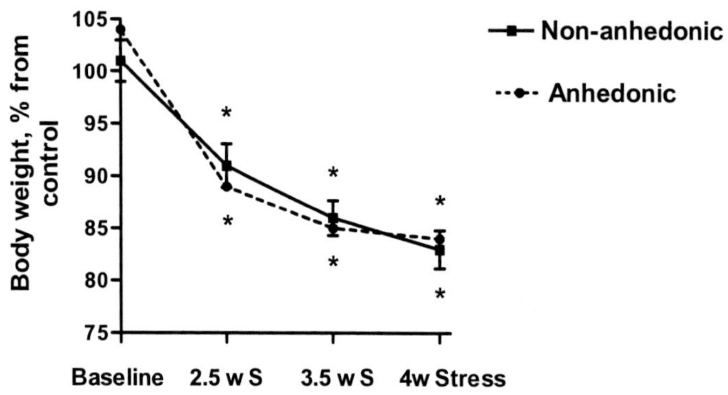

Figure 6. Similar reduction of body weight in anhedonic and non-anhedonic mice. In a course of chronic stress, at the time points 2.5, 3.5 and 4 weeks of exposure to stress, body weight is significantly decreased in both anhedonic and non-anhedonic mice in comparison to the control group (*$p < 0.05$ vs. control group; Mann-Whitney). There is no significant difference between anhedonic and non-anhedonic mice ($p > 0.05$). Data are expressed in percent from means of a control group as mean ± standard error of measurement (SEM).

The fact that the values of the body weight and behavioral parameters of anxiety and locomotion are close in non-anhedonic and anhedonic mice (see Figs.4-6), suggests a similar impact of the stress procedure in these animals. Hence, it is very unlikely that the differences described above in depressive-like behaviors in the non-anhedonic and anhedonic groups can be due to a distinct "amount" of stress perceived by the animals, but suggests rather qualitative differences between non-anhedonic and anhedonic individuals. These data also led us to consider the non-anhedonic mice to be an adequate control for the effects of chronic stress *per se*.

Dynamic of Anhedonic State After the Termination of Stress Procedure and Anhedonia Maintenance

After the termination of chronic stress, diminished sucrose preference lasts in the anhedonic group 1-3 weeks. During this period, most of the mice from this group show a sucrose preference below 65%; over time, they spontaneously recover from a hedonic deficit [116-130].

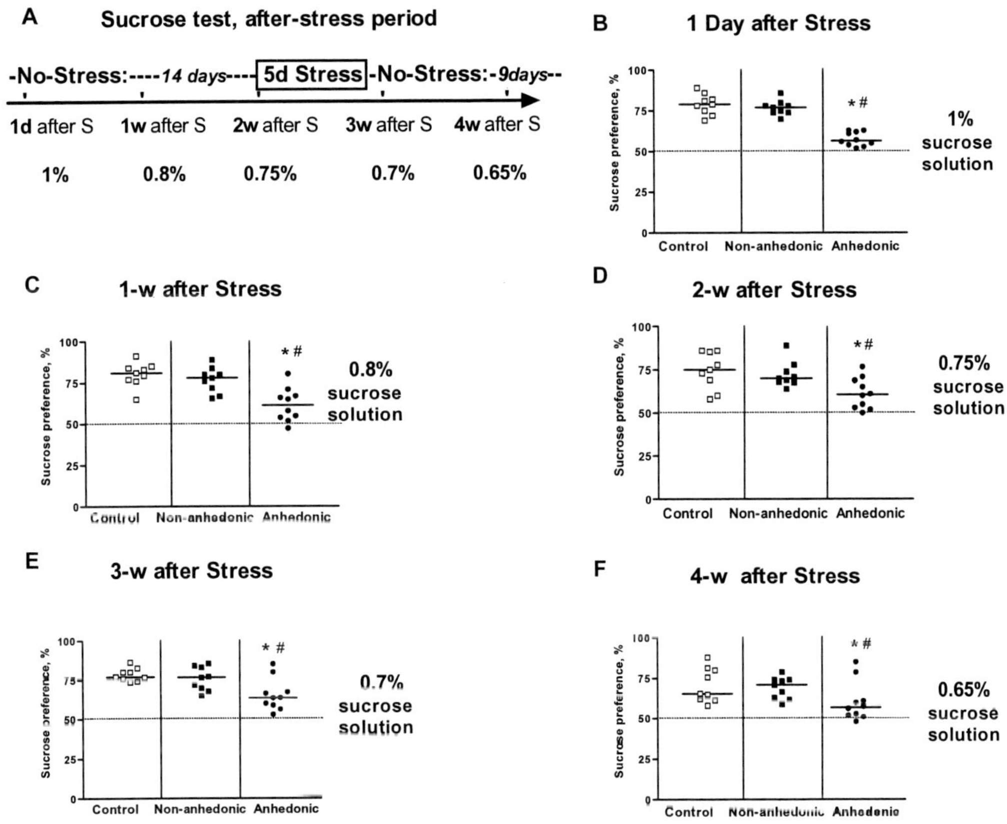

Figure 7. Post-stress dynamic of sucrose preference and maintenance of anhedonia. (A) Scheme of the experiment. After the termination of a 4-week stress procedure, chronically stressed mice were left undisturbed for 14 days; thereafter, they were submitted to a 5-day rat-exposure stress. During remaining 9 days of the 4-week after-stress period, no stress was applied. Sucrose test was performed 1 day, 1, 2, 3 and 4 weeks after the termination of chronic stress with weekly descending concentrations of sucrose solution (from 1% to 0.65%). (B) 1 day after the termination of stress, sucrose preference in the anhedonic group is significantly lower than in the control (*$p<0.05$; Mann-Whitney) and non-anhedonic (# $p < 0.05$) mice (C, D) This difference remained significant during weeks 1 and 2 (E, F), as well as after additional 5-day stress, during measurements on weeks 3 and 4. Non-anhedonic and control groups of mice do not differ in sucrose preference throughout the entire experiment ($p>0.05$). Bars indicate medians of the groups. Data on graphs are expressed as mean ± standard error of measurement (SEM).

It was shown that the anhedonic state could be reinstated/ maintained by an additional application of 5 daily exposures to the tail-suspension or rat-exposure stress [128,130]. As sucrose preference in the anhedonic group spontaneously restores to the values of the control group at a point in time between the 2nd and 3rd weeks after the termination of stress procedure [121], we applied a 5-day rat exposure during this period of the experiment to prolong a phase of anhedonia (for details, see [128]).

Therefore, mice from the stressed group were left unstressed for the first 14 days after the termination of the stress procedure, and then, between two sucrose tests on weeks 2 and 3, submitted to a 5-day rat-exposure stress (see the scheme of experiment, Fig.7A). No further stress was applied during the remaining 9 days of the experiment. Because repeated access to sweetened solutions elevates sucrose preference in mice and leads to a development of the "ceiling" effect in sucrose test that compromises its sensitivity (*see below*), sucrose preference was evaluated with solutions of weekly descending concentrations (from 1% to 0.65%).

Sucrose preference in the anhedonic group was significantly lower than in the non-anhedonic and control mice 1 day after the termination of the chronic stress and during weeks 1-4 thereafter (Fig.7B-F, adapted from [128]). There was no significant difference in the measured parameter between the non-anhedonic and control animals in the course of the entire experiment; thus, application of an additional 5-day rat exposure stress does not affect hedonic sensitivity in animals from this group.

Data from the presented study suggest that with a proposed chronic stress model, employed methods of additional stress application and gradual decrease of sucrose solution concentration during repeated testing in a sucrose test can ensure a prolonged maintenance and monitoring of a hedonic deficit in mice. This makes our paradigm useful for the evaluation of potential antidepressant effects of chronically applied treatment, which was additionally shown by the study with a 4-week post-stress administration of citalopram (*see below*).

Changes of Sucrose and Water Consumption During the Chronic Stress Experiment: Indication of Elevated Stress-Response in Anhedonic Mice?

During and after exposure to a chronic stress, mice from the anhedonic and non-anhedonic groups stably demonstrate distinct dynamics of water and sucrose consumption, as well as total liquid intake [116-130]. For example, in one of the experiments, after 2.5 weeks of stress, anhedonic mice show a short-term increase in sucrose consumption switching to a decrease of this parameter at the end of the 4-week stress period (Fig.8, adapted from [128]). Non-anhedonic animals also show elevated sucrose consumption after 2.5 and 3.5 weeks of chronic stress, which returns to the values of control on the 4th week of stress application (Fig.8B).

Similar effects were revealed in other studies with our model. We documented an increase in sucrose intake and a preference both in anhedonic and non-anhedonic animals after 3 weeks of stress. The anhedonic group showed a tendency of higher elevation of these parameters than the non-anhedonic mice [121].

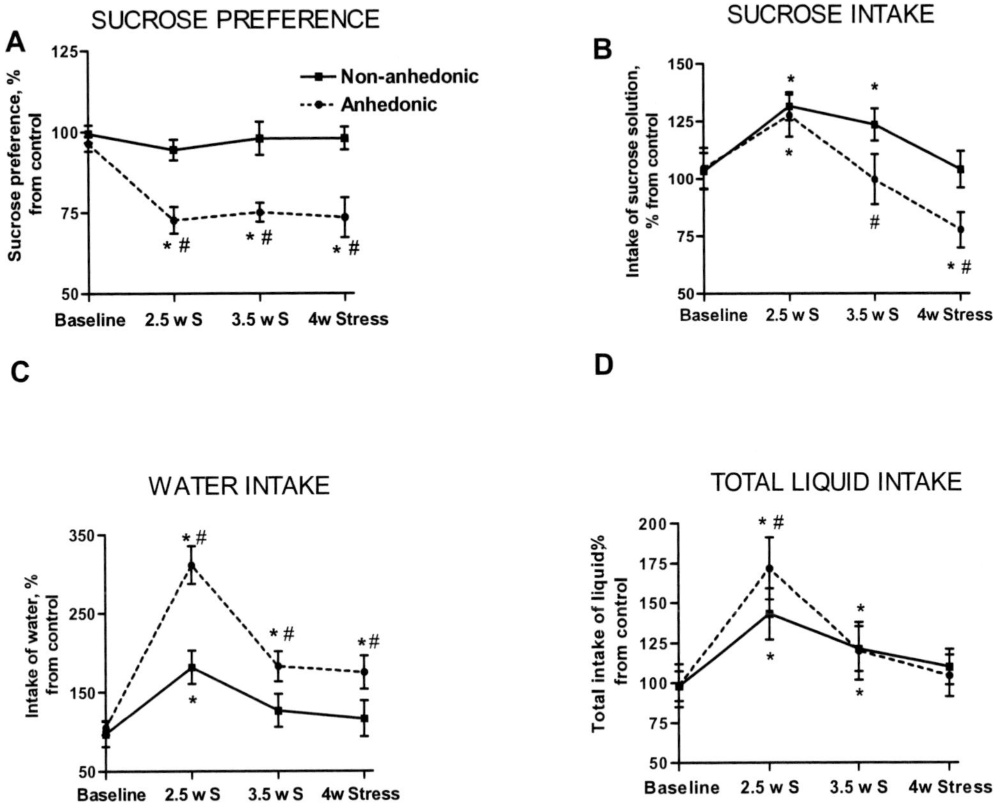

Figure 8. Dynamics of the sucrose test parameters in anhedonic and non-anhedonic mice during stress. (A) Sucrose preference in the anhedonic group is significantly lower than in the non-anhedonic and control mice after 2.5, 3.5 and 4 weeks of stress. (B) Sucrose intake in the anhedonic group is significantly increased after 2.5 weeks of stress and significantly decreased after 3.5 weeks of stress (vs. non-anhedonic group) and after 4 weeks of stress (vs. control and non-anhedonic group). Non-anhedonic mice show elevated sucrose intake after 2.5 and 3.5 weeks of stress. (C) Water intake is elevated in the anhedonic animals after 2.5 - 4 weeks of stress (vs. control and non-anhedonic group). In the non-anhedonic group; water intake is increased after 2.5 weeks of stress as compared to control. (D) Total liquid intake is elevated both in the anhedonic and in non-anhedonic animals after 2.5 and 3.5 weeks of stress (vs. control group). After 2.5 weeks, anhedonic mice show significantly higher total liquid intake then the non-anhedonic mice. Parameters of the sucrose test are expressed in percent from mean values of control group and compared between anhedonic (dashed line) and non-anhedonic (plain line) groups during 4-week stress procedure as mean ± standard error of measurement (SEM) (*$p < 0.05$ vs. control group; #$p < 0.05$ vs. non-anhedonic group; Mann-Whitney).

Anhedonic mice exhibit a profound increase in water intake after 2.5, 3.5 and 4 weeks of stress, as compared to control and non-anhedonic group, and significant elevation of total liquid consumption after 2.5 weeks (in comparison to control and non-anhedonic group) and 3.5 weeks of stress (in comparison to control mice; Fig.8C,D). In non-anhedonic mice, the magnitude and the duration of these changes are less pronounced. Significant effect of elevated water intake is observed only at the time point 2.5 weeks of stress, total intake of liquid is elevated after 2.5 and 3.5 weeks of stress exposure. Anhedonic mice show significantly higher values of water intake than the non-anhedonic mice.

Anhedonic mice demonstrate a lasting increase in water intake and total liquid consumption during the subsequent stress-free period, while non-anhedonic mice show normal values of these measurements. Both parameters are significantly elevated in anhedonic group of animals at the time points 1 and 2 weeks of the after-stress period, as compared to control and non-anhedonic groups, and as compared to controls only, on the 3rd week after the termination of chronic stress (Fig. 9; adapted from [128]).

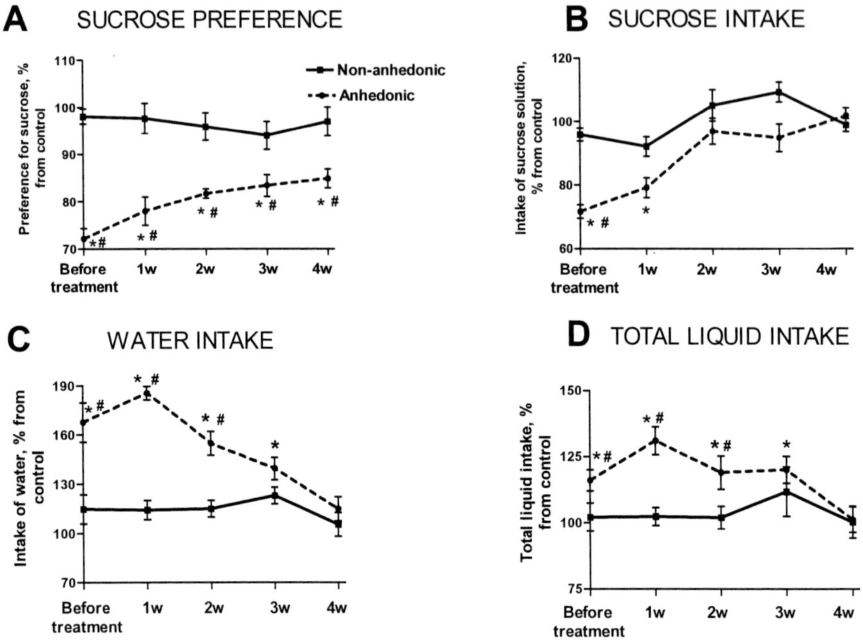

Figure 9. Dynamics of the sucrose test parameters in anhedonic and non-anhedonic mice after the termination of stress. Parameters of the sucrose test were expressed in percent from mean values of control group and compared between anhedonic (dashed line) and non-anhedonic (plain line) groups during 4-week stress procedure (*$p < 0.05$ vs. controls; # $p < 0.05$ vs. non-anhedonic group; Mann-Whitney). (A) Sucrose preference in the anhedonic group is significantly lower than in the non-anhedonic and control mice throughout the entire experiment. (B) Sucrose intake in the anhedonic group is significantly decreased after the termination of stress (vs. control and non-anhedonic group) and during week 1 of the stress-free period (vs. control group). (C, D) Water consumption and total liquid intake are elevated in the anhedonic animals up to 3rd week of a post-stress period as compared to control and up to 2nd week as compared to the non-anhedonic mice. Data on graphs are expressed as mean ± standard error of measurement (SEM).

Stress-induced increase in sucrose preference and liquid intake in mice and rats exposed to a prolonged stress were reported previously [46, 81, 90]. Parallels were made between these signs of elevated consummatory behavior and other indications of behavioral invigoration, e.g., enhanced scores of swimming behavior and increased activity during testing for anxiety-like behavior.

While the nature of the elevated intake of palatable solutions during chronic stress remains unclear, some reports suggest that consumption of sweetened solutions can evoke an antidepressant-like effect in rodents [149]. This led to speculate that possible biological meaning of "paradoxical" increase of sucrose intake and preference in chronically stressed mice might be the "adaptation" to stress. In this light, the development of anhedonia at later stages of chronic stress can be regarded as a manifestation of a "distress" state in mice. As such, large magnitude of described above increase of sucrose intake in mice from the anhedonic group, observed at early stages of stress exposure might reflect their reaction of "hyperadaptation", which might be due to enhanced stress-response of these animals.

In studies with chronic stress models, various alterations in the intake of palatable solutions were observed [148, 149]. Some research groups found no changes in sucrose/saccharin consumption and / or preference in chronically stressed rodents [85, 45, 44, 42, 9]. Our results suggest that an increase, decrease, or the lack of changes in sucrose preference and intake in rodents is a matter of the stage of chronic stress and animals' individual vulnerability to development of anhedonia.

Besides general behavioral invigoration and increase in consummatory activities, the augmentation of general liquid intake observed in our mouse model could also be explained by a stress-induced polydipsia. This phenomenon may result from general sympathetic activation and an increase of metabolic needs in water, diabetes mellitus [105], altered secretion of the hypothalamus and hypophysis [24, 108]. Thus, elevated water intake in anhedonic mice may reflect a pronounced response of these animals to stress.

According to our observations, magnitudes and dynamics of the stress-induced changes in parameters of absolute intake and preference in sucrose test are depending on characteristics of social behavior of a population, evaluated in baseline conditions. Before the onset of stress procedure, all mice in our experiments are tested for initial social behavior in a resident-intruder test (*see below*). In populations with high percentage of aggressive (dominant) individuals, we observed a significant elevation of sucrose preference and a modest transient increase of water and total liquid intake during stress [116-122]. In the populations with high percentage of subdominant (submissive) mice, we found a pronounced increase in absolute sucrose and water intake and lasting augmentation of water consumption, while sucrose preference is not elevated in a course of chronic stress [126-128]. With both profiles of stress-induced alterations of sucrose test parameters, their magnitude was more pronounced in the anhedonic group. Therefore, in our opinion, these changes may reflect animals' stress responsiveness.

Study of the Day / Night Activity During Stress and Anhedonia

Disrupted pattern of a day/night activity is one of the characteristic symptoms of depression. Therefore, we investigated a home cage locomotor activity in control, non-anhedonic and anhedonic mice at the different stages of anhedonia development during the chronic stress experiment. For this purpose, we used a recently elaborated System for Automatic Measurement of Animal Behavior (SAMAB) and specialized customized software SAMAB [31]. In this system, mice are housed individually in the SAMAB cages (35 cm x 20 cm x 15 cm). Mean duration of horizontal moving per day is continuously monitored with a spatial resolution 1.18 x 1.37 cm by two rows of infrared beams placed 2 and 7 cm above the cage bottom.

Behavioral analysis showed that during the dark phase of the day/night cycle, mean time of horizontal activity of mice from the anhedonic group is significantly elevated during 2nd-4th weeks of stress, and 1st -2nd weeks after the termination of stress procedure (Fig.10A). Animals from the non-anhedonic group show no changes in horizontal locomotion as compared to a control group. During the light phase, no differences between the groups are observed (Fig.10B).

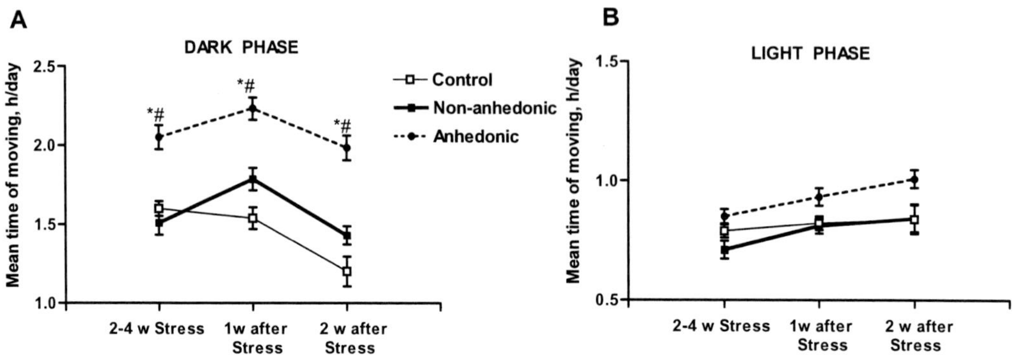

Figure 10. Anhedonic mice display lasting increase of the home cage activity during the dark phase of the day. (A) At the dark phase of a light cycle, mean time of horizontal moving of anhedonic group (dashed line) is significantly elevated during 2nd-4th weeks of stress, and during weeks 1 and 2 after the termination of stress procedure, as compared to the non-anhedonic (plain thick line) and control (plain thin line) groups (*$p < 0.05$ vs. control group and # $p < 0.05$ vs. non-anhedonic group; Mann-Whitney). This parameter does not change in the non-anhedonic group throughout the entire experiment ($p > 0.05$ vs. control group). (B) During the light phase of the day, no difference between the groups in home cage activity is observed ($p > 0.05$). Data on graphs are expressed as mean ± standard error of measurement (SEM).

Our results demonstrate that stress-induced anhedonia in mice is associated with an altered pattern of the day/night activity, which cannot be seen in stressed animals without hedonic deficit. Lasting increase of locomotion in home cages during active phase of animals' cycle in the anhedonic mice may be due to the elevated sympathetic activation and be related to their increased stress response.

Individual Variability in Susceptibility to Stress-Induced Anhedonia

Even though C57 BL/6 mouse strain is a genetically homogeneous line, we observed pronounced individual variability in animals' susceptibility to stress-induced anhedonia. Individual variability in behavioral and physiological parameters is well documented in inbred strains [13, 33, 66, 100, 143, 144]. In order to identify behaviors, which possibly predict individual predisposition of mice to the anhedonia development, we tested initial animals' behavior in baseline conditions (i.e., before the onset of chronic stress). Therefore, we used the open field test, novel cage and new object exploration paradigms, elevated Zero-maze and dark/light box. Obtained data were related to the results from the consequently performed chronic stress experiment, in which animals were assigned to non-anhedonic and anhedonic groups [121].

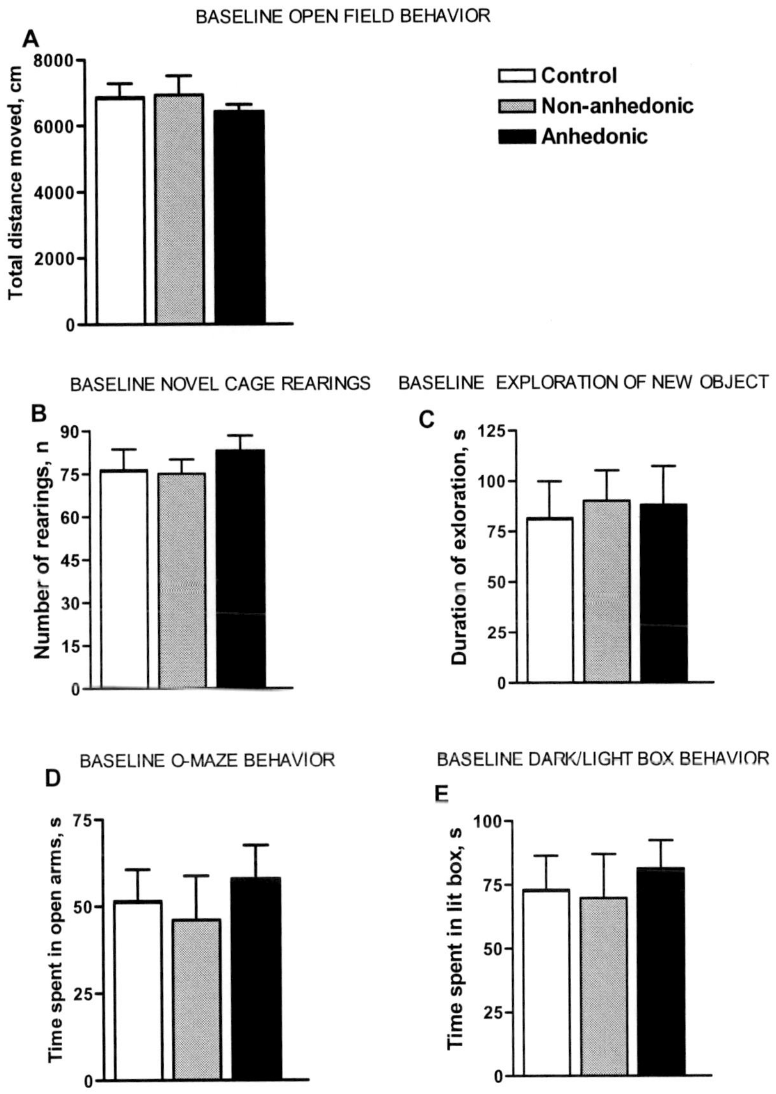

Figure 11. Similar baseline behaviors of non-anhedonic and anhedonic mice. No statistically significant differences between the non-anhedonic and non-anhedonic groups were observed (A) in the total distance moved, evaluated in the open field test, (B) number of exploratory rearings in the novel cage, (C) duration of exploration of a new object, (D) time spent in open arms scored in the elevated Zero-maze and (E) time spent in lit compartment of dark/light box ($p > 0.05$; Mann-Whitney). Data on graphs are expressed as mean ± standard error of measurement (SEM).

Mice from the non-anhedonic and anhedonic groups have similar initial total distance moved in the open field (Fig.11A), exploratory rearing activity in the novel cage and duration of new object exploration (Fig.11B,C), as well as time spent in the open arms of the elevated Zero-maze and the lit compartment of the dark/light box (Fig.11D,E).

Studies of animals' baseline behavior were repeated with slightly different environmental and procedural conditions. Generally, they replicated original result showing a lack of differences between mice with distinct vulnerability to a stress-induced anhedonia in initial characteristics of locomotion, exploration, anxiety-like behaviors and parameters of the sucrose test, such as sucrose intake and preference, water intake and total liquid consumption [116-130]. The exception was a finding of an inconsistent decrease of the open field activity in animals from the anhedonic group observed during winter [117].

Clinical studies indicate that timid behavior is a pre-existing feature, which predisposes to the development of depression and anhedonia. Low social rank in rats in repeated defeat paradigm has been associated with a decrease in sucrose consumption [29, 149]. Ethological studies revealed similarities in the behavioral characteristics of animals with a subdominant type of social behavior on one hand, and in behavioral models of depression of helplessness and social defeat, on another [70, 133,13]. These and other facts suggest a rationale for the analysis of animals' initial social traits in chronic stress depression paradigm.

For analysis of the social behavior in mice we used a resident-intruder test (adapted from [67]). In this paradigm, animals are classified as submissive (subdominant or non-aggressive), non-submissive (dominant or aggressive) and socially neutral individuals. 3.5 months old male CD1 mice, used as intruders in a resident-intruder test, are kept in groups of five with their littermates for 2 weeks before the experiment. During the experiment, the C57 BL/6 behaviorally naive mice of the same age, single housed during 2 weeks prior testing, are placed as a resident in an observation cage for 30 min; thereafter, male CD1 mice are introduced to the same cage for 8 min period. Complete lack of attacks towards the partner, accompanied by specific "submissive" postures, escape and defense are regarded as submissive (subdominant) type of social behavior in this test. Initiation of attacks towards the partner and fighting back in response to attacks are categorized as non-submissive (dominant or aggressive) behavior. Social interaction and/or exploratory behavior instead of confrontation are interpreted as socially neutral behaviors.

Our studies with different variants of the 4-week stress procedure showed, that individual predisposition of C57 BL/6 mice to a stress-induced anhedonia is associated with their features of social interaction. Individuals with submissive social traits are found to be more vulnerable to stress-induced anhedonia [134, 121, 122, 126-128]. In populations with initially low percentage of submissive animals (15-20%), all submissive mice develop anhedonia (Fig.12A; [121]). In populations with high percentage of submissive individuals (over 50%), the percentage of non-submissive (dominant or aggressive) animals susceptible to anhedonia is significantly lower in comparison to a submissive cohort of mice (Fig.12B,C; [128]).

Notably, while in populations with initially low percentage of submissive animals, statistically significant decrease of sucrose preference in the stressed group occurs at 3.5 weeks of stress time point [121], in experimental groups containing high percentage of submissive individuals (over 50%) the drop in sucrose preference is detected already after 2.5 weeks of stress [126-128].

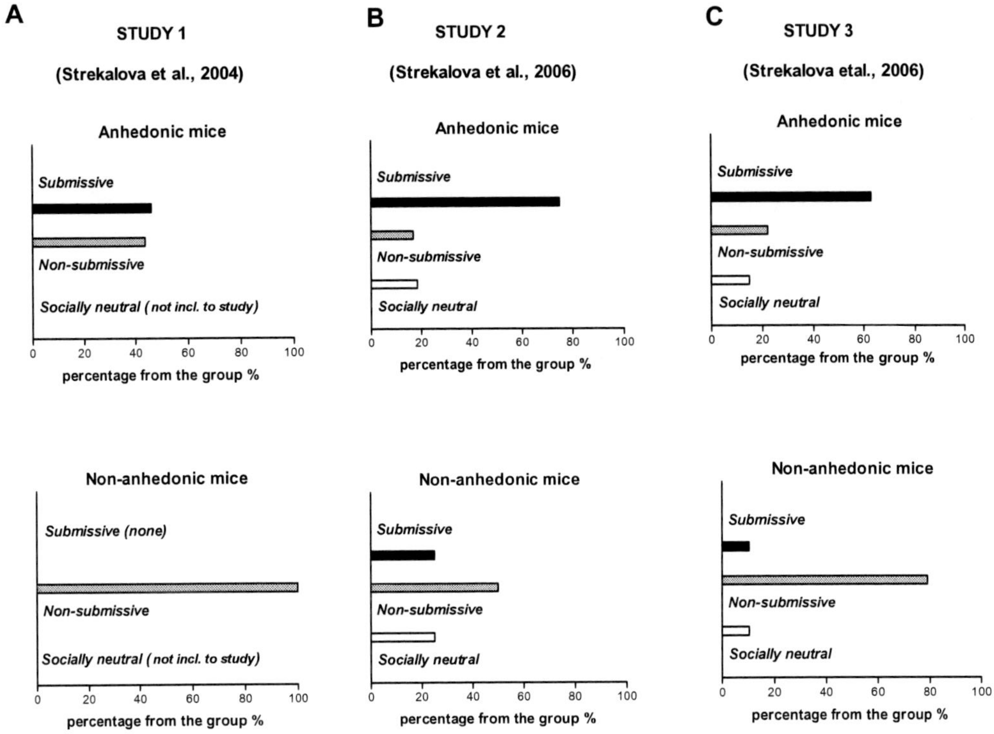

Figure 12. Individual vulnerability to stress-induced anhedonia and initial social traits. In three experiments: (A) Study 1 [121], (B) Study 2 [128] and (C) Study 3 [128], percentage of individuals with submissive, non-submissive and socially neutral initial social status was calculated from a total number of animals in anhedonic or non-anhedonic groups. In all experiments, percentage of non-submissive animals was significantly lower in anhedonic than in the non-anhedonic groups (Fisher' exact test).

In a course of 4-week stress, anhedonia was found to occur earlier in submissive animals. In one of the experiments, at the time point of 3 ½ weeks of stress, 100% of submissive mice showed sucrose preference below 65% and matched a taken criterion of anhedonia, while only 16.6% of aggressive mice from the anhedonic group had a hedonic deficit at this stage of the chronic stress [121]. Furthermore, in a resident-intruder test, non-submissive animals from the anhedonic group are characterized by longer latency of the first attack (mean 328.3 ± 74.2 s) and diminished number of attacks (mean 6.9 ± 1.8) in comparison to the animals with the same type of social behavior from the non-anhedonic group (14.5 ± 18.6 s and 13.3 ± 3.9, respectively; [121]).

Detailed ethological analysis of more than thirty parameters of social behavior in a resident-intruder test revealed reduced scores of aggressive behavior and social contacts of a dominant-like type in mice predisposed to a stress-induced anhedonia, thus, confirming our original finding [134]. This study employed a variant of our 4-week stress procedure and a sucrose test paradigm in the C57 BL6/N mice. Thus, a body of evidence obtained in our chronic stress model suggests that submissive social traits correlate with vulnerability to a stress-induced anhedonia.

CHARACTERISTICS OF STRESSORS EMPLOYED IN A 4-WEEK STRESS PROCEDURE

Presumably, unpredictable uncontrollable chronic stress excludes animals' adaptation to applied stressors and represents itself the only type of chronic stress procedures, which is thought to fit the construct validity criteria of animal models of depression [18, 35, 57]. We believe that our protocol matches these criteria also. Procedural variations of a proposed chronic stress model comprise of several stressors (three to five different stressors per experiment), which are switched 7-18 times over a four-week period i.e., they resemble unpredictable chronic stress (Fig.13).

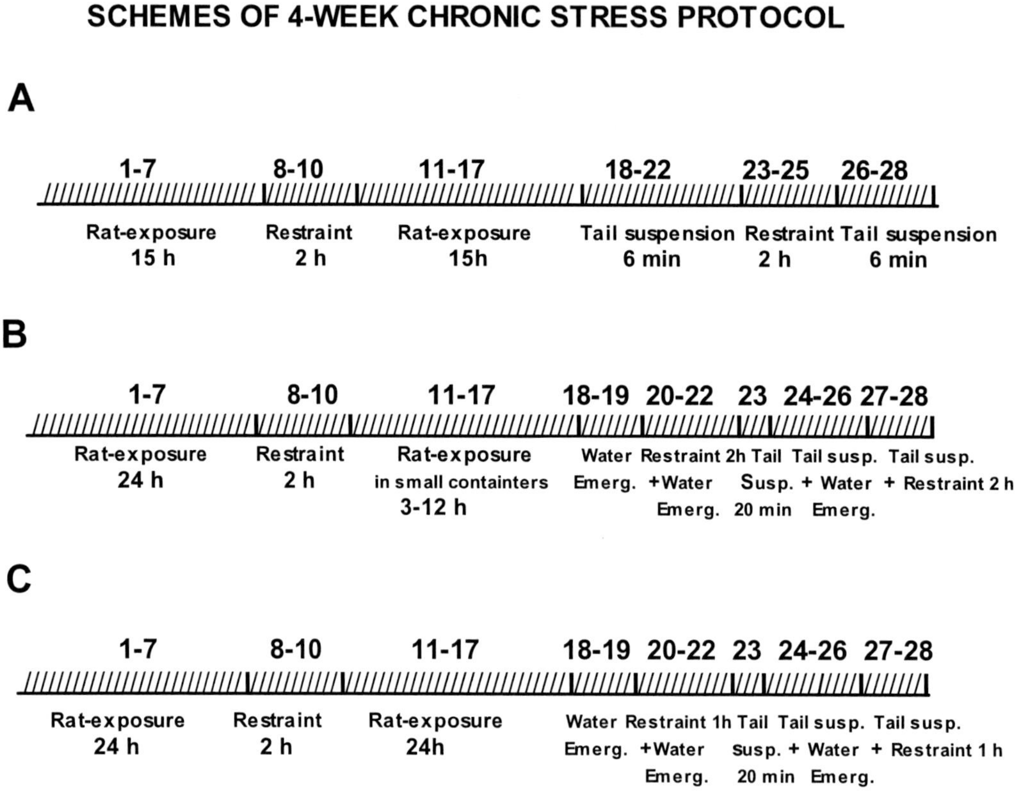

Figure 13. Variants of the 4-week chronic stress protocol. (A) Protocol of chronic stress comprised of a periodic 15h rat-exposure, 2h restraint and 6-min tail suspension stress [121]. (B) A variant of chronic stress comprised of a continuous 24h rat-exposure, water emergency combined with restraint stress, 6-min tail suspension stress and 1 or 2h restraint stress [125]. (C) A modification of the chronic stress procedure comprised of continuous 24-h and periodic 3-12h rat-exposure, water emergency combined with a restraint stress, 20-min tail suspension stress and 2h restraint stress [128].

Rotation of stressors in our model prevents the animals' habituation to them similarly to the unpredictable stress: experimental data evidence reduced body weight and decreased sucrose intake and preference in both paradigms [116-130, 148, 149]. Repeated and monotonous exposures to the same stressor over a prolonged period, which are accompanied by signs of animals' adaptation to stress, most of the time lead to a number of substantial physiological alterations, but not to a diminished sensitivity to reward and other features of depressive-like state in rodents [21].

In order to monitor animals' habituation to the chronic stress in a course of the 4-week experiment, we analyzed weekly-measured body weight; lack of significant differences between control and stressed groups was interpreted as an indicator of adaptation to a chronic stress. To control habituation to single stress sessions during chronic stress, we evaluated the immediate post-stress locomotor activity of mice; lack of behavioral inhibition right after the exposure to a stressor was taken as a sign of their habituation to a stressor. Adjustment of original protocol with defined stress intensity, as described elsewhere [121], to the changing characteristics of habituation and stress response in different batches of C57 BL/6 mice resulted in several procedural variations of the 4-week chronic stress model.

Similar reduction of body weight and sucrose preference with unpredictable stress procedures and a proposed 4-week chronic stress protocol suggest that these paradigms result to comparable physiological effects in rodents and respond the construct validity criteria of animal models of depression.

Repeated stress of limited duration of application used in our model might be of advantage over the procedures based on the unpredictable uncontrollable stress, as an application of the same stressor during several days in a row is likely to induce a state of learned helplessness, a pathogenetic factor of depression. Besides, in comparison to the models of unpredictable chronic stress, our stress paradigm with repeated stressors application might be closer to the modeling of life stress in human, which typically consists in recurrent repetition of several stress situations, rather than in daily experiencing of new stressors.

Another important feature of the construct validity criteria of the depression models is the stressor uncontrollability, a factor of development of learned helplessness in animals [141, 2, 36]. In our 4-week procedure, we use exclusively uncontrollable stressors.

Altogether, several variations of the 4-week chronic stress protocol were tested. We employed a tail suspension and a restraint stress of various durations, water emergency stress (as a variant of the restraint stress) and three variants of the rat-exposure stress (for experimental details, see [116-130]. In one of the protocols, tail suspension was omitted from a stress procedure in order to use this paradigm for behavioral characterization of stressed animals at the end of the experiment [118]. As it was mentioned above, duration and type of stressors were adjusted to specific characteristics of tested population, as significant diversity in behavioral stress response of mice from different batches was observed. This variability is supposively due to environmental and seasonal factors. With all variants of the 4-week chronic stress, a significant reduction in body weight and sucrose preference were reliably induced in the stress group.

Definition of Anhedonia and Sensitivity to Reward with a Sucrose Test

Several paradigms have been proposed to measure sensitivity to reward in rodents, such as consumption of palatable solutions, intracranial self-stimulation, progressive ratio responding, novel-object place conditioning and conditioned place preference [145, 68, 12, 102, 72]. Free access paradigm of the sucrose / saccharine consumption test is thought to be a useful method of assessing the hedonic state in the chronic stress depression models, as its experimental procedure is not demanding for the animals and characterized by relatively high through output. In comparison to the other tests, evaluation of hedonic state with a sucrose test excludes involvement of learning component and minimally implicates factors of anxiety and locomotion, which alteration, known to be typical for a depressive-like state, can affect a measurement of sensitivity to reward in animal depression paradigms.

Decreased intake of sucrose and other palatable solutions was proposed to be a measure of hedonic deficit in chronic stress depression models (for a review, see [149]). Meanwhile, absolute intake of sucrose solution in mice and rats can be influenced by many factors, not related to their hedonic status. It can be affected by subtle stressors occurring during the sucrose test and by prolonged effects of acutely applied stress procedures, deprivation of food and water, stress-induced changes in consummatory behaviors, as well as by altered metabolic needs in calories and liquids [96, 73, 37, 97]. In our studies, we modified the protocol of the two-bottle sucrose test (*see below*) and used preference to a 1%-sucrose solution over water, calculated as a percentage of consumed sweetened solution of the total amount of liquid drunk, as a parameter of hedonic state in mice.

A decrease of preference for 1%-sucrose solution below 65%, measured after 4-week stress, is taken in our model as criterion for anhedonia. Importantly, none of the control mice exhibit ≤ 65% preference for sucrose at the end of the 4-week stress experiment. In other models, animals with a preference to 1%-sucrose solution ≤ 65% had shown features of anhedonia [146, 149, 93, 85], as well as an increased threshold of intracranial self-stimulation [78] and a decreased latency and increased duration of REM sleep [137, 79, 23]; these changes are considered to be physiological features of a depressive syndrome in rodents. Stress-induced decrease of sucrose preference below 65% was also accompanied by a reduction of sexual activity [30, 15] and alterations of circadian rhythms [113, 111]. Our studies revealed pronounced differences in depressive-like features between the groups of chronically stressed mice formed upon the 65%-sucrose preference criterion, thus, confirming a physiological validity of chosen definition of anhedonia applied in our model [116-130].

Sucrose Test: Identification of Major Artifacts in Mice and Modification for the Analysis of Inter-Individual Differences

Various protocols of the sucrose test were initially elaborated in rat models of depression and later were extrapolated, often without changes, to the mouse models [58, 145, 110]. In contrast to the experimental situation in rats, most of them reveal differences in the hedonic state between the groups of mice, but not between individual animals [14, 131].

Insufficient resolution of the mouse sucrose test can be due to the species-specific physiological features of drinking behavior in mice and physical artifacts in its evaluation. For example, in comparison to rats, mice generally demonstrate lower values of sucrose preference and sucrose intake [146, 26, 87], a pronounced neophobic behavior during the very first access to a sucrose solution [65, 40, 109], large individual variability in sucrose preference and, especially, in absolute values of liquid intake [77, 1, 87]. Preference and intake of sucrose solutions were found to increase substantially with repeated sessions of the sucrose test, suggesting that the results of testing in this paradigm depend on the animals' previous experience of sucrose consumption [121, 3]. As mentioned above features of the sucrose test in mice may limit its accuracy, we analyzed their potential contribution to the outcome from this test.

When given a choice between two drinking bottles placed on the left and right corners of the cage, mice show a preference to drinking either at one or another position. Side preference in drinking position is not related to any obvious external factors, such as sources of noise and lighting, localization of a home cage in a room, etc. In order to evaluate a contribution of this phenomenon in a measurement of preference to 2.5%-sucrose solution, we first estimated a side preference either to a left or right corner for each mouse used in the experiment in a 10-h two-bottle water test. Then, in a two-bottle paradigm of choice between water and sucrose solution, mice were allowed to consume 2.5% sucrose solution either from the preferable (Day 1) or non-preferable (Day 2) sides of the cage during two consequent 10-h tests (Fig.14A). On the third day, the bottles with sucrose solution and water were switched in a midway of 10-h testing. We found that sucrose preference was significantly higher when a bottle with 2.5% sucrose solution was placed in a preferable side of the cage than when it was situated on the non-preferable corner. Switching of the bottles resulted in an intermediate mean value of sucrose preference, as compared to the values obtained during the testing on Day 1 and Day 2. This experiment suggested that a side preference essentially affects an outcome from the sucrose test. Switching of the bottles with water and sucrose solution in a midway of the sucrose test led to minimize this confounding factor.

In another experiment, mice were first housed with one drinking bottle; their side preference was evaluated in a 10 h two-bottle water consumption test (Fig.14B). Thereafter, they were housed with two drinking bottles for a time period of 2 weeks and their preference in water consumption for initially preferable side of the cage was re-evaluated in a 10-h drinking test. The housing with water on both sides of the cage abolished side preference in drinking behavior in some but not all animals. Thus, this method cannot reliably prevent confounding effects of side preference in evaluation of a preference for palatable solutions with a two-bottle sucrose test.

As measurement of absolute sucrose intake can be confounded by individual differences in animals' daily patterns of drinking behavior, we studied dynamics of water consumption in a one-bottle 10-h test during 4 measurements spaced by 2.5 h in a group of mice (Fig.14C). Three patterns of drinking behavior were revealed: 36.4% and 13.6% of mice showed their peaks in water intake 5 h and 10 h after the onset of the test (and the dark phase of the day), respectively, and 50% of animals had a peak in water intake 7.5 h thereafter. The latter group demonstrated significantly higher values of total water intake, as compared to the other two groups of mice. These data evidence a large variability of daily dynamics of water intake, thus, pointing to importance of a prolonged testing of drinking behavior in mice. Obtained results also suggest a convenience of the application of relative parameters of the sucrose /

saccharine test in evaluation of reward sensitivity in mice, rather than of absolute values of consumption of palatable solutions.

When given a choice between two drinking bottles placed on the left and right corners of the cage, mice show a preference to drinking either at one or another position. Side preference in drinking position is not related to any obvious external factors, such as sources of noise and lighting, localization of a home cage in a room, etc. In order to evaluate a contribution of this phenomenon in a measurement of preference to 2.5%-sucrose solution, we first estimated a side preference either to a left or right corner for each mouse used in the experiment in a 10-h two-bottle water test. Then, in a two-bottle paradigm of choice between water and sucrose solution, mice were allowed to consume 2.5% sucrose solution either from the preferable (Day 1) or non-preferable (Day 2) sides of the cage during two consequent 10-h tests (Fig.14A). On the third day, the bottles with sucrose solution and water were switched in a midway of 10-h testing. We found that sucrose preference was significantly higher when a bottle with 2.5% sucrose solution was placed in a preferable side of the cage than when it was situated on the non-preferable corner. Switching of the bottles resulted in an intermediate mean value of sucrose preference, as compared to the values obtained during the testing on Day 1 and Day 2. This experiment suggested that a side preference essentially affects an outcome from the sucrose test. Switching of the bottles with water and sucrose solution in a midway of the sucrose test led to minimize this confounding factor.

In another experiment, mice were first housed with one drinking bottle; their side preference was evaluated in a 10 h two-bottle water consumption test (Fig.14B). Thereafter, they were housed with two drinking bottles for a time period of 2 weeks and their preference in water consumption for initially preferable side of the cage was re-evaluated in a 10-h drinking test. The housing with water on both sides of the cage abolished side preference in drinking behavior in some but not all animals. Thus, this method cannot reliably prevent confounding effects of side preference in evaluation of a preference for palatable solutions with a two-bottle sucrose test.

As measurement of absolute sucrose intake can be confounded by individual differences in animals' daily patterns of drinking behavior, we studied dynamics of water consumption in a one-bottle 10-h test during 4 measurements spaced by 2.5 h in a group of mice (Fig.14C). Three patterns of drinking behavior were revealed: 36.4% and 13.6% of mice showed their peaks in water intake 5 h and 10 h after the onset of the test (and the dark phase of the day), respectively, and 50% of animals had a peak in water intake 7.5 h thereafter. The latter group demonstrated significantly higher values of total water intake, as compared to the other two groups of mice. These data evidence a large variability of daily dynamics of water intake, thus, pointing to importance of a prolonged testing of drinking behavior in mice. Obtained results also suggest a convenience of the application of relative parameters of the sucrose / saccharine test in evaluation of reward sensitivity in mice, rather than of absolute values of consumption of palatable solutions.

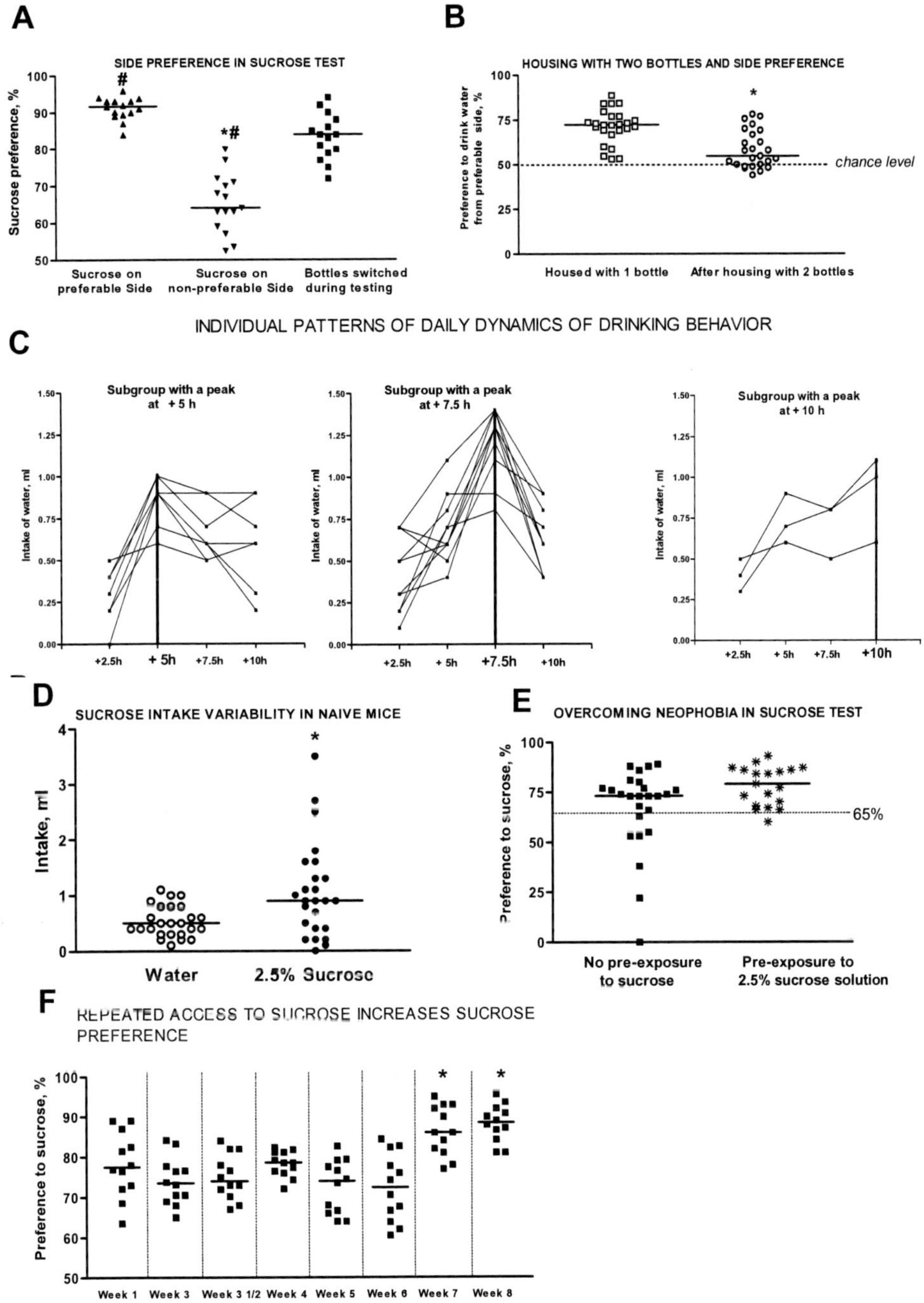

Figure 14. Identification of major confounding factors of sucrose test in mice. (A) Mice, allowed to consume 2.5% sucrose solution either from preferable or non-preferable sides of the cage in the two-bottle choice paradigm, had the means of sucrose preference, which were significantly different from each other (n=15; $*p < 0.05$; Wilcoxon) and from the value obtained in the test, when the bottles were switched in a midway of testing ($\#p < 0.05$; Wilcoxon). (B) In a choice test, preference to intake water from the bottle placed on originally preferable side of the cage was significantly reduced after 2-week housing with two drinking bottles (n=25; $*p < 0.05$; Wilcoxon). However, individual data show that

part of mice kept original side preference in drinking behavior. (C) 10-h water consumption test revealed three patterns of drinking behavior in mice: with peaks in water intake 5 h and 10 h after the onset of the dark phase of the light cycle and with a peak at 7.5 h thereafter. The latter group had significantly higher values of total water intake, as compared to the other two groups (n=20; *$p<0.05$, Mann-Whitney). (D) Comparison of absolute intake of water and 2.5% sucrose solution in mice naive for a sucrose taste evidences a pronounced individual variability of sucrose solution consumption during the first access to a sucrose solution in a 3-h one-bottle test. Intake of sucrose solution was significantly higher than intake of water (n=25; *$p<0.05$, Wilcoxon). (E) Mice pre-exposed to a 2.5% sucrose solution (n=19) showed decreased variability in preference to a 1% sucrose solution during a consequent testing in comparison to mice naive for a taste of sucrose (n=25). Thus, pre-exposure of mice to sucrose solution allows overcome neophobia in sucrose test. (F) Repeated weekly testing with 1% sucrose solution results to increase of sucrose preference (n=12; *$p<0.05$, versus measurement on week 1, Wilcoxon).

Measurement of intake of water and 2.5% sucrose solution, which was performed during two consequent days in a 3-h test, evidenced a pronounced individual variability in sucrose consumption of mice, naive for a sucrose taste. Individual data show diversity in animals' reactivity to a sweet taste, which is varying from neophobia to an exceeded consumption of sucrose solution (Fig.14D). Overall, mice showed a higher intake of sucrose solution than water intake.

Mice inexperienced with a taste of sucrose, showed lower sucrose preference to 1% solution and significantly larger variability of this parameter than mice pre-exposed for 2 h to a 2.5% sucrose solution on a day preceding the sucrose test experiment (Fig.14E). Thus, in a 1%-sucrose preference test, pre-exposure of mice to a concentrated sucrose solution allows to reduce variability in sucrose preference; most likely, this effect is due to an inhibition of neophobic reaction to a novel taste.

In another study, weekly repeated testing with 1% sucrose solution of a group of mice resulted in a significant increase of sucrose preference on the 7th week (Fig.14F). This phenomenon may also to be due to a decrease of the neophobic reaction of mice towards a sucrose solution, and a sensitization to a taste of sucrose.

Thus, our data suggest that side preference, neophobia, inter-individual differences in circadian patterns of liquid consumption, and previous experience of sucrose intake essentially affect an outcome from the sucrose test. Taking into account these and other results, we modified a sucrose test protocol that allowed us to overcome identified confounding factors of this procedure and assess the hedonic state of chronically stressed mice on the individual basis.

First, in contrast to the majority of other paradigms, the evaluation of a hedonic state in our model was based on the sucrose preference, not the absolute sucrose intake [19, 42, 149]. Apart from described above experimental arguments for that, we found that values of absolute intake of sucrose solution are characterized by greater variability, in comparison to those of sucrose preference, as significant differences of variances were revealed by the F test. This defines sucrose preference as a more stable parameter, which is apparently less influenced by animals' individual patterns in drinking behavior.

Second, we used refined testing conditions of the sucrose test procedure, consisting in the following. 1) Duration of the sucrose test was 10 or 24 h, instead of 1-2 h; in order to prevent a sensitization to sucrose taste and avoid an interruption too long of the chronic stress procedure, it was limited by 24 h. 2) To diminish the effect of neophobia, mice naive to a sucrose taste were allowed to consume 2.5%-sucrose solution for 2 h a day before the first sucrose test. 3) In order to minimize the effect of side preference, position of the bottles was switched in the middle of testing. 4) Sucrose test was started with the offset of the light and carried out during the dark (active) phase of animal's cycle. When 24-h protocol was employed, presence of preferable and non-preferable solutions on each side of the cage was equilibrated with respect to a circadian cycle. 5) Temperatures of the drinking solutions and air in the lab rooms were balanced. 6) No food and water deprivation was permitted before the test for at least 24 h. 7) No stressors are applied for at least 12 h prior the test. 8) For multiple repeated testing, gradually descending sucrose concentrations were used.

Methodological modifications of the sucrose test listed above increased its sensitivity that enabled the individual mice with and without stress-induced "qualitative" changes in a sucrose preference to be differentiated, according to a taken criterion of anhedonia. This differentiation correlated with appearance of other depressive-like pathological alterations, as it was shown in our model with various methods [10, 41, 31, 123, 130]. Together, these data speak in favor of the validity of the sucrose test paradigm as a method of evaluation of depressive-like state in rodents.

Hyperlocomotion as a Confounding Factor in Behavioral Analysis of Chronically Stressed Mice

Majority of studies with models of stress-induced anhedonia in rodents failed to define consistent behavioral correlates of stress-induced decrease in sucrose intake and preference and resulted to controversial data obtained in various behavioral paradigms, in particular, in tests for locomotion, anxiety, exploration and behavioral despair / coping traits. Our recent results suggest that discrepancies in the behavioral testing of chronically stressed mice might be due to the phenomenon of hyperlocomotion, an unspecific consequence of prolonged exposure to stress in rodents, which is triggered by a stressful procedure of behavioral testing [122, 125]. According to these findings, we modified protocols of behavioral analysis of chronically stressed mice by using "mild" conditions of testing with low stress impact that precluded an occurrence of stress-induced hyperlocomotion and enabled a characterization of behavioral correlates of anhedonia in our chronic stress model without artifacts (*see above*).

Previous studies with chronic stress models revealed some paradoxical changes in rodents' behavior, which contradict generally known behavioral profiles of states of stress and depression. Chronic mild stress procedure was shown to increase time spent in anxiety-related areas of the elevated plus maze and dark/light box in mice that was regarded as "anxiolytic-like" changes [30, 20, 64, 99]. Mice, subjected to chronic tail-suspension stress and crowded housing had decreased parameters of anxiety in the dark/light paradigm; no changes in anxiety-related behavior was found after chronic foot-shock stress [101]. Alternations of lighting rhythm, restraint stress, environmental temperature, and acute cold stress did not affect mouse behavior in the elevated plus maze but resulted in reduced immobility time in forced swimming in mice [47-49]. Diminished immobility during forced swimming associated with increased plasma corticosterone levels was detected after neonatal handling stress in mice [88]. Two-week stress, consisting of restraint stress and food shortage increased sensitivity to amphetamine-induced hyperlocomotion and diminished immobility in the forced swim test [17]. Application of uncontrollable foot shock in mice induced prolongation of swimming, which could be prevented by an injection of diazepam [94].

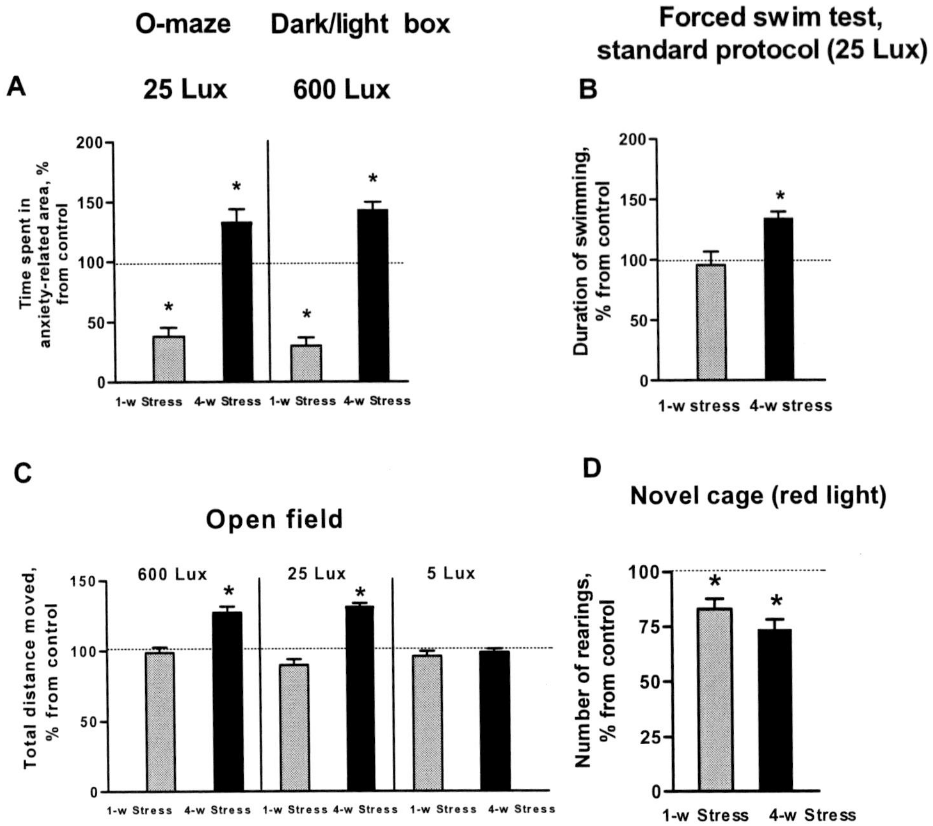

Figure 15. Light triggers hyperlocomotion in chronically stressed mice that confounds their behavioral testing. (A) Time spent by mice in the open arms of the Zero-maze (lighting intensity 25 Lux) is increased after a 4-week stress and decreased after a 1-week stress, as compared to a control group. Time spent in the lit compartment of the dark/light box (lighting intensity 600 Lux) is significantly elevated in chronically stressed mice and significantly decreased in mice subjected to a 1-week stress. (B) Mean duration of swimming evaluated in a "standard" procedure of the forced swim test (lighting intensity 25 Lux) is significantly increased after a 4-week stress and does not change after a 1-week stress as compared to controls. (C) 4-week chronic stress induces hyperlocomotoion triggered by bright and moderate, but not weak lighting. In the open field lit with light of an intensity of 600 and 25 Lux, mean total distance moved is significantly increased in mice submitted to a 4-week stress, but not in animals exposed to a 1-week stress, as compared to the non-stressed controls. Mean total distance moved is not different from that of control in both chronically and subchronically stressed mice in the open field lit with light of an intensity of 5 Lux. (D) Both 1-week subchronic and 4-week chronic stress inhibit novel cage rearings scored under red lighting (*$p < 0.05$ with respect to the control group; Mann-Whitney). Data are expressed in percent from means of a control group as mean ± standard error of measurement (SEM).

We tested a hypothesis that chronic stress in mice induces unspecific changes in locomotion, which interfere with their behavior in the forced swim and anxiety tests. Therefore, we applied a 4-week chronic stress protocol and 1-week subchronic stress in C57BL/ 6N mice. Both protocols were consisting of the same stressors: rat exposure, restraint stress and tail suspension. After stress, animals were studied in elevated Zero-maze, dark/light box and forced swim tests under lighting of strong, modest and subtle intensities. Locomotor activity in the open field and rearing in the novel cage were scored in animals under lighting conditions, respective to those, which have been used in anxiety and swimming tests (for experimental details, see [125]).

Time spent by mice in the open arms of the Zero-maze (lit with modest lighting of 25 Lux) and the lit compartment of the dark/light box (lighting intensity was 600 Lux) increased after a 4-week stress and decreased after a 1-week stress (Fig.15A, adapted from [125]). Thus, chronically stressed mice tested under modest and strong illumination exhibit an "anxiolytic-like" behavior under conditions, when subchronically stressed animals show elevated scores of anxiety. Mean duration of swimming evaluated in a "standard" protocol of the forced swim test (lighting intensity of 25 Lux was employed) increased after a 4-week stress and did not change after a 1-week stress (Fig.15B). In the open field lit with 600 and 25 Lux light, mean total distance moved was elevated in mice submitted to a 4-week stress, but not in animals exposed to a 1-week stress. No differences in the locomotion scores were detected between chronically and subchronically stressed mice tested in the open field under dimmed illumination of 5 Lux (Fig.15C). Both 1-week and 4-week stress inhibited novel cage rearings scored under red lighting (Fig.15D). Thus, 4-week chronic stress induces a hyperlocomotoion triggered by bright and moderate illumination, which does not occur under weak and red lighting.

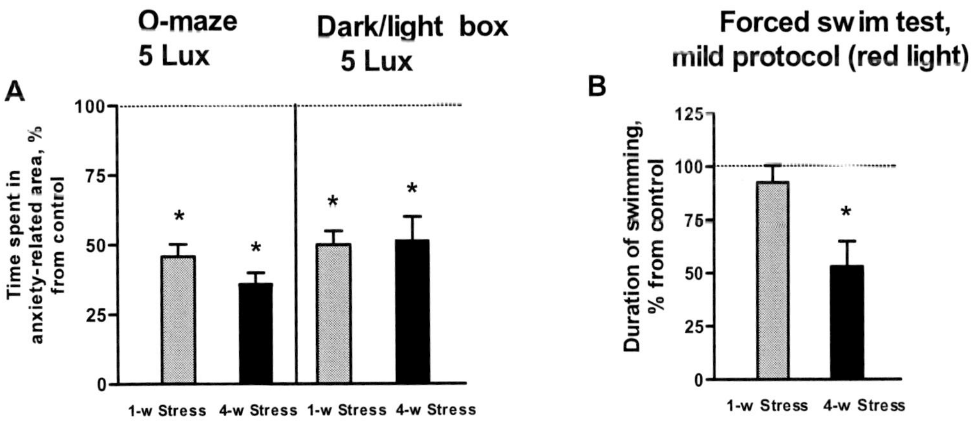

Figure 16. Testing under weak (5 Lux) and red illumination precludes hyperlocomotion of chronically stressed mice in anxiety and forced swim tests. (A) Time spent in the open arms in the Zero-maze (lighting intensity 5 Lux) and in the lit compartment of the dark/light box (lighting intensity 5 Lux) is significantly decreased after 4-week and 1-week stress when compared to controls. (B) Mean duration of swimming in the forced swim test (red lighting was employed) is significantly decreased after a 4-week chronic stress and remain unchanged in mice stressed during 1 week (*$p < 0.05$ with respect to the control group; Mann-Whitney). Data are expressed in percent from means of a control group as mean ± standard error of measurement (SEM).

Weakening of lighting employed during testing of chronically stressed mice in the anxiety and forced swim tests revealed increased anxiety-like behavior and inhibited swimming in these animals. Time spent in the open arms in the Zero-maze and in the lit compartment of the dark/light box both illuminated with 5 lux light was significantly decreased in mice subjected either to a 4-week or a 1-week stress (Fig.16A). Mean duration of swimming scored in the forced swim test under red lighting was significantly decreased after a 4-week chronic stress and remained unchanged in mice stressed during 1 week (Fig.16B, adapted from [125]).

Together, a 4-week chronic stress induces hyperlocomotion in mice, which is triggered by subtle stressors like light and confounds evaluation of anxiety and forced swimming behaviors. Mice, subjected to a 1-week stress did not change parameters of locomotion and forced swimming, and showed elevated scores of anxiety-lime behavior irrespectively to illumination, thus validating anxiogenic effects of employed stressors and paradigms of anxiety-like behavior that were used in the study. Obtained data demonstrate that only prolonged stress could evoke a phenomenon of hyperlocomotion in C57BL/ 6N mice.

A single injection of low dose of diazepam abolished hyperlocomotion, "anxiolytic-like" pattern and prolongation of swimming observed during testing under strong illumination [125] that is in line with previous observations [94]. Because the main pharmacological activity of benzodiazepines is a blockade of the acute stress reaction, the fact that the administration of diazepam elicited the same effect as a decrease in the illumination intensity suggests that a stress-induced hyperlocomotion, documented in various tests here, is due to the stress impact of lighting used during testing.

Our data are in line with results of other groups. On one hand, various effects of chronic stress on general locomotion in rodents: hyperactivity, a hypoactivity, and lack of changes, were described [101, 29, 46, 52, 82, 16]. On another hand, lighting conditions employed during testing were reported to be a significant factor of general locomotor activity in the stressed animals [54, 135, 11]. Systematic analysis of locomotion in stressed mice with respect to the characteristics of employed lighting and duration of stress performed with our study revealed a relationship between these two factors, and let to define lighting conditions, which preclude occurrence of behavioral artifacts in testing of chronically stressed C57BL/ 6N mice. We believe that our finding with stress-induced hyperactivity in mice can explain previously reported contradictions resulting from behavioural testing in chronic stress models.

Chronic Citalopram Administration Counteracts Anhedonia Development

Demonstration of the preventive action of antidepressant treatment on the development of stress-induced anhedonia is regarded as an indication of construct validity of the chronic stress depression models [77, 39, 44, 150]. Apart from that, premedication with antidepressants in animal models simulates preventive pharmacotherapy of patients with high risk of iatrogenic depression [132].

In order to study whether antidepressant treatment counteracts the development of anhedonia and depressive-like phenotype in our model, we employed selective serotonin reuptake inhibitor (SSRI) citalopram, a commonly used antidepressant [64, 136]. The selection of citalopram as a pharmacological reference was based on the preceding experiments, in which this compound delivered orally in a dose 15 mg/kg/day during 4 weeks did not alter basal parameters of the sucrose test and locomotion in naive mice. In contrast, another standard antidepressant drug imipramine, used under the same conditions, significantly affected these behaviors in normal mice.

In a so-called protocol of "pre-stress" antidepressant treatment, administration of citalopram with drinking water was started 1 week before the onset of stress procedure and continued throughout the entire 4-week period of stress. Parameters of the sucrose test and floating behavior in the forced swim test were assessed after the termination of chronic stress. Citalopram solution was replaced with tap water during 12 h-sucrose test sessions (for details, see [128]).

Five-week administration of citalopram decreased the percentage of mice defined as anhedonic in 2.5 times and delayed a decrease of sucrose preference in the stressed group of mice from 2.5th to the 4th week (Fig.17A,B, adapted from [128]). Citalopram rescued normal floating, which was elevated in non-treated stressed mice (Fig.17C).

Effects of chronic treatment with citalopram on sucrose and forced swim tests observed in the study with oral administration of antidepressant were further confirmed in the chronic stress experiments with 5-week "pre-stress" citalopram delivery (15 mg/kg/day) performed via i.p. injections and via continuous subcutaneous administration with the osmotic mini-pumps [126, 127].

Irrespectively to the method of administration, citalopram diminished elevated water consumption and total liquid intake in chronically stressed mice. Citalopram-treated stressed group showed higher values of body weight than the non-treated stressed group [126-128]. Our results suggest that apart from the antidepressant-like behavioral effects, chronic treatment with citalopram counteracts with the development of other consequences of chronic stress in mice, such as elevated water consumption and reduction of body weight.

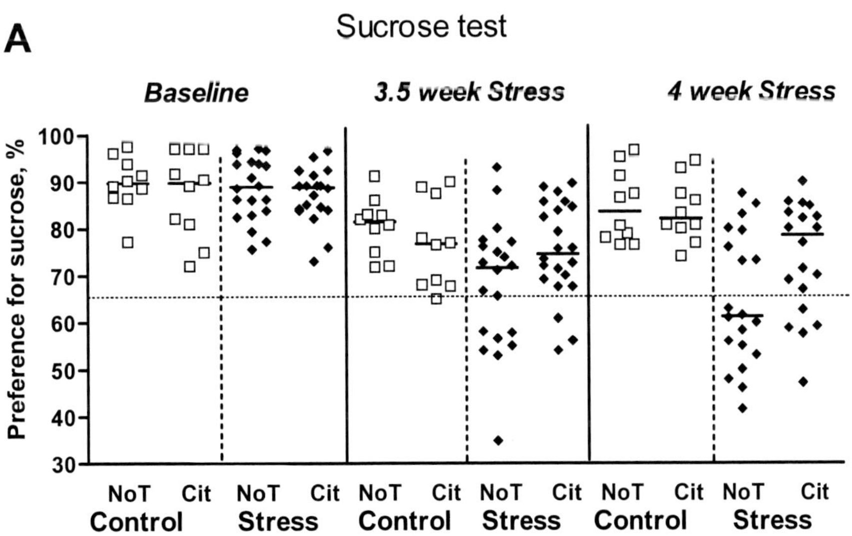

Figure 17. Continued on next page.

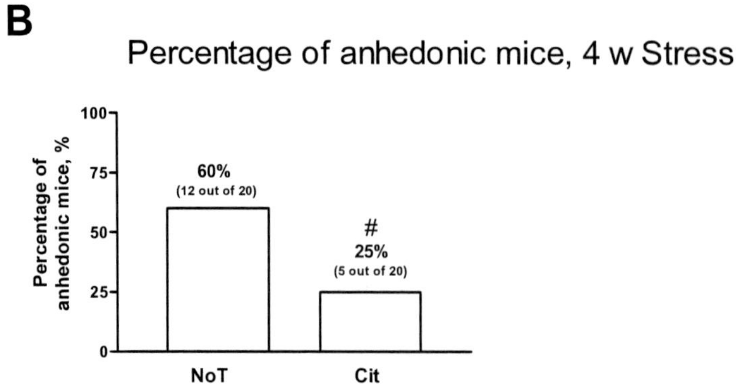

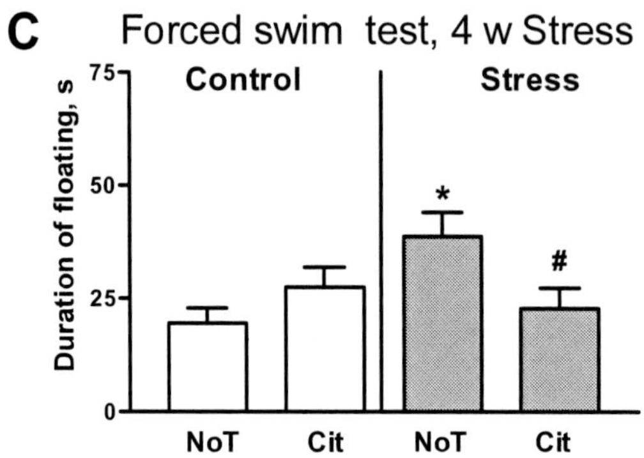

Figure 17. Administration of citalopram counteracts with a development of depressive-like syndrome. (A) Individual data show that in the course of stress, the citalopram-treated stressed group contain fewer anhedonic mice in comparison to non-treated animals. (B) After 4 weeks of stress, the percentage of anhedonic mice is significantly lower in the citalopram-treated stressed group, as compared to non-treated stressed animals (#$p < 0.05$; Fisher exact test). (C) Floating behavior is significantly elevated in the non-stressed non-treated group, as compared to a control group (*$p < 0.05$; Mann-Whitney) and stressed citalopram-treated mice (#$p < 0.05$). Data are expressed as mean ± standard error of measurement (SEM). NoT: non-treated group; Cit: citalopram-treated group. Bars indicate medians of the groups.

Thus, stress-induced depressive-like behavioral features, a decrease in sucrose preference and increase of floating behavior in our model can be counteracted by the treatment with SSRIs, applied predominantly simultaneously with chronic stress. These data validate pharmacological sensitivity of stress-induced decrease in sucrose preference in the employed model to a treatment with antidepressants of SSRI class.

Selective Effects of Citalopram on Stressed Mice with and without Hedonic Deficit

The administration of the antidepressant in rodents after the induction of a depressive-like syndrome mimics the clinical situation with treatment of depressed patients. With this type of chronic stress experiments, restoration of normal sucrose intake and preference are generally taken as an indicator of the antidepressant effect. Using a protocol of a "post-stress" treatment with citalopram, we further assessed a construct validity of our chronic stress paradigm and addressed the question, of whether the effects of the antidepressant treatment in stressed mice are selective with respect to their hedonic state.

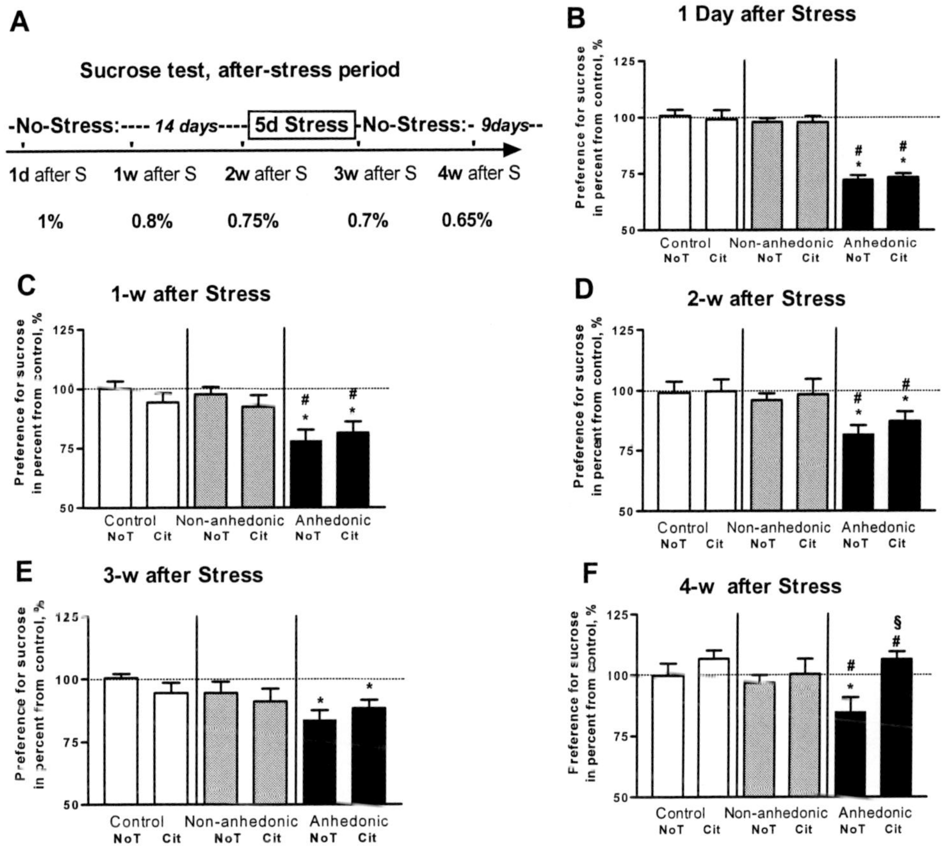

Figure 18. Administration of citalopram during post-stress period restores sucrose preference in anhedonic mice on the 4th week of treatment. (A) Scheme of the experiment with a post-stress treatment with citalopram. After the termination of a 4-week stress procedure, mice were left undisturbed for 2 weeks and thereafter submitted to a 5-day rat-exposure stress in order to maintain hedonic deficit in anhedonic group; they remained unstressed till the end of the 4th week of the after-stress period. Sucrose test was performed weekly with descending concentrations of sucrose solution (1% - 0.65% (B) Groups of animals assigned for planned treatment had similar means of sucrose preference before the beginning of citalopram administration. (C, D, E) Both non-treated and citalopram-treated anhedonic animals have a decreased sucrose preference versus the control group during the 1st - 3rd weeks of treatment and in comparison to the non-anhedonic group, during the 1st-2nd weeks. (F) Anhedonic mice treated with citalopram show a significant increase in sucrose preference as compared to the non-treated anhedonic group on the 4th week of treatment (*p < 0.05 vs. control group; # $p < 0.05$ vs. non-anhedonic group; § $p < 0.05$ vs. non-treated anhedonic mice; Mann-Whitney). NoT: non-treated group; Cit: citalopram-treated group. Data on graphs are expressed sucrose in percent from the means of the control group, as mean ± standard error of measurement (SEM).

In this study, we first applied a chronic stress procedure; immediately thereafter citalopram delivery via drinking solution (15 mg/kg/day) was begun in control, anhedonic and non-anhedonic mice. Parameters of the sucrose test and body weight were weekly monitored during a consequent 4-week treatment period. Citalopram solution was replaced with tap water for 12 h during sucrose test sessions. Additionally, floating behavior was evaluated in the forced swim test after three weeks of antidepressant treatment (for details, see [128]).

In order to maintain a hedonic deficit for the entire period of antidepressant delivery, we applied a 5-day rat exposure stress between 3rd and 2nd weeks of the treatment (Fig.18A, see also above). Citalopram restored sucrose preference in the anhedonic group on the 4th week of the post-stress drug administration, while non-treated mice from anhedonic group showed a diminished sucrose preference throughout the entire experiment (Fig.18B-F, adapted from [128]). Citalopram changed sucrose preference selectively in animals with anhedonia, but not in the stressed non-anhedonic and non-stressed control groups.

Multiple regression analysis showed that restoration of a sucrose preference in the anhedonic group was due to a pronounced increase of sucrose consumption, a phenomenon typical for prolonged treatment with antidepressants [34, 55, 150]. Importantly, while at this time point of the study, mean sucrose consumption in the non-treated anhedonic group was not different from the values of non-treated control and non-anhedonic groups, citalopram affected sucrose intake exclusively in anhedonic animals [128]. This suggests that stress experience alone does not make animals susceptible to the action of antidepressants. More studies are required to elucidate the physiological basis of selective effects of citalopram on sucrose intake in stressed mice from the anhedonic group.

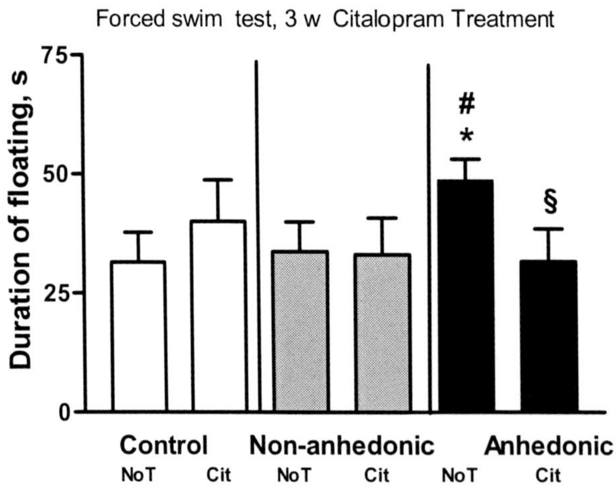

Figure 19. Post-stress treatment with citalopram decreases floating in anhedonic mice. Anhedonic mice treated with citalopram demonstrate significantly reduced duration of floating, in comparison to non-treated anhedonic group. In the non-treated anhedonic mice this parameter is significantly elevated versus the control group. $*p < 0.05$ vs. control group; # $p < 0.05$ vs. non-anhedonic group; § $p < 0.05$ vs. non-treated anhedonic mice. NoT: non-treated group; Cit: citalopram-treated group. Data on graphs are expressed as mean ± standard error of measurement (SEM).

In the forced swim test, anhedonic mice treated with an antidepressant, showed a decrease in duration of floating behavior, which was elevated in the non-treated anhedonic group. Non-stressed citalopram-treated control mice demonstrated a tendency of increased duration of floating. Chronic administration of citalopram did not affect floating of the non-anhedonic animals (Fig.19).

Chronic administration of citalopram significantly increased body weight in anhedonic mice, but not in the non-anhedonic animals; among non-treated animals, there was no significant difference in body weight between the non-anhedonic and anhedonic groups [128]. Restoration of body weight is known to accompany recovery from a depressive-like syndrome, induced by antidepressant treatment, in particular, with citalopram [71]. In anhedonic animals, chronic administration of citalopram reduced water intake, which was increased in this group of mice, and did not affect this parameter in the non-anhedonic animals.

In our study, we found selective efficacy of citalopram in changing several variables in stressed mice with, but not without a hedonic deficit. Formally, a possibility remains that citalopram modifies the sucrose preference, water consumption and the floating behavior exclusively in anhedonic mice because of the ceiling effects of stress in this group of animals. While it is not feasible to rule out this possibility, our findings and experimental approach may be compared to the clinical situation of the application and therapeutic effects of antidepressants. In patients, depressive symptoms can be seen as ceiling changes of a number of biological parameters selectively treatable by the antidepressant therapy; these variables are normal in healthy subjects and remain unchanged by antidepressants. In addition, a principle of ceiling effect does not explain selective effects of citalopram in anhedonic mice on sucrose intake and body weight, since their values did not differ between non-treated anhedonic and non-anhedonic groups.

As the effects of restoration of sucrose intake and preference, as well as normal floating behavior in rodents are considered to be the signs of antidepressant-like changes [145-133], presented findings argue for a construct validity of a proposed paradigm of a depressive-like state in mice. Our results evidence the late onset of the antidepressant effects of citalopram, which is consistent with the clinical experience of SSRIs application [147, 139].

Together, obtained data suggest that the proposed mouse model of stress-induced anhedonia allows for the separation of the effects of treatment with SSRI in animals with and without a hedonic deficit. Hence, it provides a possibility for a more refined analysis of the effects of antidepressant treatment with respect to the states of stress and anhedonia.

CONCLUSIONS

It is our conclusion that the results obtained, in a proposed model of stress-induced anhedonia with an internal control for stress, speak in favor for the validity of the chronic stress and sucrose test paradigms in modeling depressive states in rodents. The data suggest that chronic stress can induce a decrease in sucrose preference, which is sensitive to antidepressant treatment and cannot be attributed merely to the general consequences of prolonged stress. Inter-individual variability, in a stress-induced decrease of sucrose preference in our model, parallels the diversity in the development of other behavioral and physiological features of a depressive-like syndrome in mice, thus, further arguing for the validity of the sucrose test and chronic stress methods.

Our views are based on the data collected with a renewed experimental approach. Essential modifications in experimental design and behavioral testing enabled us to overcome some of the methodological and conceptual disadvantages of previously elaborated versions of the chronic stress depression paradigm in mice. In our opinion, the primary modifications consist of: 1) providing an internal control for the effects of stress alone by a separate analysis of stressed non-anhedonic animals; 2) increasing the sucrose test accuracy to the extent that it allows for the detection of inter-individual differences in a stressed mouse population; 3) identification of a phenomenon of stress-induced hyperlocomotion in chronically stressed mice and determination of testing conditions precluding its confounding effect on the animals' behavioral analysis.

The following facts obtained in a novel mouse paradigm of stress-induced anhedonia argue for its improved face and construct validity. First, the analysis of stressed mice demonstrated that anhedonia, but not chronic stress alone, is accompanied by key depressive-like behaviors, such as increased floating in the forced swim test, decreased exploration of novelty, prolonged immobilization in the tail suspension test, disrupted hippocampal plasticity, an increase in the percentage of REM sleep, disturbed day/night activity and altered gene expression [116-130, 10, 31, 41]. Other behavioral alterations, like stress-induced changes in locomotion, increased anxiety, as well as loss of body weight, develop in chronically stressed mice independently from a hedonic deficit. Hence, a stress-induced decrease in sucrose intake and preference is accompanied by a number of pathological alterations that cannot be observed in stressed individuals without a hedonic deficit.

Second, stress-induced changes specific for anhedonic animals in our model, parallel symptoms in patients with depression. Behavioral despair in the forced swim and tail suspension tests, found during anhedonia, is regarded as an analogue to the coping deficits in depressed humans. In rodents, these behaviors are well documented to be reduced by antidepressant treatment (for review, see [92, 36]. Diminished motivation to explore novelty is another behavioral characteristic of patients with depression. In some animal models, decreased novelty exploration was taken as a criterion of depressive-like status [103, 12]. One of the major features of depression observed in our model is a cognitive impairment, which is thought to be associated with deficient hippocampal function [75, 106, 6]. An increase in the percentage of REM sleep and disturbed day/night activity documented in a proposed mouse paradigm are well known characteristics of a depressive state [25, 69].

Third, we found that the above described depressive-like changes that were tested for sensitivity to antidepressants can be reversed and / or prevented by SSRI administration [123, 126-128]. Application of citalopram, in our paradigm with an internal control for chronic stress, suggests a selective activity of the SSRI antidepressant treatment in stressed animals with anhedonia. This suggestion originates from the fact that while the values of sucrose intake and body weight in animals from the non-anhedonic and anhedonic groups are similar, only the latter one displayed significant changes of these parameters after treatment with citalopram. This finding is in line with human studies demonstrating a lack of antidepressant effects in healthy volunteers.

Furthermore, we believe that the proposed chronic stress depression model has resemblance with several aspects of the depressive disorder in human, which so far was not achieved with other paradigms. First, we showed an animals' individual predisposition to develop stress-induced anhedonia in our model. To our knowledge, a percentage of individuals not developing a hedonic deficit and depressive-like phenotype is significant in other stress paradigms as well (R. Rygula, E. Fuchs; personal communications), but as non-anhedonic animals are typically removed from the analysis as outliers, the issue as a whole is largely underrepresented in the literature. Clinical studies evidence a pronounced inter-individual diversity in a susceptibility to depression precipitated by stress and suggest that congenital and environmental factors underlie this phenomenon [104, 8, 140]. Recently published data from an extensive study of Nestler's group [66] support our findings on large inter-individual variability in susceptibility to a stress-induced depressive state in mice. The fact that our model reflects an aspect of individual predisposition to depression additionally speaks in favor of its face validity. The proposed mouse paradigm can be regarded as the first model that addresses the biological basis of inter-individual differences in vulnerability to develop a depressive phenotype.

Second, we demonstrated that submissive traits in the social behavior of male mice can predict individual susceptibility to stress-induced anhedonia. Similarly, humans (men) with timid behavior and lower self-esteem were found to have a higher risk to develop clinical depression [98, 51]. Thus, our paradigm can be a useful tool for the modeling of a depressive profile in this cohort of patients.

Third, in our model we found that elevated anxiety accompanies a stress-induced hedonic deficit but does not itself represent a specific correlate. While in clinical practice, anxiety is well documented to be associated with the depressive syndrome, depression and anxiety disorders are regarded as two separate co-morbid pathologies, i.e., elevated anxiety is not regarded as a part of the depressive syndrome [112, 22, 38, 57]. Thus, our model mimics a co-morbidity of depression and anxiety diseases that can be used for the investigation of its biological basis.

Together, a demonstration of the specific physiological correlates of stress-induced anhedonia, as compared to the stress alone, and their sensitivity to the antidepressant treatment with SSRIs, speaks in favor of the face and construct validity of the initially proposed chronic stress depression model, which defines a stress-induced reduction in sucrose intake and preference as a sign of the hedonic deficit and depressive-like state in rodents [58, 145]. In our view, the presented data suggests that a new mouse model of stress-induced anhedonia does reflect significant pathological aspects of clinical depression. Generally, this increases the potential of our and other chronic stress paradigms for the investigation of the mechanisms of depressive disorders. Latest findings using social defeat paradigm [66] confirm these expectations.

ACKNOWLEDGEMENTS

I would like to thank the authorities of the Central Institute of Mental Health in Mannheim, the University of Aachen and United Technologies in Aachen/Zurich, for providing excellent research possibilities, Brandon Cline and Natalia Budygina for their helpful corrections of English, and Dr Roman Zolotowizki for his kind assistance in my work. I am especially thankful to Drs Natalia Gorenkova and Oleg Dolgov for their devotion and irreplaceable contribution to the project.

Reproduction of the material presented in this review is permitted by Macmillan Publishers Ltd: Neuropsychopharmacology, Strekalova T, Spanagel R, Bartsch D, Henn FA, Gass P. Stressed-induced anhedonia in mice is associated with deficits in forced swimming and exploration. Neuropsychopharm 2004; 11:2007-2017, and Lippincott Williams & Wilkins, Behavioral Pharmacology, 1) Strekalova T, Spanagel R, Dolgov O, Bartsch D. Stress-induced hyperlocomotion as a confounding factor in anxiety and depression models in mice. Behav Pharmacol 2005; 16:171-180; 2) Strekalova T, Gorenkova N, Schunk E, Dolgov O, Bartsch D. Selective effects of citalopram in the mouse model of stress-induced anhedonia with control effects for chronic stress. 2006;17:271-287.

The work on this chapter was supported by Deutsche Forschungsgemeinschaft (grant 699/1-2 to T.S.), EuroMedInnovation and OwnResearch group.

References

[1] Adriani W, Macri S, Pacifici R, Laviola G. Restricted daily access to water and voluntary nicotine oral consumption in mice: methodological issues and individual differences. *Behav Brain Res* 2002;134: 21-30.

[2] Amat J, Baratta MV, Paul E, Bland ST, Watkins LR, Maier SF. Medial prefrontal cortex determines how stressor controllability affects behavior and dorsal raphe nucleus. *Nat Neurosci* 2005;8: 365-371.

[3] Amico JA, Vollmer RR, Cai HM, Miedlar JA, Rinaman L. Enhanced initial and sustained intake of sucrose solution in mice with an oxytocin gene deletion. *Am J Physiol Regul Integr Comp Physiol* 2005;289: R1798-1806.

[4] Anokhin KV, Sudakov KV. Genome of brain neurons in organization of systemic mechanisms of behavior. *Bull Exp Biol Med* 2003;135:107-113.

[5] Ashmarin IP, Levitskaya NG, Antonova LV, Nezavibatko VN, Alfeeva LYu, Dubinin VA, Golubovich VP, Ponomareva-Stepnaya MA, Kamensky AA. The neurotropic activity of a structural analog of ACTH(5-7). *Regul Pept* 1994;51: 49-54.

[6] Atre-Vaidya N, Taylor M, Seidenberg M, Reed R, Perrine A, Glick-Oberwise F. Cognitive deficits, psychopathology and psychosocial functioning in bipolar mood disorder. *Neuropsychiatry Neuropsychol Behav Neurol* 1998;11: 120-126.

[7] Auriacombe M, Reneric JP, Le Moal M. Animal models of anhedonia. *Psychopharmacol* 1997;134: 337-338.

[8] Avenevoli S, Merikangas KR. Implications of high-risk family studies for prevention of depression. *Am J Prev Med* 2006; 6 Suppl 1:S126-135.

[9] Baker SL, Kentner AC, Konkle AT, Santa-Maria Barbagallo L, Bielajew C. Behavioral and physiological effects of chronic mild stress in female rats. *Physiol Behav* 2006;87: 314-322.

[10] Berger S, Strekalova T, Schoenig K, Beck C, Weber T, Leimer U, Li L, Gretz N, Bartsch D. Gene expression profiling in a mouse model of stress induced anehdonia with control for chronic stress. Abstracts of 34th Annual meeting of American Society for Neuroscience. 2004; http://sfn.abstractcentral.com/main.html.

[11] Bertoglio LJ, Carobrez AP. Behavioral profile of rats submitted to session 1-session 2 in the elevated plus-maze during diurnal/nocturnal phases and under different illumination conditions. *Behav Brain Res* 2002;132: 135-143.

[12] Bevins RA, Besheer J. Novelty reward as a measure of anhedonia. *Neurosci Biobehav* Rev. 2005;29: 707-714.

[13] Bolivar VJ, Walters SR, Phoenix JL. Assessing autism-like behavior in mice: variations in social interactions among inbred strains. *Behav Brain Res* 2007;176: 21-26.

[14] Brennan K, Roberts DC, Anisman H, Merali Z. Individual differences in sucrose consumption in the rat: motivational and neurochemical correlates of hedonia. *Psychopharmacology* (Berl) 2001;157: 269-276.

[15] Brotto LA, Gorzalka BB, LaMarre AK. Melatonin protects against the effects of chronic stress on sexual behaviour in male rats. *Neuroreport* 2001;16: 3465-3469.

[16] Burne TH, O'Loan J, McGrath JJ, Eyles DW. Hyperlocomotion associated with transient prenatal vitamin D deficiency is ameliorated by acute restraint. *Behav Brain Res* 2006; 174:119-124.

[17] Cabib S, Zochi A, Puglisi-Allegra S. A comparison of the behavioral effects of minaprine, amphetamine and stress. *Psychopharm* (Berl) 1995;121: 73-80.

[18] Cabib S. What is mild in mild stress? *Psychopharmacology* (Berl) 1997;134: 344-346.

[19] Calvo-Torrent A, Brain PF, Martinez M. Effect of predatory stress on sucrose intake and behavior on the plus-maze in male mice. *Physiol Behav* 1999;67: 189-196.

[20] Cancela LM, Bregonzio C, Molina VA. Anxiolytic-like effect induced by chronic stress is reversed by naloxone pretreatment. *Brain Res Bull* 1995; 36: 209-213.

[21] Cespuglio R, Marinesco S, Baubet V, Bonnet C, el Kafi B. Evidence for a sleep-promoting influence of stress. *Adv Neuroimmunol* 1995;5: 145-154.

[22] Charlton BG. Dialog on depression. *Science.* 2000;288(5468):975.

[23] Cheeta S, Ruigt G, van Proosdij J, Willner P. Changes in sleep architecture following chronic mild stress. *Biol Psychiatry* 1997;41: 419-427.

[24] Cole BJ, Koob GF. Corticotropin-releasing factor and schedule-induced polydipsia. *Pharmacol Biochem Behav* 1994;47: 393-398.

[25] Costa e Silva JA Sleep disorders in psychiatry. *Metabolism* 2006;55: S40-44.

[26] Coudereau JP, Stain F, Drion N, Sandouk P, Monier C, Debray M, Scherrmann JM, Bourre JM, Frances H. Effect of social isolation on the metabolism of morphine and its passage through the blood-brain barrier and on consumption of sucrose solutions. *Psychopharmacology* (Berl) 1999;144: 198-204.

[27] Cryan JF, Markou A, Lucki I. Assessing antidepressant activity in rodents: recent developments and future needs. *Trends Pharmacol Sci* 2002;23: 238-245.

[28] Cryan JF, Holmes A. The ascent of mouse: advances in modelling human depression and anxiety. *Nat Rev Drug Discov* 2005;4: 775-790.

[29] D'Amato FR, Rizzi R, Moles A. A model of social stress in non-submissive mice: effects on sociosexual behaviour. *Physiol Behav* 2001;73: 421-426.

[30] D'Aquila PS, Brain P, Willner P. Effects of chronic mild stress on performance in behavioural tests relevant to anxiety and depression. *Physiol Behav* 1994;56: 861-867.

[31] Dolgov O, Bartsch D, Strekalova T. Continuous automatic measurement of behavior in a new model of stress-induced anhedonia in mice. 2005. Proceedings of the 6th Meeting of the German Neuroscience Society http://www.neuro.uni-goettingen. de

[32] Ducottet C, Aubert A, Belzung C. Susceptibility to subchronic unpredictable stress is related to individual reactivity to threat stimuli in mice. *Behav Brain Res* 2004;155: 291-299.

[33] Ducottet C, Belzung C. Correlations between behaviours in the elevated plus-maze and sensitivity to unpredictable subchronic mild stress: evidence from inbred strains of mice. *Behav Brain Res* 2005;156: 153-162.

[34] Einat H, Manji HK, Belmaker RH. New approaches to modelling bipolar disorder. *Psychopharmacol Bull* 2003;371: 47-63.

[35] Einat H. Establishment of a battery of simple models for facets of bipolar disorder: a practical approach to achieve increased validity, better screening and possible insights into endophenotypes of disease. *Behav Genet* 2007;37: 244-255.

[36] El Yacoubi M, Vaugeois JM. Genetic rodent models of depression. *Curr Opin Pharmacol* 2007;7: 3-7.

[37] Forbes NF, Stewart CA, Matthews K, Reid IC. Chronic mild stress and sucrose consumption: validity as a model of depression. *Physiol Behav* 1996;60: 1481-1484.

[38] Freeman MP, Freeman SA, McElroy SL. The comorbidity of bipolar and anxiety disorders: prevalence, psychobiology, and treatment issues. *J Affect Disord* 2002; 68: 1-23.

[39] Gambarana C, Scheggi S, Tagliamonte A, Tolu P, De Montis M. Animal models for the study of antidepressant activity. *Brain Res Brain Res Protoc* 2001;7: 11-20.

[40] Glendinning JI, Gresack J, Spector AC. A high-throughput screening procedure for identifying mice with aberrant taste and oromotor function. *Chem Senses* 2002;27: 461-274.

[41] Gorenkova N, Kunert J, Berger S, Strekalova T, Bartsch D. Gene expression profiling in stressed mice with and without depressive-like syndrome. Proceedings of EMBO conference of Young Investigator Awardees 2005. http://www.embo.org/about_embo/EMBO_AR_2005.pdf.

[42] Gronli J, Murison R, Fiske E, Bjorvatn B, Sorensen E, Portas CM, Ursin R. Effects of chronic mild stress on sexual behavior, locomotor activity and consumption of sucrose and saccharine solutions. *Physiol Behav* 2005;84: 571-577.

[43] Hamilton M. Development of a rating scale for primary depressive illness. *Br J Soc Clin Psychol* 1967;6: 278-296.

[44] Harkin A, Houlihan DD, Kelly JP. Reduction in preference for saccharin by repeated unpredictable stress in mice and its prevention by imipramine. *J Psychopharmacol* 2002; 16: 115-123.

[45] Harro J, Tonissaar M, Eller M, Kask A, Oreland L.Chronic variable stress and partial 5-HT denervation by parachloroamphetamine treatment in the rat: effects on behavior and monoamine neurochemistry. *Brain Res* 2001;899: 227-239.

[46] Harris RB, Zhou J, Youngblood BD, Smagin GN, Ryan DH. Failure to change exploration or saccharin preference in rats exposed to chronic mild stress. *Physiol Behav* 1997;63: 91-100.

[47] Hata T, Itoh E, Nishikawa H. Behavioral characteristics of SART-stressed mice in the forced swim test and drug action. *Pharmacol Biochem Behav* 1995;51: 849-853.

[48] Hata T, Nishikawa H, Itoh E, Watanabe A. Depressive state with anxiety in repeated cold-stress mice in forced swimming tests. *Jpn J Pharmacol* 1999;79: 243-249.

[49] Hata T, Nishikawa H, Itoh E, Funakami Y. Anxiety-like behavior in elevated plus-maze in repeatedly cold stressed mice. *Jpn J Pharmacol* 2001;85: 189-196.

[50] Hatcher JP, Bell DN, Reed TJ, Hagan JJ. Chronic mild stress-induced reductions in saccharin intake depend upon feeding status. *J Psychopharmacol* 1997;11: 331-338.52.

[51] Hayward C, Wilson KA, Lagle K, Kraemer HC, Killen JD, Taylor CB. The developmental psychopathology of social anxiety in adolescents. *Depress Anxiety* 2007 Mar 8; [Epub ahead of print].

[52] Heiderstadt KM, McLaughlin RM, Wright DC, Walker SE, Gomez-Sanchez CE. The effect of chronic food and water restriction on open-field behavior and serum corticosterone levels in rats. *Lab Anim* 2000;34: 20-28.

[53] Holmes A, le Guisquet AM, Vogel E, Millstein RA, Leman S, Belzung C. Early life genetic, epigenetic and environmental factors shaping emotionality in rodents. *Neurosci Biobehav Rev* 2005;29:1335-1346.

[54] Igarashi E, Takeshita S. Effects of illumination and handling upon rat open field activity. *Physiol Behav* 1995;57: 699-703.

[55] Jayatissa MN, Bisgaard C, Tingstrom A, Papp M, Wiborg O. Hippocampal cytogenesis correlates to escitalopram-mediated recovery in a chronic mild stress rat model of depression. *Neuropsychopharmacology* 2006;31: 2395-2404.

[56] Kalueff AV, Gallagher PS, Murphy DL. Are serotonin transporter knockout mice 'depressed'? Hypoactivity but no anhedonia. *Neuroreport* 2006;17: 1347-1351.

[57] Kalueff AV, Wheaton M, Murphy DL. What's wrong with my mouse model? Advances and strategies in animal modelling of anxiety and depression. *Behav Brain Res* 2007;179: 1-18.

[58] Katz RJ. Animal models and human depressive disorders. *Neurosci Biobehav Rev* 1981; 5: 231-246.

[59] Katz RJ. Animal model of depression: pharmacological sensitivity of a hedonic deficit. *Pharmacol Biochem Behav* 1982;16: 965-968.

[60] Kendler K, Neale M, Kessler R, Heath A, Eaves L. Major depression and generalized anxirty disorder. Same genetics (partly) different environments? *Arch Gen Psychiatry* 1992;9: 716-722.

[61] Kessler RC, Chiu WT, Demler O, Merikangas KR, Walters EE. Prevalence, severity and comorbidity of 12-month DSM-IV disorders in the National Comorbidity Survey Replication. *Arch Gen Psychiatry* 2005;62: 617-627.

[62] Kessler RC, Akiskal HS, Ames M, Birnbaum H, Greenberg PE, Hirschfeld RM, Wang PS. Considering the costs of bipolar depression. *Behav Health* 2007;27: 45-47.

[63] Klein DF. Endogenomorphic depression. A conceptual and terminological revision. *Arch Gen Psychiatry* 1974;31: 447-454.

[64] Kopp C, Vogel E, Rettori MC, Delagrange P, Misslin R. The effects of melatonin on the behavioural disturbances induced by chronic mild stress in C3H/He mice. *Behav Pharmacol* 1999;10: 73-83.

[65] Krimm RF, Nejad MS, Smith JC, Miller IJ Jr, Beidler LM. The effect of bilateral sectioning of the chorda tympani and the greater superficial petrosal nerves on the sweet taste in the rat. *Physiol Behav* 1987;41: 495-501.

[66] Krishnan V, Han MH, Graham DL, Berton O, Renthal W, Russo SJ, Laplant Q, Graham A, Lutter M, Lagace DC, Ghose S, Reister R, Tannous P, Green TA, Neve RL, Chakravarty S, Kumar A, Eisch AJ, Self DW, Lee FS, Tamminga CA, Cooper

DC, Gershenfeld HK, Nestler EJ. Molecular adaptations underlying susceptibility and resistance to social defeat in brain reward regions. *Cell* 2007;131:391-404.

[67] Krsiak M, Podhorna J, Miczek KA. Aggressive and social behavior after alprazolam withdrawal: experimental therapy with Ro 19-8022. *Neurosci Biobehav Rev* 1998;23: 155-161.

[68] Le Pen G, Gaudet L, Mortas P, Mory R, Moreau JLDeficits in reward sensitivity in a neurodevelopmental rat model of schizophrenia. *Psychopharmacology* (Berl). 2002;161: 434-441.

[69] Leistedt S, Dumont M, Lanquart JP, Jurysta F, Linkowski P. Characterization of the sleep EEG in acutely depressed men using detrended fluctuation analysis. *Clin Neurophysiol* 2007;118: 940-950.

[70] Lipina TV, Mikhnevich NV, Kudriavtseva NN. The development of catatonic reactions in male mice of CBA/Lac strain: the effect of repeated experience of aggression and submission. *Zh Vyssh Nerv Deiat Im I P Pavlova* 2003;53: 88-93.

[71] Maina G, Albert U, Salvi V, Bogetto F. Weight gain during long-term treatment of obsessive-compulsive disorder: a prospective comparison between serotonin reuptake inhibitors. *J Clin Psychiatry* 2004;65: 1365-1371.

[72] Malkesman O, Braw Y, Zagoory-Sharon O, Golan O, Lavi-Avnon Y, Schroeder M, Overstreet DH, Yadid G, Weller A.Reward and anxiety in genetic animal models of childhood depression. *Behav Brain Res*. 2005;164: 1-10.

[73] Matthews K, Forbes N, Reid IC. Sucrose consumption as an hedonic measure following chronic unpredictable mild stress. *Physiol Behav* 1995;57: 241-248.

[74] McArthur R, Borsini F. Animal models of depression in drug discovery: a historical perspective. *Pharmacol Biochem Behav* 2006;84: 436-452.

[75] Miller G. Informational processing deficits in anhedonia and perceptual aberration: a psychophysiological analysis. *Biol Psychiatry* 1986;21: 100-115.

[76] Mineur YS, Belzung C, Crusio WE. Effects of unpredictable chronic mild stress on anxiety and depression-like behavior in mice. *Behav Brain Res* 2006;175: 43-50.

[77] Monleon S, D'Aquila P, Parra A, Simon VM, Brain PF, Willner P. Attenuation of sucrose consumption in mice by chronic mild stress and its restoration by imipramine. *Psychopharmacology* (Berl) 1995;117: 453-457.

[78] Moreau JL, Jenck F, Martin JR, Mortas P, Haefely WE. Antidepressant treatment prevents chronic unpredictable mild stress-induced anhedonia as assessed by ventral tegmentum self-stimulation behavior in rats. *Eur Neuropsychopharmacol* 1992;2:43-49.

[79] Moreau JL, Scherschlicht R, Jenck F, Martin JR. Chronic mild stress-induced anhedonia model of depression; sleep abnormalities and curative effects of electroshock treatment. *Behav Pharmacol* 1995;6: 682-687.

[80] Murray J L and Lopez A D. Evidence-based health policy-lessons from global burden of disease study. *Science* 1996;274: 740-743.

[81] Murison R, Hansen AL. Reliability of the chronic mild stress paradigm: implications for research and animal welfare. *Integr Physiol Behav Sci* 2001;36: 266-274.

[82] Negroni J, Venault P, Pardon MC, Perez-Diaz F, Chapouthier G, Cohen-Salmon C. Chronic ultra-mild stress improves locomotor performance of B6D2F1 mice in a motor risk situation. *Behav Brain Res* 2004;155: 265-273.

[83] Nestler EJ, Gould E, Manji H, Buncan M, Duman RS, Greshenfeld HK, Hen R, Kester S, Ledehendleer I, Meaney M, Robbins T, Winsky L, Zalcman, S. Preclinical models: Status of basic research in depression. *Biol Psychiatry* 2002;52: 503-528.

[84] Newport DJ, Stowe ZN, Nemeroff CB. Parental depression: animal models of an adverse life event. *Am J Psychiatry*. 2002;159(8):1265-1283.

[85] Nielsen CK, Arnt J, Sanchez C. Intracranial self-stimulation and sucrose intake differ as hedonic measures following chronic mild stress: interstrain and interindividual differences. *Behav Brain Res* 2000;107: 21-33.

[86] Nutt DJ, Ballenger JC, Sheehan D, Wittchen HU. Generalized anxiety disorder: comorbidity, comparative biology and treatment. *Int J Neuropsychopharmacol* 2002;5: 315-325.

[87] O'Callaghan MJ, Croft AP, Little HJ. Effects of intraperitoneal injections of saline on the alcohol and sucrose consumption of C57/BL10 mice. *Psychopharmacology* (Berl) 2002; 160: 206-212.

[88] Papaioannou A., Gerozissis K., Prokopiou A., Bolaris S. and Stylianopoulou F. Sex differences in the effects of neonatal handling on the animal's response to stress and the vulnerability for depressive behavior. *Behav Brain Res* 2002;129: 131-139.

[89] Papp M, Nalepa I, Vetulani J. Reversal by imipramine on serotonergic and beta-adrenergic receptor binding in a chronic mild stress model of depression. *Eur J Pharmacol* 1994;261: 141-147.

[90] Pijlman FT, Wolterink G, Van Ree JM. Physical and emotional stress has differential effects on preference for saccharine and open field behaviour in rats. *Behav Brain Res* 2003;139: 131-138.

[91] Porsolt RD, Bertin A, Jalfre M. Behavioral despair in mice: a primary screening test for antidepressants. *Arch Int Pharmacodyn Ther* 1977;229: 327-336.

[92] Porsolt R, Papp M. CNS – Psychiatric models of disease: Depression. In: *Current Protocols in Pharmacology*. Enna S, Williams M, Ferkany J, Kenakin T, Porsolt R and Sullivan J (editors). New York: Wiley; 1998; pp. 591-598.

[93] Pothion S, Bizot JC, Trovero F, Belzung C. Strain differences in sucrose preference and in the consequences of unpredictable chronic mild stress. *Behav Brain Res* 2004;155: 135-146.

[94] Prince CR, Collins C, Anisman H. Stressor-provoked response patterns in a swim task: modification by diazepam. *Pharmacol Biochem Behav* 1986;24: 323-328.

[95] Prisciandaro JJ, Roberts JE. A taxometric investigation of unipolar depression in the national comorbidity survey. *J Abnorm Psychol.* 2005;114: 718-728.

[96] Pucilowski O, Overstreet DH, Rezvani AH, Janowsky DS. Chronic mild stress-induced anhedonia: greater effect in a genetic rat model of depression. *Physiol Behav* 1993;54: 1215-1220.

[97] Reid I, Forbes N, Stewart C, Matthews K. Chronic mild stress and depressive disorder: a useful new model? *Psychopharmacology* (Berl) 1997;134: 365-567.

[98] Romney DM, Bynner JM. A re-examination of the relationship between shyness, attributional style, and depression. J Genet Psychol 1997;158: 259-270.

[99] Rossler AS, Joubert C, Chapouthier G. Chronic mild stress alleviates anxious behaviour in female mice in two situations. *Behav Processes* 2000;49: 163-165.

[100] Ruiz-Durantez E, Hall SK, Steffen C, Self DW. Enhanced acquisition of cocaine self-administration by increasing percentages of C57BL/6J genes in mice with a nonpreferring outbred background. *Psychopharmacology* (Berl) 2006;186: 553-560.

[101] Sanchez C. Acute stress enhances anxiolytic-like drug responses of mice tested in a black and white test box. *Eur Neuropsychopharmacol* 1997;7: 283-288.

[102] Sanchis-Segura C, Spanagel R, Henn FA, Vollmayr B. Reduced sensitivity to sucrose in rats bred for helplessness: a study using the matching law. *Behav Pharmacol* 2005;16: 267-270.

[103] Santarelli L, Saxe M, Gross C, Surget A, Battaglia F, Dulawa S, Weisstaub N, Lee J, Duman R, Arancio O, Belzung C, Hen R. Requirement of hippocampal neurogenesis for the behavioral effects of antidepressants. *Science* 2003;301: 805-809.

[104] Sapolsky R. Why stress is bad for your brain. *Science* 1996;273: 749-750.

[105] Schoenecker B, Heller KE, Freimanis T. Development of stereotypies and polydipsia in wild caught bank voles (Clethrionomys glareolus) and their laboratory-bred offspring. Is polydipsia a symptom of diabetes mellitus? *Appl Anim Behav Sci* 2000;68: 349-357.

[106] Sheline Y, Wany P, Gado M, Csernansky J, Vannier M. Hippocampal athrophy in recurrent major depression. *Proc Natl Acad Sci USA* 1996;93: 3908-3913.

[107] Shelton CI. Long-term management of major depressive disorder: are differences among antidepressant treatments meaningful? *J Clin Psychiatry* 2004;65: 29-33.

[108] Skuse D, Albanese A, Stanhope R, Gilmour J, Voss LA. New stress-related syndrome of growth failure and hyperphagia in children, associated with reversibility of growth-hormone insufficiency. *Lancet* 1996;348: 353-358.

[109] Slattery DA, Markou A, Cryan JF. Evaluation of reward processes in an animal model of depression. *Psychopharmacology* (Berl) 2007;190: 555-568.

[110] Smith GP, Schneider LH. Relationships between mesolimbic dopamine function and eating behavior. *Ann N Y Acad Sci* 1988;537: 254-261.

[111] Solberg LC, Horton TH, Turek FW. Circadian rhythms and depression: effects of exercise in an animal model. *Am J Physiol* 1999;276: R152-R161.

[112] Stahl SM. Phenomenology of anxiety disorders: clinical heterogeneity and comorbidity. In: Westenberg HGM, Den Boer JA, Murphy DL (editors). *Advances in the Neurobiology of Anxiety Disorders*. John Willey & Sons, 1996; pp. 21-38.

[113] Stewart KT, Rosenwasser AM, Hauser H, Volpicelli JR, Adler NT. Circadian rhythmicity and behavioral depression: I. Effects of stress. *Physiol Behav* 1990;48: 149-155.

[114] Strekalova TV. The characteristics of the defensive behavior of rats in accordance with their resistance to emotional stress. *Zh Vyssh Nerv Deiat Im I P Pavlova* 1995;45: 420-422.

[115] Strekalova T, Wotjak C, Schachner M. Intrahippocampal administration of an antibody against the HNK-1 carbohydrate impairs memory consolidation in an inhibitory learning task in mice. *Mol Cell Neurosci* 2001;17: 1102-1113.

[116] Strekalova T, Bartsch D, Spanagel R, Gass P. Behavioral correlates of chronic stress-induced anhedonia in C57BL/6N mice. *Eur Arch Psych Clin Neurosci* 2002;252: 39 S161.

[117] Strekalova T, Gass P. Behavioral strategies in mice C57BL/6N with different individual vulnerability to stress-induced anhedonia. In: *Mechanisms of stress-resistance*. Moscow NIINF, Sudakov (editor) 2002; pp.160-168.

[118] Strekalova T, Spanagel R, Bartsch D, Henn FA, Gass P. Stressed-induced anhedonia in mice is associated with coping and exploration deficits. *Behav Pharm* 2003;14: Suppl 1 S 70.

[119] Strekalova T, Berger S, Leimer U, Beck C, Pesold B, Weber D, Bartsch D. Model of stress-induced anhedonia in mice with control for chronic stress: features of stress and anhedonia are distinct. Proceedings of the 33rd Annual Meeting of American Society for Neuroscience. 2003b; http://sfn.abstractcentral.com/main.html.

[120] Strekalova T, Frynta D, Berger S, Henn F, Bartsch D. Social behavior in a resident-intruder test predicts individual vulnerability to stress-induced anhedonia in C57BL/6N mice. *Abstracts of the International Behavioral Neuroscience Society*. 2004;13: 36.

[121] Strekalova T, Spanagel R, Bartsch D, Henn FA, Gass P. Stressed-induced anhedonia in mice is associated with deficits in forced swimming and exploration. *Neuropsychopharm* 2004;11: 2007-2017.

[122] Strekalova T, Spanagel R, Bartsch D, Gass P. Chronic stress induces hyperactivity in anxiety and forced swim tests in mice: implications for depression model. *Psychopharm Biol Narcology* 2004;4: 763.

[123] Strekalova T, Bartsch D. Distinct physiological, behavioral and cognitive features of chronically stressed mice with and without hedonic deficit. Proceedings of the IX International Conference “Stress and Behavior”, St. Petersburg, 2005: 25.

[124] Strekalova T, Dolgov O, Gorenkova N, Bartsch D. Distinct behavioral and cognitive correlates of hedonic deficit and chronic stress in a new model of stress-induced anehdonia in mice. 2005a. Proceedings of the 6th Meeting of the German Neuroscience Society http://www.neuro.uni-goettingen.de.

[125] Strekalova T, Spanagel R, Dolgov O, Bartsch D. Stress-induced hyperlocomotion as a confounding factor in anxiety and depression models in mice. *Behav Pharmacol* 2005; 16: 171-180.

[126] Strekalova T, Gorenkova N, Dolgov O, Kutscherjawj S, Bartsch D. Model of anhedonia with control for stress: effects of citalopram. *J Biol Psych* 2005;6S1: 306.

[127] Strekalova T, Gorenkova N, Kutscherjawj S, Schunk E, Dolgov O, Bartsch D. Mouse model of anhedonia with internal control for chronic stress: effects of pre- and post-stress treatment with antidepressant citalopram. *Behav Pharm* 2005;16 S1: S44.

[128] Strekalova T, Gorenkova N, Schunk E, Dolgov O, Bartsch D. Selective effects of citalopram in the mouse model of stress-induced anhedonia with control effects for chronic stress. *Behav Pharm* 2006;17: 271-287.

[129] Strekalova T, Dolgov O. Sucrose test method in C57BL/6 mice: facts, artifacts and application in depression models. Proceedings of Xth International Conference of Biological Psychiatry “Stress and Behavior”, St. Petersburg, May 16-20, 2007. P. 26.

[130] Strekalova T, Dolgov O, Valatx J-L, Cespuglio R. Impairment of sleep parallels deficits in cognitive functions: evidence for relationship from a mouse depression model. Proceedings of the International Conference “Sleep as a window to the world of wakefulness”, Moscow, June 8-9; 2007: 106.

[131] Tonissaar M, Herm L, Rinken A, Harro J. Individual differences in sucrose intake and preference in the rat: circadian variation and association with dopamine D2 receptor function in striatum and nucleus accumbens. *Neurosci Lett* 2006;403: 119-124.

[132] Trask PC, Esper P, Riba M, Redman B. Psychiatric side effects of interferon therapy: prevalence, proposed mechanisms, and future directions. *J Clin Oncol* 2000;18: 2316-2326.

[133] Tsankova NM, Berton O, Renthal W, Kumar A, Neve RL, Nestler EJ. Sustained hippocampal chromatin regulation in a mouse model of depression and antidepressant action. *Nat Neurosci* 2006;9: 519-525.

[134] Vachova H. Chronic stress depression model in mice. In: Social characteristics in mice and their physiological correlates: two studies. Diploma work at the department of zoology, Charle's University in Prague. 2004; 2-32.

[135] Valentinuzzi VS, Buxton OM, Chang AM, Scarbrough K, Ferrari EA, Takahashi JS, Turek FW. Locomotor response to an open field during C57BL/6J active and inactive phases: differences dependent on conditions of illumination. *Physiol Behav* 2000;69: 269-275.

[136] Venault P, Rudrauf D, Lepicard EM, Berthoz A, Jouvent R, Chapouthier G. Balance control and posture in anxious mice improved by SSRI treatment. *Neuroreport* 2001;12: 3091-3094.

[137] Vogel G, Neill D, Hagler M, Kors D. A new animal model of endogenous depression: a summary of present findings. *Neurosci Biobehav Rev* 1990;14: 85-91.

[138] Von Frijtag JC, Reijmers LG, Van der Harst JE, Leus IE, Van den Bos R, Spruijt BM. Defeat followed by individual housing results in long-term impaired reward- and cognition-related behaviours in rats. *Behav Brain Res* 2000;117: 137-146.

[139] Wade AG, Toumi I, Hemels ME. A pharmacoeconomic evaluation of escitalopram versus citalopram in the treatment of severe depression in the United Kingdom. *Clin Ther* 2005;27: 486-496.

[140] Wasserman D, Sokolowski M, Rozanov V, Wasserman J. The CRHR1 gene: a marker for suicidality in depressed males exposed to low stress. *Genes Brain Behav* 2008; 7: 14-19.

[141] Weiss JM, Simson PG. Depression in an animal model: focus on the locus ceruleus. Ciba Found Symp 1986;123:191-215.

[142] Weiss JM. Does decreased sucrose intake indicate loss of preference in CMS model? *Psychopharmacology* (Berl) 1997;134: 368-377.

[143] Willis-Owen SA, Flint J. Identifying the genetic determinants of emotionality in humans; insights from rodents. *Neurosci Biobehav Rev* 2007;31: 115-124.

[144] Willis-Owen SA, Flint J. The genetic basis of emotional behaviour in mice. *Eur J Hum Genet* 2006;14: 721-728.

[145] Willner P, Towell A, Sampson D, Sophokleous S, Muscat R. Reduction of sucrose preference by chronic unpredictable mild stress, and its restoration by a tricyclic antidepressant. *Psychopharmacology* (Berl) 1987;93: 358-364.

[146] Willner P. Animal models as simulations of depression. *Trends Pharmacol Sci* 1991;12: 131-136.

[147] Willner P, D'Aquila PS, Coventry T, Brain P. Loss of social status: preliminary evaluation of a novel animal model of depression. *J Psychopharmacol* 1995;9: 207-213.

[148] Willner P. Validity, reliability and utility of the chronic mild stress model of depression: a 10-year review and evaluation. *Psychopharmacology* 1997;134: 319-329.

[149] Willner P. Chronic mild stress (CMS) revisited: consistency and behavioural-neurobiological concordance in the effects of CMS. *Neuropsychobiology* 2005;52: 90-110.

[150] Yalcin I, Aksu F, Belzung C. Effects of desipramine and tramadol in a chronic mild stress model in mice are altered by yohimbine but not by pindolol. *Eur J Pharmacol* 2005;514:165-174.

In: Behavioral Models in Stress Research
Editors: A.V. Kalueff and J.L. LaPorte, pp. 159-183 ISBN: 978-1-60456-361-0

Chapter 7

DOMAIN-ORIENTED ANALYSIS FOR UNDERSTANDING MOLECULAR INTERACTIONS AND TRANSLATING ANIMAL GENETIC MODELS INTO NEUROPSYCHIATRIC DISORDERS: FOCUS ON SEROTONIN TRANSPORTER AND BDNF

Allan V. Kalueff*[*], *Renee F. Ren-Patterson, Justin L. LaPorte and Dennis L. Murphy

Laboratory of Clinical Science, National Institute of Mental Health, NIH, Bethesda, MD, USA

INTRODUCTION

Animal experimental models of brain disorders represent a valuable tool in today's biological psychiatry [23, 35, 77, 78, 142, 153, 154, 172, 173]. Various genetic animal models, based on selectively bred, hybrid, gene-targeted or transgenic animals, are widely used in neuroscience research for screening neuroactive drugs, testing neurobiological hypotheses and finding candidate genes for stress-related brain disorders [35, 70, 76, 162, 163].

Numerous data indicate that brain disorders in many cases represent overlapping pathogenetic pathways with common genetic determinants and clinical manifestations (Table 1) [71, 76, 78]. Together, this raises the possibility that a *combination* of several distinct but interacting domains, and their *interplay*, may contribute to clinical and experimental phenotypes. This also implies that a closer in-depth analysis of different domains may be needed for developing new, clinically relevant genetic experimental modeling of neuropsychiatric disorders.

[*] E-mail address: avkalueff@inbox.ru. Phone: 1-301-594-0393. Laboratory of Clinical Science, Building 10, Room 3D41, National Institute of Mental Health, Bethesda, Maryland 20892-1264, USA

Table 1. Clinical comorbidity (gray fields) and common candidate SERT (s) and BDNF (b) genetic determinants (white fields) of several common psychiatric disorders.

Disorders	GAD	OCD	PD	PTSD	SA	AU	UD	BD	SC	ED	SD	TS. RS	OCSD	ADHD
	Comorbidity													
Generalized anxiety disorder (GAD)	sb	++	++	++	++	++	++	+	+	+	+	+	+	+
Obsessive-compulsive disorders (OCD)		sb				++	++			++	++	++	++	++
Panic disorder (PD)			s	++	++	+								
Post-traumatic stress disorder (PTSD)				s	+	+	++		+				+	
Social anxiety (SA)					s	++	+							
Autism (AU)						sb	++	++	++	++		++	++	++
Unipolar depression (UD)							sb	++	+	++	++	++	++	++
Bipolar depression (BD)								sb	++	+	+	+	+	++
Schizophrenia (SC)									sb					
Eating disorders (ED)										sb				
Sleep disorders (SD)														
Tourette and Rett syndromes (TS, RS)												sb		+
OCD spectrum disorders (OCSD)													sb	+
Attention deficit/hyperactivity (ADHD)														sb

In this chapter, we will discuss how analysis of different behavioral and physiological domains in genetic animal models may optimize further experimental research in this field. We will also argue that these approaches can be used to pursue another far-reaching goal: to elucidate complex brain processes and their potential molecular underpinnings. To illustrate the utility of this approach, we will specifically focus on two key brain molecules – serotonin transporter (SERT) and brain-derived neurotrophic factor (BDNF), both implicated in multiple neuropsychiatric disorders (see further).

Serotonin (5-HT), SERT and BDNF

5-HT plays a key role in the regulation of human and animal behavior throughout their lifespans. In addition to its well-recognized role as a neurotransmitter [3-6, 96, 161, 167, 168], 5-HT is an important morphogenetic contributor to the developing brain [16, 25, 26, 164-166, 168, 169]. Disrupted signaling of this neurotransmitter during early development produces lasting changes in the morphology and function of the central nervous system [48, 49, 51]. Altered developmental and post-natal 5-HT effects numerous facets of cognition and emotional regulation [47-51, 105, 126], as evidenced by its implication in the pathogenesis of many brain disorders, such as anxiety, depression, mania, addiction, schizophrenia, autism and obsessive-compulsive disorder (OCD) [41, 83, 97-100, 115, 144, 146, 177]. In order to treat these disorders, we must first understand their biological underpinnings and comprehend how the mechanisms that regulate 5-HT can contribute to brain pathogenesis.

The uptake of synaptic 5-HT into nerve terminals - the most important mechanism of serotonergic regulation - is mediated by SERT, a high-affinity plasma membrane transporter [98, 120-122, 140, 141, 182]. In humans, a SERT common polymorphism in the promoter region, a variable-number tandem repeat in intron 2, and a coding region mutation have been reported to be associated with a variety of neuropsychiatric diseases, including anxiety, autism, obsessive-compulsive disorder and depression (see [41, 44, 45, 57, 60, 74, 122, 127] for details).

BDNF is another key brain molecule- a member of the nerve-growth factor family of neurotrophic factors [46, 69, 90, 110, 123, 147, 148, 150, 160]. It is the most abundant brain neurotrophin that regulates neuronal survival, migration, axon and dendrite growth and the activity-dependent synaptic development [10, 29, 32, 89, 91, 92, 110, 111]. BDNF is an important modulator of dopaminergic, cholinergic, and serotonergic neurons, implicated in synaptic vesicle function and synaptic plasticity, leading to specific alterations in behaviors, including memory, activity, eating behavior, depression and anxiety [9, 12, 13, 15, 17, 19, 24, 34, 84, 91, 114, 116, 120, 131, 176].

As can be seen in Table 1, numerous clinical and animal studies have implicated 5-HT, SERT and BDNF in pathogenesis of anxiety, depression, autism, schizophrenia and other brain disorders. Moreover, mounting clinical and experimental evidence (summarized in Tables 2 and 3) indicates that these molecules not only are both involved in the regulation brain normal and pathological mechanisms [111], but directly interact at many levels in determining complex neuropsychiatric phenotypes [14, 80, 82, 112, 155, 156, 159].

Table 2. Clinical data showing interactions between SERT and BDNF.

Models/subjects	Data	Refs
Genetic		
Maltreated children	5-HTLPR x BDNF (variant val^{66}met) x maltreatment interaction	[82]
Children	5-HTLPR x BDNF (variant val^{66}met) x depression interaction	[171]
Elderly	5-HTLPR x BDNF (variant val^{66}met) x life stressors interaction	[86]
5-HTTLPR* S allele carriers	↓ structural connectivity in amygdala and cortex in BDNF val^{66}met vs. val^{66}val genotype ↓ cortical volume modulated in BDNF val^{66}met	[129]
Pharmacological		
Depressed adults	↑ serum BDNF after 8-week SSRI** or SNRI*** treatment	[179]
	↓ serum BDNF after SSRI treatment, compared to controls	[130]
	↑ serum BDNF after SSRI treatment in female respondents	[67]
Pharmacogenetic		
Depressed adults	BDNF polymorphism modulates efficacy of SERT-active antidepressants	[33, 178]

* 5-HTLPR – SERT promoter region polymorphism, ** SSRI – selective serotonin (5-HT) reuptake inhibitor, *** SNRI – selective serotonin/norephinephrine reuptake inhibitor.

Table 3. Experimental data showing interactions between 5-HT, SERT and BDNF (footnote as in Table 2)

Models/subjects	Data	Refs
Genetic		
BDNF -/- mice	↑ 5-HT neurons and SERT in raphe nucleus (vs. +/-mice)	[38]
BDNF+/- mice	↓ 5-HT axons in aged BDNF+/- (vs. +/+) mice	[107]
	↑ extracellular 5-HT in ventral hippocampus	[52]
	↓ hippocampal 5-HT clearance in older adult mice	[37]
	↑ age-related degeneration of 5-HT forebrain innervation	[155]
	↓ hippocampal 5-HT axons in aged mutant (vs. +/+) mice	[106]
	↓ extracellular 5-HT in aged BDNF+/- mice	[106]
	5-HT clearance varies in age/BDNF genotype-dependent manner	[37]
	↓ sensitivity to the 5-HT clearance-inhibiting effect of an SSRI	[37]
	↓ 5-HT in cortex and hippocampus (vs. +/+ mice)	[87]
SERT-/- x BDNF +/- mice	↓BDNF in SERT[+/- and -/-] x BDNF+/- mice vs. +/+ mice	[134]
	↑ anxiety-like behaviors	[134]
	↓ number and size of cortical neurons, dendritic branching and axon projections vs SERT[+/+ or +/-] x BDNF+/- mice	[134]
	↓ 5-HT in hypothalamus, anxiety-like behaviors, ↑ Trk B receptors in hypothalamus (in females vs. males)	[133]
	↓ 5-HT in some brain areas (vs. SERT-/- x BDNF+/+ mice)	[122]
trkB.TK+ mice	↓ 5-HT metabolism in hippocampus of trkB.TK+ (vs. +/+) mice	[88]
Pharmacological		
Rats	↑ 5-HT after the infusion of BDNF into midbrain	[148]
	↓ protection by BDNF of 5-HT neurons after toxic axon damage	[109]
Gerbils (ischemia)	↑ hippocampal BDNF following SSRI treatment	[85]
Pharmacogenetic		
TrkB.T1 mice	↓ ability of SSRI to activate Trk B after monoamine depleting agents in cortex of transgenic (vs. +/+) mice after tyrosine treatment	[132]
	↑ BDNF-associated genes in cortex, hippocampus and gyrus after chronic SSRI treatment	[8]
Mesolimbic-specific BDNF knockout mice	Chronic SSRI ↓ social avoidance caused by chronic defeat stress	[19]
Rats	↑ hippocampal BDNF mRNA after chronic tesofensine, a triple monoamine reuptake inhibitor	[95]
Neurochemical		
Bulbectomy, mice	↓ 5-HT turnover and ↑ BDNF in hippocampus and cortex	[58]
Rats	↑ BDNF by 5-HT-ergic activation of cAMP/cAMP response element-binding protein in hippocampus	[124]
	↓ SERT and ↑ BDNF after subdural engraftment of 5-HT neurons into hemisectioned spinal tissues	[53]
	↑ Protection of neurotoxin-treated 5-HT-ergic terminals by BDNF-producing fibroblasts administered in striatum	[42]

Models/subjects	Data	Refs
In-vitro models		
Mouse raphe cultures	↑ neuronal 5-HT-ergic phenotype by BDNF through ↑ 5-HT neurons and arborization	[143]
SERT -/- mouse raphe cultures	↑ neuronal 5-HT-ergic phenotype by BDNF	[143]
	↓ effects of BDNF on 5-HT neuron number and neurite extension	[143]
Human lymphoblasts	↓ 5-HT uptake in-vitro by BDNF	[118]
RN46A cells	↑ 5-HT cells by BDNF and partial depolarization	[170]

Animal Genetic Models of 5-HT/SERT and BDNF Dysregulation

SERT and BDNF are especially interesting genes for experimental modeling neuropsychiatric disorders [79, 101]. For example, human variants at both the SERT and BDNF gene loci have been implicated in affective disorders, OCD, autism and poly-substance abuse liability [1, 20, 30, 31, 54, 68, 81, 125, 145, 180] (Table 1), strengthening the importance of studying these two genes and their interactions [37, 122] using animal experimental and genetic models.

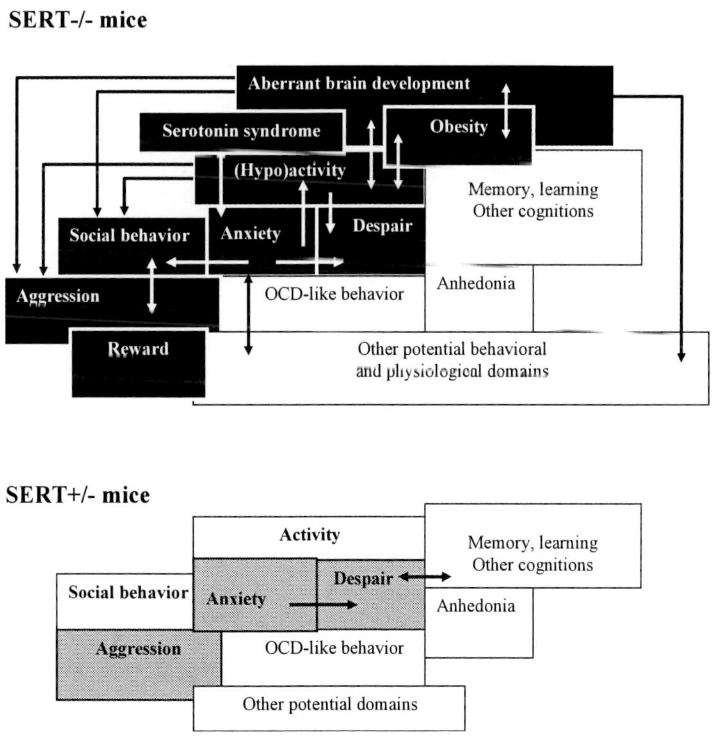

Figure 1. Summary of behavioral and other domains, and their possible interplay [indicated by arrows], in 5-HT transporter knockout (SERT-/-) and heterozygous knockout (SERT+/-) mice. White; unaltered, black ;disordered, gray ; intermediate domains/phenotypes.

Differences in animal SERT expression and function affect many quantitative traits, including aggression, anxiety- and depression-related behaviors, locomotor activity, sleep and circadian behaviors, gut function, body weight, nociceptive responses and preferences for drugs like cocaine and alcohol [27, 28, 36, 72, 120-122, 157]. Several genetic animal models based on targeting SERT gene in mice [18, 181] and rats [65] have been reported in the literature. SERT knockout (-/-) mice demonstrate elevated extracellular 5-HT, up- and down-regulation of some 5-HT receptors, and numerous behavioral abnormalities [2, 7, 61-64, 102, 103, 183], with multiple affected domains that seem to interact with each other in determining the mouse phenotype (Fig. 1).

Here we will use domain-oriented analysis to dissect different behavioral and physiological phenotypes in these mice. For example, reduced activity (hypolocomotion domain), increased body weight (obesity domain) and skeletal anomalies in SERT-/- mice are likely to affect their performance in various behavioral tests, such as locomotion in the open field (anxiety domain) or the forced swim or tail suspension tests (depression-like "despair" domain) [74]. While these multiple factors, acting together, may affect our interpretations of animal behavioral phenotypes, a focus on domain interplay may help elucidate this problem. Another example of the importance of examining *several* domains simultaneously is the relation between obesity, anxiety and inactivity in SERT-/- mice. Since SERT-/- mouse obesity develops later in life than anxiety and hypoactivity (own systematic observations across three different strains), it is possible to conclude that hypoactivity or anxiety may lead to obesity in this model, and not vise versa. Moreover, the fact that heterozygous SERT+/- mice do not have overt obesity or hypoactivity, but still develop anxiety-like behaviors, permits a clearer focus on SERT-related anxiety and dissects it from obesity of activity domains [74].

Similarly, while reduced aggression in SERT-/- mice [63] may be partially explained by their hypoactivity, social deficits or anxiety, SERT+/- mice (with normal activity and social behaviors) also display reduced aggression, confirming the link between low aggression and reduced SERT function [74]. In line with relatively normal cognitive domain (as evidenced by unimpaired short- and long-term spatial memory [73]) in SERT+/- and -/-mice, these findings confirm that aggression domain is indeed affected by SERT mutation, and not due to altered activity, cognitive or anxiety domains (see similar data on reduced aggression and social play in SERT-/- rats: [66]).

Likewise, data are conflicting on depression domain in SERT-/- mice across different tests and genetic backgrounds. Using two popular "behavioral despair" models of depression (the tail suspension and the forced swim tests), several groups have examined depressiveness in SERT-/- mice on different genetic backgrounds. These studies have shown that SERT-/- mice on 129S6 background exhibit increased despair (immobility) in the forced swim test but produce a surprising antidepressant-like reduction of immobility in the tail suspension test [64, 104]. In contrast, mutants on C57BL/6 background displayed unaltered tail suspension and forced swim behavior [55, 64, 104].

This situation is not uncommon in behavioral research, as genetic background and the nature of behavioral tasks can often influence the phenotype in question [119, 174, 175]. In the case of SERT-/- mice, given the lack of a consistent phenotype across different backgrounds, and apparently incompatible depression-related findings in the tail suspension and forced swim tests, we need an explanation that may reconcile these results. One explanation, in our opinion, may be the presence of motor/musculo-skeletal abnormalities in

SERT-/- mice (see [74] for details), coinciding with, and therefore strengthened by, similar problems in 129 background strains [75, 119], and accompanied by a non-depressive (or even anti-depressant?) phenotype of SERT mutant mice. Thus, motor/musculo-skeletal abnormalities would be consistent with low activity levels seen in SERT-/- mice in the novelty tests and other tasks (such as the forced swim test) that require motor coordination and physical activity, while allowing the mouse to still have reduced immobility in the tail suspension test, which does not require a high degree of activity and motor coordination. The reduced immobility in the latter task indicates that describing the SERT mutant strain as "depressive" phenotype does not adequately explain their behavioral profile, making the motor/muscular impairment a reasonable alternative explanation. In contrast, a well-known tail-climbing phenotype of C57BL/6 background strain [113] greatly compromises the validity of the tail suspension test results using mice on this background, with mice of all genotypes (own unpublished observations) producing this behavior frequently during a 6-min trial. Therefore, caution is needed when considering SERT-/- mice on C57BL/6 background as "unaffected" in the tail suspension test. In fact, it is the combination of lower overall motor activity in SERT-/- mutants, and strain-specific tail climbing in C57BL/6 background that may contribute to this phenotype, most likely confounded by the use of inappropriate test for this background.

Again, given high dependence of some "despair" tests upon motor activity [152, 153], hypoactivity (consistently replicated in many studies using other tests) would seem to limit the utility of these tests in SERT-/- mice. In addition, high baseline anxiety in SERT-/- mice may non-specifically affect the tail suspension and forced swim performance in these mice, further complicating interpretation of these data.

Therefore, a depression phenotype in SERT-/- animals cannot be solidly established by tests oriented on despair domain alone (such as forced swim and tail suspension tests), and may require other depression-related, and less dependent on activity or anxiety, domains [72]. To accomplish this goal, we have recently tested SERT-/- mice in a well-validated, anhedonia-based depression paradigm - the sucrose preference test [151, 152], in combination with parallel assessment of activity/anxiety domains using marble burying and digging tests [72]. Overall, SERT+/+, +/- and -/- mice on C57BL/6 genetic background showed similar sucrose preference, suggesting that depression is unaltered in SERT-/- mice. In contrast, marble-burying and bedding tests revealed a marked reduction in activity of SERT-/- mice, dissecting activity domain (which is affected) from depression. Notably, increased digging (indicative of anxiety) was not seen in SERT-/- mice in the marble burying or digging tests [72]. These findings again demonstrate the interplay between anxiety and activity domains, confirming the extent to which in some paradigms (e.g., forced swim, tail suspension despair tests, anxiety tests) multiple domains in SERT-/- mice may be confounded by their hypoactivity domain.

In contrast, adding an additional domain (such as cognitive functions) enabled a better focus on behavioral abnormalities in these mice, as assessed in recent studies showing increased despair after repeated forced swim testing [28]. Notably, recent studies have revealed strain-specific alterations in cortical morphology in SERT-/- mice, suggesting of complex epistatic interactions with other genes in determining the mouse morphological and behavioral phenotypes [11]. Taken together, these data confirm that careful comparative analyses of domain interplay is crucial for a reliable dissection and correct interpretation of phenotypes in genetic models of brain disorders (Fig. 2, 3).

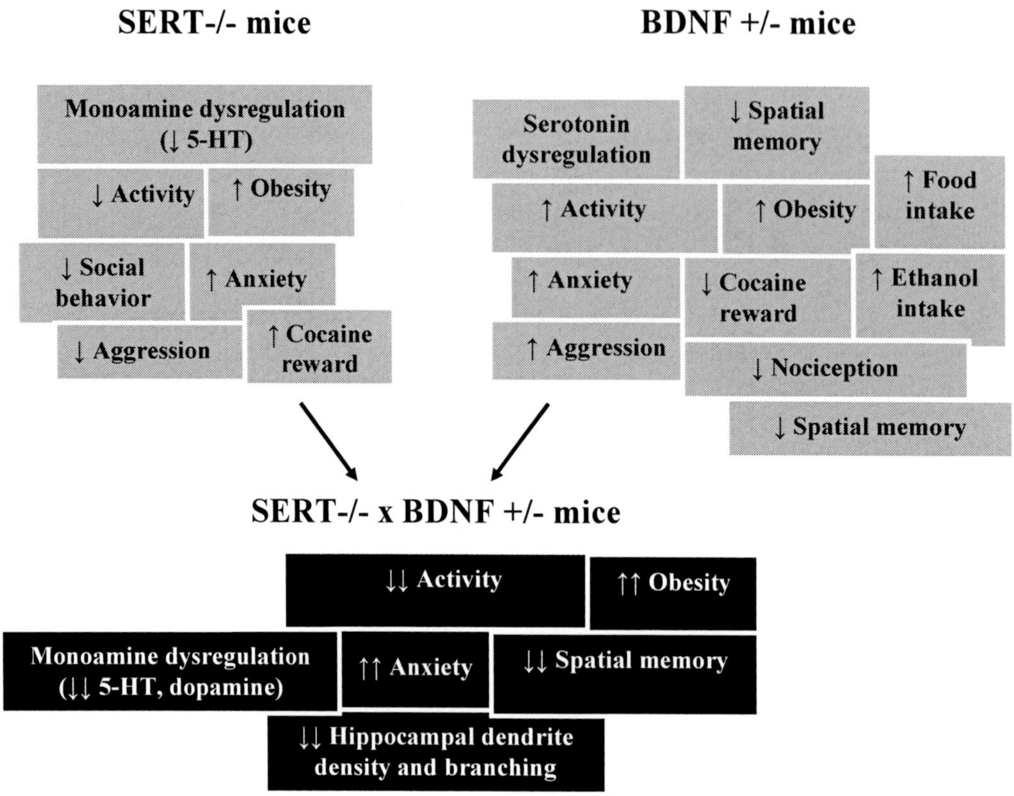

Figure 2. Domain architectonics in 5-HT transporter knockout (SERT-/-), brain derived neurotrophic factor heterozygous knockout (BDNF +/-), and double mutant (SERT-/- x BDNF +/-) mice. Note that only selected disordered domains are presented (gray) for each genetic model (↑ - increased, ↓ - decreased profile). Exacerbation of the respective known domains in the double knockout (SERT-/- x BDNF +/-) model, as a result of genetic interactions, is marked with black color and double arrows.

BDNF mutant mice have also been a useful tool in neurobehavioral research [17, 34, 39, 155]. For example, BDNF conditional knockout mice display severe dysregulation of postsynaptic 5-HT receptors as well as hyperactivity, anxiety and obesity [136, 137]. Likewise, BDNF+/- mice exhibit several phenotypic features, including 5-HT dysregulation, increased motor activity, aggression, anxiety, obesity, ethanol/food intake, and reduced nociception, cocaine reward and hippocampal learning [17, 39, 52, 56, 59, 87, 107, 108, 117, 133, 134, 139] (Fig. 2). Collectively, this makes BDNF+/- mice an indispensable model for biomedical research (also see Table 3).

However, most pronounced physiological and behavioral changes, strongly supporting the importance of SERT-BDNF interactions, have been observed in a double SERT-/- x BDNF+/- knockout mouse model (Fig. 2) developed by Ren-Patterson and colleagues [133-135]. Several results based on this model and reflecting SERT-BDNF interplay will illustrate the utility of dissecting individual domains and studying them as a system of interacting endophenotypes (Fig. 3).

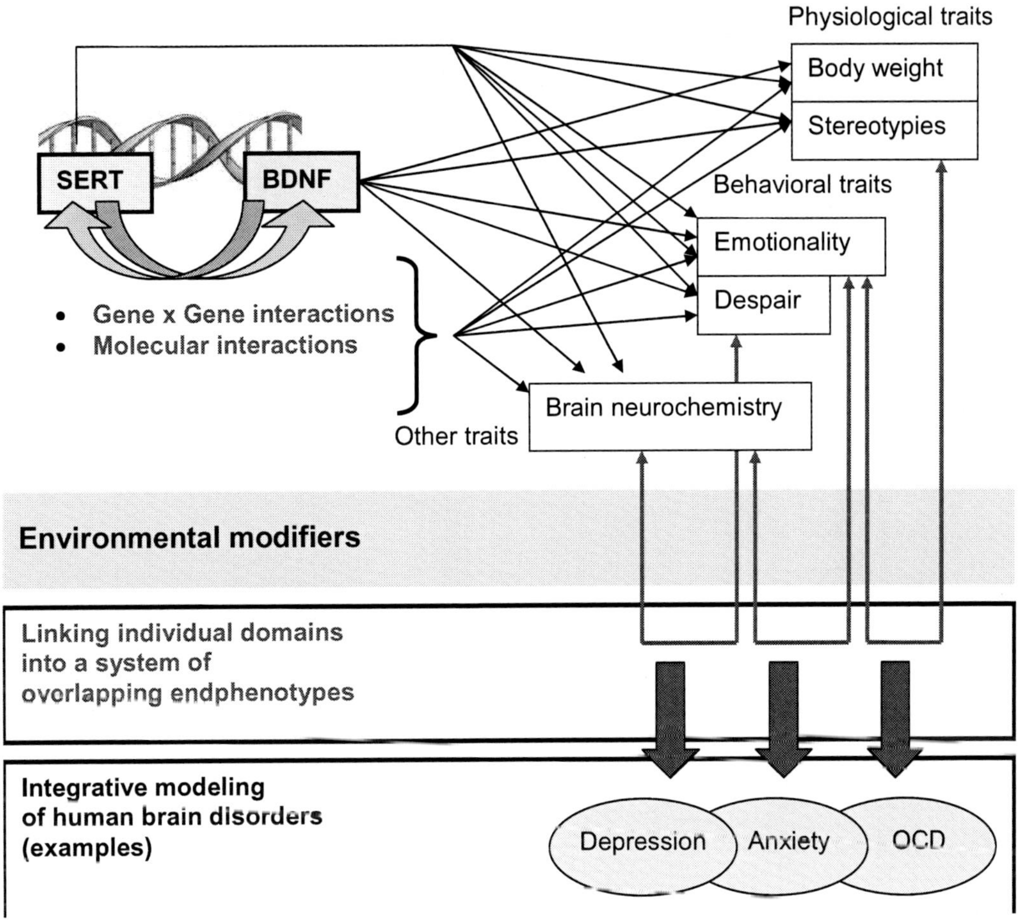

Figure 3. Network-based modeling of brain disorders (based on models developed in [122], with modifications), see Fig. 2 for details.

Figure 2 compares several altered domains in SERT-/- and BDNF+/- gene-targeted mice, outlining their possible interplay in SERT-/- x BDNF+/- double mutant mice. For example, SERT-/- x BDNF+/- mutant mouse data show that reduced BDNF availability during development exaggerates the consequences of absent SERT function, leading to increased anxiety and obesity [122, 134] (see increased body weight and larger heart in Fig. 4). Interestingly, using neonatal models, Garoflos et al. [43] examined the effects of early experience on spatial learning and memory, food intake, hippocampal glucocorticoid, mineralocorticoid and 5-HT1A receptors, and BDNF. They found that neonatal handling has a beneficial effect in the male mice, improving their cognitive ability, accompanied by increased hippocampal gluco/mineralocorticoid receptors ratio and BDNF. Another pathway underlying anti-stress effects of handling may involve up-regulated 5-HT1A receptors that prevent stress-induced hyperphagia, obesity and resistance to leptin [43, 128]. These findings are in line with our observations that SERT x BDNF double-mutant mice have greater stress-induced increases in plasma adrenocorticotropic hormone (than any single-knockout mice) [122], confirming that the multiple gene interactions affect many systems (including the neuroendocrine system) co-modulating the animal behavioral and physiological phenotypes.

Importantly, BDNF, SERT and 5-HT are present not only in the brain, but also in peripheral tissues involved in metabolic functions and responses to stress [157, 158]. Thus, both central and peripheral 5-HT/BDNF-mediated mechanisms are affected in the double-mutant SERT x BDNF mice. One of the mechanisms for this may be mediated by corticotropin releasing hormone that originates from hypothalamus paraventricular nucleus, which in turn results in the release of adrenocorticotropic hormone from the pituitary into general circulation. Furthermore, stress-induced obesity [21, 22, 138] is believed to be associated with glucocorticoid-induced resistance to leptin [149], although other important neuroendocrine mechanisms [40, 93, 94] (potentially associated with 5-HT, SERT and BDNF) have recently been reported.

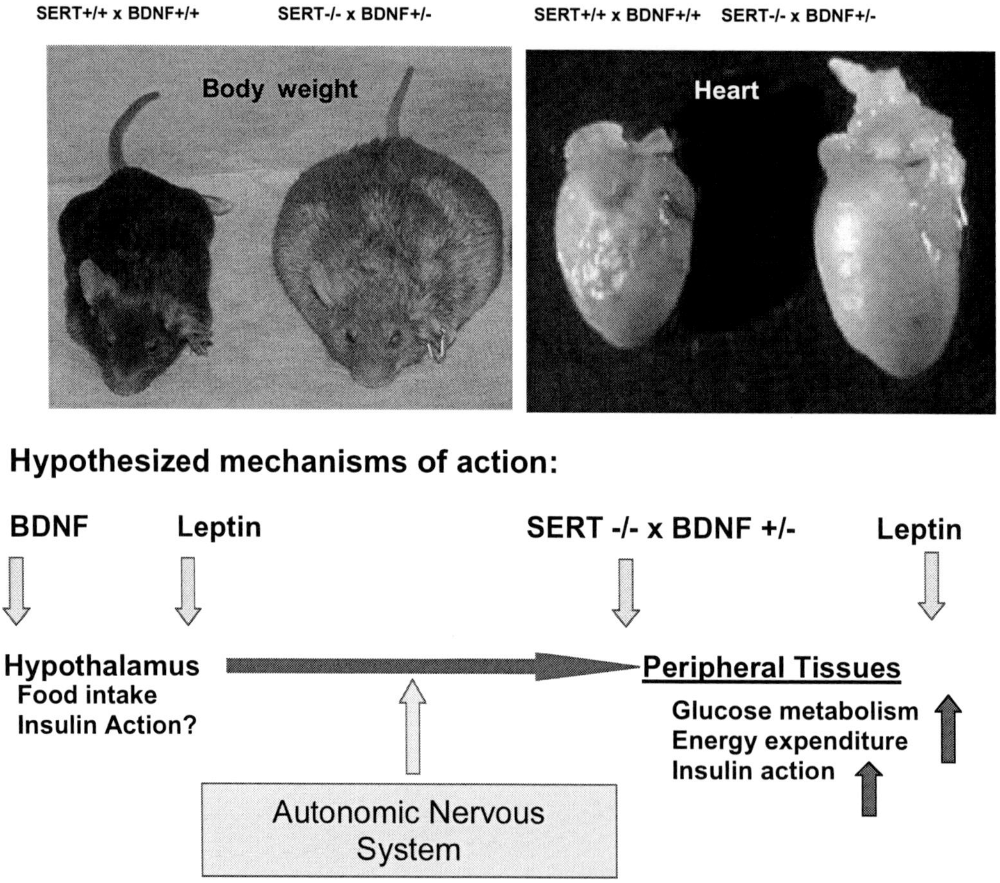

Figure 4. Life long SERT and BDNF deficiency and stress result in adult-onset obesity in SERT-/- x BDNF+/- mice. These double-mutant mice show more significant increase in body weight than other (including single mutant mice) genotypes.

Likewise, as BDNF plays a crucial role in the development and plasticity of neuronal circuits in the nervous system, analysis of neuronal morphology showed that hypothalamus and hippocampus neurons exhibited 25-30% reductions in dendrites (especially in multiple, highly-ordered dendrites branches) in double-mutant mice compared with BDNF +/- mice

(Fig. 5). These morphological changes suggest that the deletion of BDNF x SERT genes significantly affects the development and growth of dendrites - the structural elements that are crucial for synaptogenesis. Furthermore, the double-mutant mice showed poorer performance in the radial maze (compared with any single-mutant mice; R. Ren-Patterson et al., unpublished data). This may indicate aberrant hippocampal memory caused by abnormal hippocampal morphology (but also does not exclude other hippocampal abnormalities, such as impaired navigation and exploration).

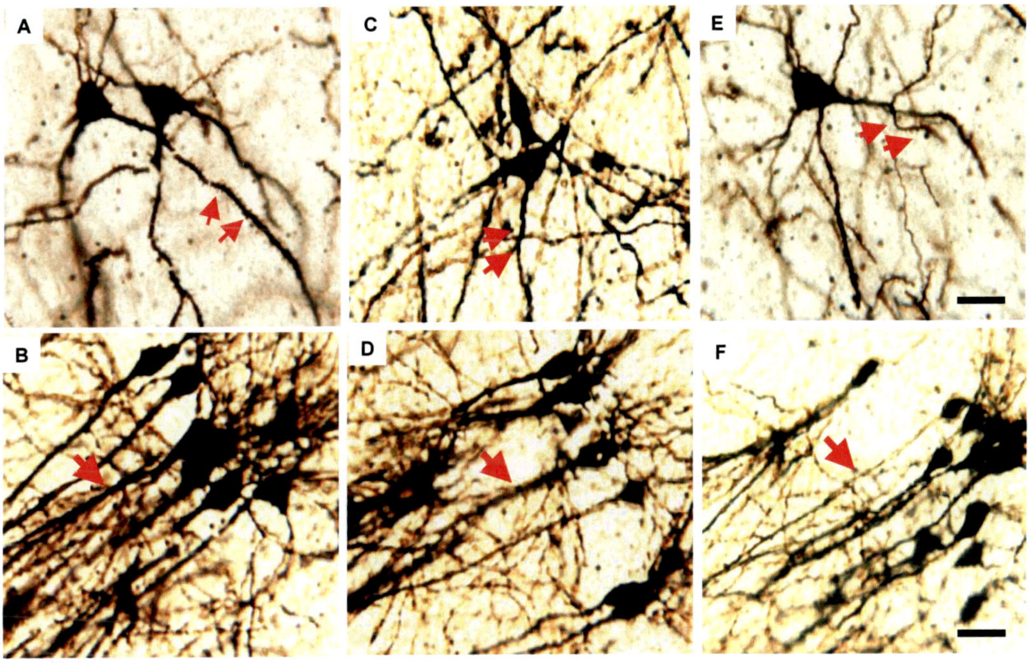

Figure 5. Reductions of 5-HT and BDNF affect development of neuronal dendritic branches in double-mutant mice (E, D) compared to the wild type (A, B) +/+ mice and single SERT-/- knockout mice (C, D). The morphology of brain neuronal hippocampal near dentrate gyrus dentrities and spines was evaluated in 20 fields (scale bars = 10 μm). The quantity of dendrites in brain sections with Golgi impregnation. Double mutant (SERT-/- x BDNF+/-) mice had significant reductions ($p<0.0001$) compared to other genotypes by two-way ANOVA test.

Modelling Genetic Effects in SERT and BDNF Mutants

The other interesting aspect of this genetic model is in the ways that the different domains interact, and in the richness of different types of such interactions. For example, activity is reduced in SERT-/- mice, increased in BDNF+/- mice, and markedly reduced in SERT-/- x BDNF+/- mice (Fig. 2). This suggests a *non-additive* nature ($-1 + 1 = 2$) of genetic influences for this specific phenotype. In contrast, spatial memory is relatively normal in SERT-/- mice, reduced in BDNF+/- mice, and markedly impaired in double mutant mice (Fig. 2), suggesting *potentiation-like* ($0 + 1 = 2$) co-modulation from SERT and BDNF genes of this phenotype.

Also, although SERT-/- mice and BDNF+/- mice do not have marked dopaminergic abnormalities, the double mutant mice display robust reduction in dopaminergic tone in striatum [133, 134], thus showing another type of interactions – *generative* (0 + 0 = 2). Finally, while 5-HT dysregulations, anxiety and obesity are seen in both single mutant strains, they are exaggerated in the double mutant model, showing an *additive* nature (1 + 1 = 2) of SERT and BDNF genetic influences for these three phenotypes.

Clearly, a further dissection of different domains may be possible in this model, elucidating the role of the two genes in their regulation and co-modulation. For example, as mentioned above, we seem to have non-additive genetic influences from SERT and BDNF on the motor activity domain, while observing additive influences on anxiety and obesity domains. This observation further supports the idea that specific physiological abnormalities in SERT-/- x BDNF+/- mice (Fig. 4), rather that simply inactivity per se, may be responsible for high-obesity phenotype in these double mutant mice. While different *other* mechanisms may be involved, analyses of individual domains such as those presented here, will better our understanding of complex abnormalities in SERT-/- x BDNF+/- mutant mice.

Concluding Remarks

In addition to the vital roles that SERT and BDNF play independently of each other, it is clear that SERT and BDNF interact at numerous levels and play an integral part in the regulation of physiological and behavioral functions [19, 82, 133-135] (Tables 1-3). This evidence, as summarized in this chapter, powerfully demonstrates that this linked involvement allows for the effective co-modulation of a range of neural mechanisms. However, genetic interactions also play an active part in this regulatory process, adding another interesting dimension to the interplay between SERT and BDNF (Fig. 3). The elucidation of such mechanisms offers encouraging potential for novel avenues of investigation into the pathogenesis of common and devastating brain maladies. With the new possibilities for innovative exploration, there arises the necessity for developing relevant animal models that foster treatment-oriented research. Given the importance that genetic interactions have on the development and perpetuation of many disorders, genetic models based on mutant or transgenic mice are ideal candidates for this task. SERT, BDNF, and SERT x BDNF mutant mice are particularly promising models because of they have been shown to be pertinent analogues of many prevalent human disorders.

The behavioral research conducted in these animal models has revealed a complex spectrum of relevant phenotypes that interact with each other in a variety of ways. Just as human disorders are often characterized by an intricate set of signs and symptoms, the interaction of these mutant animal phenotypes creates very rich and multifaceted models. The combined alterations of SERT and BDNF in mutant mice produce distinct phenotypical qualities that are different from either SERT or BDNF mutants alone. Based on the research, these mutants can display additive, non-additive, generative or potentiated phenotypes that are highly relevant to human clinical scenarios, bringing them added validity. This domain-interplay approach is absolutely necessary for accurately conceptualizing and dissecting the phenotypes, which will promote additional translational research and support the development of clinical treatments.

ACKNOWLEDGEMENTS

This study is supported by the Intramural Research Program of the National Institute of Mental Health (NIMH/NIH, USA) and NARSAD YI Award (to AVK).

REFERENCES

[1] H.M. Abdolmaleky, S. Thiagalingam, M. Wilcox, Genetics and epigenetics in major psychiatric disorders: dilemmas, achievements, applications, and future scope, *Am J Pharmacogenomics* 5 (2005) 149-160.

[2] R. Adamec, P. Burton, J. Blundell, D.L. Murphy, A. Holmes, Vulnerability to mild predator stress in serotonin transporter knockout mice, *Behav Brain Res* 170 (2006) 126-140.

[3] T. Adayev, B. Ranasinghe, P. Banerjee, Transmembrane signaling in the brain by serotonin, a key regulator of physiology and emotion, *Bioscience reports* 25 (2005) 363-385.

[4] G.K. Aghajanian, Serotonin-induced inward current in rat facial motoneurons: evidence for mediation by G proteins but not protein kinase C, *Brain Res* 524 (1990) 171-174.

[5] G.K. Aghajanian, G.J. Marek, Serotonin induces excitatory postsynaptic potentials in apical dendrites of neocortical pyramidal cells, *Neuropharmacology* 36 (1997) 589-599.

[6] G.K. Aghajanian, G.J. Marek, Serotonin, via 5-HT2A receptors, increases EPSCs in layer V pyramidal cells of prefrontal cortex by an asynchronous mode of glutamate release, *Brain Res* 825 (1999) 161-171.

[7] C. Alexandre, D. Popa, V. Fabre, S. Bouali, P. Venault, K.P. Lesch, M. Hamon, J. Adrien, Early life blockade of 5-hydroxytryptamine 1A receptors normalizes sleep and depression-like behavior in adult knock-out mice lacking the serotonin transporter, *J Neurosci* 26 (2006) 5554-5564.

[8] M.N. Alme, K. Wibrand, G. Dagestad, C.R. Bramham, Chronic fluoxetine treatment induces brain region-specific upregulation of genes associated with BDNF-induced long-term potentiation, *Neural plasticity* 2007 (2007) 1-24.

[9] L. Aloe, A. Iannitelli, F. Angelucci, G. Bersani, M. Fiore, Studies in animal models and humans suggesting a role of nerve growth factor in schizophrenia-like disorders, *Behav Pharmacol* 11 (2000) 235-242.

[10] B. Alsina, T. Vu, S. Cohen-Cory, Visualizing synapse formation in arborizing optic axons in vivo: dynamics and modulation by BDNF, *Nat Neurosci* 4 (2001) 1093-1101.

[11] C. Altamura, M.L. Dell'Acqua, R. Moessner, D.L. Murphy, K.P. Lesch, A.M. Persico, Altered neocortical cell density and layer thickness in serotonin transporter knockout mice: a quantitation study, *Cereb Cortex* 17 (2007) 1394-1401.

[12] F. Angelucci, L. Aloe, P.J. Vasquez, A.A. Mathe, Mapping the differences in the brain concentration of brain-derived neurotrophic factor (BDNF) and nerve growth factor (NGF) in an animal model of depression, *Neuroreport* 11 (2000) 1369-1373.

[13] F. Angelucci, S. Brene, A.A. Mathe, BDNF in schizophrenia, depression and corresponding animal models, *Mol Psychiatry* 10 (2005) 345-352.

[14] F. Angelucci, A.A. Mathe, L. Aloe, Brain-derived neurotrophic factor and tyrosine kinase receptor TrkB in rat brain are significantly altered after haloperidol and risperidone administration, *J Neurosci Res* 60 (2000) 783-794.

[15] F. Angelucci, A.A. Mathe, L. Aloe, Neurotrophic factors and CNS disorders: findings in rodent models of depression and schizophrenia, *Progress in brain research* 146 (2004) 151-165.

[16] M.S. Ansorge, M. Zhou, A. Lira, R. Hen, J.A. Gingrich, Early-life blockade of the 5-HT transporter alters emotional behavior in adult mice, *Science* 306 (2004) 879-881.

[17] A. Bartoletti, L. Cancedda, S.W. Reid, L. Tessarollo, V. Porciatti, T. Pizzorusso, L. Maffei, Heterozygous knock-out mice for brain-derived neurotrophic factor show a pathway-specific impairment of long-term potentiation but normal critical period for monocular deprivation, *J Neurosci* 22 (2002) 10072-10077.

[18] D. Bengel, D.L. Murphy, A.M. Andrews, C.H. Wichems, D. Feltner, A. Heils, R. Mossner, H. Westphal, K.P. Lesch, Altered brain serotonin homeostasis and locomotor insensitivity to 3, 4-methylenedioxymethamphetamine ("Ecstasy") in serotonin transporter-deficient mice, *Mol Pharmacol* 53 (1998) 649-655.

[19] O. Berton, C.A. McClung, R.J. Dileone, V. Krishnan, W. Renthal, S.J. Russo, D. Graham, N.M. Tsankova, C.A. Bolanos, M. Rios, L.M. Monteggia, D.W. Self, E.J. Nestler, Essential role of BDNF in the mesolimbic dopamine pathway in social defeat stress, *Science* 311 (2006) 864-868.

[20] J. Beuten, J.Z. Ma, T.J. Payne, R.T. Dupont, P. Quezada, W. Huang, K.M. Crews, M.D. Li, Significant association of BDNF haplotypes in European-American male smokers but not in European-American female or African-American smokers, *Am J Med Genet B Neuropsychiatr Genet* 139 (2005) 73-80.

[21] P. Bjorntorp, Do stress reactions cause abdominal obesity and comorbidities?, *Obes Rev* 2 (2001) 73-86.

[22] P. Bjorntorp, R. Rosmond, Obesity and cortisol, *Nutrition* (Burbank, Los Angeles County, Calif 16 (2000) 924-936.

[23] S.J. Bonasera, L.H. Tecott, Mouse models of serotonin receptor function: toward a genetic dissection of serotonin systems, *Pharmacology & therapeutics* 88 (2000) 133-142.

[24] T. Bonhoeffer, Neurotrophins and activity-dependent development of the neocortex, *Current opinion in neurobiology* 6 (1996) 119-126.

[25] A. Bonnin, W. Peng, W. Hewlett, P. Levitt, Expression mapping of 5-HT1 serotonin receptor subtypes during fetal and early postnatal mouse forebrain development, *Neuroscience* 141 (2006) 781-794.

[26] A. Bonnin, M. Torii, L. Wang, P. Rakic, P. Levitt, Serotonin modulates the response of embryonic thalamocortical axons to netrin-1, *Nat Neurosci* 10 (2007) 588-597.

[27] J.M. Boyce-Rustay, L.M. Wiedholz, R.A. Millstein, J. Carroll, D.L. Murphy, L.C. Daws, A. Holmes, Ethanol-related behaviors in serotonin transporter knockout mice, *Alcohol Clin Exp Res* 30 (2006) 1957-1965.

[28] J.C. Carroll, J.M. Boyce-Rustay, R. Millstein, R. Yang, L.M. Wiedholz, D.L. Murphy, A. Holmes, Effects of mild early life stress on abnormal emotion-related behaviors in 5-HTT knockout mice, *Behav Genet* 37 (2007) 214-222.

[29] A.R. Carter, C. Chen, P.M. Schwartz, R.A. Segal, Brain-derived neurotrophic factor modulates cerebellar plasticity and synaptic ultrastructure, *J Neurosci* 22 (2002) 1316-1327.

[30] A. Caspi, T.E. Moffitt, Gene-environment interactions in psychiatry: joining forces with neuroscience, *Nature reviews* 7 (2006) 583-590.

[31] A. Caspi, K. Sugden, T.E. Moffitt, A. Taylor, I.W. Craig, H. Harrington, J. McClay, J. Mill, J. Martin, A. Braithwaite, R. Poulton, Influence of life stress on depression: moderation by a polymorphism in the 5-HTT gene, *Science* 301 (2003) 386-389.

[32] P. Celada, J.A. Siuciak, T.M. Tran, C.A. Altar, J.M. Tepper, Local infusion of brain-derived neurotrophic factor modifies the firing pattern of dorsal raphe serotonergic neurons, *Brain Res* 712 (1996) 293-298.

[33] M.J. Choi, R.H. Kang, S.W. Lim, K.S. Oh, M.S. Lee, Brain-derived neurotrophic factor gene polymorphism (Val66Met) and citalopram response in major depressive disorder, *Brain Res* 1118 (2006) 176-182.

[34] S. Chourbaji, R. Hellweg, D. Brandis, B. Zorner, C. Zacher, U.E. Lang, F.A. Henn, H. Hortnagl, P. Gass, Mice with reduced brain-derived neurotrophic factor expression show decreased choline acetyltransferase activity, but regular brain monoamine levels and unaltered emotional behavior, *Brain Res Mol Brain Res* 121 (2004) 28-36.

[35] J.N. Crawley, Behavioral phenotyping of transgenic and knockout mice: experimental design and evaluation of general health, sensory functions, motor abilities, and specific behavioral tests, *Brain Res* 835 (1999) 18-26.

[36] L.C. Daws, S. Montanez, J.L. Munn, W.A. Owens, N.L. Baganz, J.M. Boyce Rustay, R.A. Millstein, L.M. Wiedholz, D.L. Murphy, A. Holmes, Ethanol inhibits clearance of brain serotonin by a serotonin transporter-independent mechanism, *J Neurosci* 26 (2006) 6431-6438.

[37] L.C. Daws, J.L. Munn, M.F. Valdez, T. Frosto-Burke, J.G. Hensler, Serotonin transporter function, but not expression, is dependent on brain-derived neurotrophic factor (BDNF): in vivo studies in BDNF-deficient mice, *J Neurochem* 101 (2007) 641-651.

[38] S. Djalali, M. Holtje, G. Grosse, T. Rothe, T. Stroh, J. Grosse, D.R. Deng, R. Hellweg, R. Grantyn, H. Hortnagl, G. Ahnert-Hilger, Effects of brain-derived neurotrophic factor (BDNF) on glial cells and serotonergic neurones during development, *J Neurochem* 92 (2005) 616-627.

[39] D.E. Dluzen, X. Gao, G.M. Story, L.I. Anderson, J. Kucera, J.M. Walro, Evaluation of nigrostriatal dopaminergic function in adult +/+ and +/- BDNF mutant mice, *Experimental neurology* 170 (2001) 121-128.

[40] M.A. Dutton, E.W. Lee, Z. Zukowska, NPY and extreme stress: lessons learned from posttraumatic stress disorder, *Exs* (2006) 213-222.

[41] C. Firk, C.R. Markus, Review: Serotonin by stress interaction: a susceptibility factor for the development of depression?, *J Psychopharmacol* 21 (2007) 538-544.

[42] D. Frechilla, R. Insausti, P. Ruiz-Golvano, A. Garcia-Osta, M.P. Rubio, J.M. Almendral, J. Del Rio, Implanted BDNF-producing fibroblasts prevent neurotoxin-induced serotonergic denervation in the rat striatum, *Brain Res Mol Brain Res* 76 (2000) 306-314.

[43] E. Garoflos, T. Panagiotaropoulos, S. Pondiki, A. Stamatakis, E. Philippidis, F. Stylianopoulou, Cellular mechanisms underlying the effects of an early experience on cognitive abilities and affective states, *Annals of general psychiatry* [electronic resource] 4 (2005) 8.

[44] C.E. Glatt, J.A. DeYoung, S. Delgado, S.K. Service, K.M. Giacomini, R.H. Edwards, N. Risch, N.B. Freimer, Screening a large reference sample to identify very low frequency sequence variants: comparisons between two genes, Nat Genet 27 (2001) 435-438.

[45] C.E. Glatt, V.I. Reus, Pharmacogenetics of monoamine transporters, *Pharmacogenomics* 4 (2003) 583-596.

[46] J.A. Gorski, S.R. Zeiler, S. Tamowski, K.R. Jones, Brain-derived neurotrophic factor is required for the maintenance of cortical dendrites, *J Neurosci* 23 (2003) 6856-6865.

[47] F.G. Graeff, F.S. Guimaraes, T.G. De Andrade, J.F. Deakin, Role of 5-HT in stress, anxiety, and depression, *Pharmacol Biochem Behav* 54 (1996) 129-141.

[48] C. Gross, R. Hen, The developmental origins of anxiety, *Nature reviews* 5 (2004) 545-552.

[49] C. Gross, R. Hen, Genetic and environmental factors interact to influence anxiety, *Neurotox Res* 6 (2004) 493-501.

[50] C. Gross, L. Santarelli, D. Brunner, X. Zhuang, R. Hen, Altered fear circuits in 5-HT(1A) receptor KO mice, *Biol Psychiatry* 48 (2000) 1157-1163.

[51] C. Gross, X. Zhuang, K. Stark, S. Ramboz, R. Oosting, L. Kirby, L. Santarelli, S. Beck, R. Hen, Serotonin1A receptor acts during development to establish normal anxiety-like behaviour in the adult, *Nature* 416 (2002) 396-400.

[52] B.P. Guiard, D.J. David, T. Deltheil, F. Chenu, E. Le Maitre, T. Renoir, I. Leroux-Nicollet, P. Sokoloff, L. Lanfumey, M. Hamon, A.M. Andrews, R. Hen, A.M. Gardier, Brain-derived neurotrophic factor-deficient mice exhibit a hippocampal hyperserotonergic phenotype, *Int J Neuropsychopharmacol* (2007) 1-14.

[53] B.C. Hains, S.D. Fullwood, M.J. Eaton, C.E. Hulsebosch, Subdural engraftment of serotonergic neurons following spinal hemisection restores spinal serotonin, downregulates serotonin transporter, and increases BDNF tissue content in rat, *Brain Res* 913 (2001) 35-46.

[54] D. Hall, A. Dhilla, A. Charalambous, J.A. Gogos, M. Karayiorgou, Sequence variants of the brain-derived neurotrophic factor (BDNF) gene are strongly associated with obsessive-compulsive disorder, *American journal of human genetics* 73 (2003)370-376.

[55] F.S. Hall, The forced swim test in dopamine, serotonin and norepinephrine transporter knockout mice. *IBANGS*, Vol. 8, 2006, p. 46.

[56] F.S. Hall, J. Drgonova, M. Goeb, G.R. Uhl, Reduced behavioral effects of cocaine in heterozygous brain-derived neurotrophic factor (BDNF) knockout mice, *Neuropsychopharmacology* 28 (2003) 1485-1490.

[57] A.R. Hariri, A. Holmes, Genetics of emotional regulation: the role of the serotonin transporter in neural function, *Trends Cogn Sci* 10 (2006) 182-191.

[58] R. Hellweg, M. Zueger, K. Fink, H. Hortnagl, P. Gass, Olfactory bulbectomy in mice leads to increased BDNF levels and decreased serotonin turnover in depression-related brain areas, *Neurobiology of disease* 25 (2007) 1-7.

[59] J.G. Hensler, E.E. Ladenheim, W.E. Lyons, Ethanol consumption and serotonin-1A (5-HT1A) receptor function in heterozygous BDNF (+/-) mice, *J Neurochem* 85 (2003) 1139-1147.

[60] A. Holmes, A.R. Hariri, The serotonin transporter gene-linked polymorphism and negative emotionality: placing single gene effects in the context of genetic background and environment, *Genes Brain Behav* 2 (2003) 332-335.

[61] A. Holmes, Q. Lit, D.L. Murphy, E. Gold, J.N. Crawley, Abnormal anxiety-related behavior in serotonin transporter null mutant mice: the influence of genetic background, *Genes Brain Behav* 2 (2003) 365-380.

[62] A. Holmes, D.L. Murphy, J.N. Crawley, Abnormal behavioral phenotypes of serotonin transporter knockout mice: parallels with human anxiety and depression, *Biol Psychiatry* 54 (2003) 953-959.

[63] A. Holmes, D.L. Murphy, J.N. Crawley, Reduced aggression in mice lacking the serotonin transporter, *Psychopharmacology* (Berl) 161 (2002) 160-167.

[64] A. Holmes, R.J. Yang, D.L. Murphy, J.N. Crawley, Evaluation of antidepressant-related behavioral responses in mice lacking the serotonin transporter, *Neuropsychopharmacology* 27 (2002) 914-923.

[65] J.R. Homberg, J.D. Olivier, B.M. Smits, J.D. Mul, J. Mudde, M. Verheul, O.F. Nieuwenhuizen, A.R. Cools, E. Ronken, T. Cremers, A.N. Schoffelmeer, B.A. Ellenbroek, E. Cuppen, Characterization of the serotonin transporter knockout rat: a selective change in the functioning of the serotonergic system, *Neuroscience* 146 (2007) 1662-1676.

[66] J.R. Homberg, O.J. Schiepers, A.N. Schoffelmeer, E. Cuppen, L.J. Vanderschuren, Acute and constitutive increases in central serotonin levels reduce social play behaviour in peri-adolescent rats, *Psychopharmacology* (Berl) (2007).

[67] T.L. Huang, C.T. Lee, Y.L. Liu, Serum brain-derived neurotrophic factor levels in patients with major depression: Effects of antidepressants, *J Psychiatr Res* (2007).

[68] R. Hunnerkopf, A. Strobel, L. Gutknecht, B. Brocke, K.P. Lesch, Interaction between BDNF Val66Met and Dopamine Transporter Gene Variation Influences Anxiety-Related Traits, *Neuropsychopharmacology* (2007).

[69] X. Jin, H. Hu, P.H. Mathers, A. Agmon, Brain-derived neurotrophic factor mediates activity-dependent dendritic growth in nonpyramidal neocortical interneurons in developing organotypic cultures, *J Neurosci* 23 (2003) 5662-5673.

[70] A. Kaiser, M. Fedrowitz, U. Ebert, E. Zimmermann, H.J. Hedrich, D. Wedekind, W. Loscher, Auditory and vestibular defects in the circling (ci2) rat mutant, *Eur J Neurosci* 14 (2001) 1129-1142.

[71] A. Kalueff, D.J. Nutt, Role of GABA in memory and anxiety, *Depress Anxiety* 4 (1996) 100-110.

[72] A.V. Kalueff, P.S. Gallagher, D.L. Murphy, Are serotonin transporter knockout mice 'depressed'?: hypoactivity but no anhedonia, *Neuroreport* 17 (2006) 1347-1351.

[73] A.V. Kalueff, C.L. Jensen, D.L. Murphy, Locomotory patterns, spatiotemporal organization of exploration and spatial memory in serotonin transporter knockout mice, *Brain Res* 1169 (2007) 87-97.

[74] A.V. Kalueff, R.F. Ren-Patterson, D.L. Murphy, The developing use of heterozygous mutant mouse models in brain monoamine transporter research, *Trends Pharmacol Sci* 28 (2007) 122-127.

[75] A.V. Kalueff, P. Tuohimaa, Contrasting grooming phenotypes in C57Bl/6 and 129S1/SvImJ mice, *Brain Res* 1028 (2004) 75-82.

[76] A.V. Kalueff, P. Tuohimaa, Experimental modeling of anxiety and depression, *Acta Neurobiol Exp* (Wars) 64 (2004) 439-448.

[77] A.V. Kalueff, P. Tuohimaa, The Suok ("ropewalking") murine test of anxiety, Brain *Res Brain Res Protoc* 14 (2005) 87-99.

[78] A.V. Kalueff, M. Wheaton, D.L. Murphy, What's wrong with my mouse model? Advances and strategies in animal modeling of anxiety and depression, *Behav Brain Res* 179 (2007) 1-18.

[79] A.V. Kalueff, M. Wheaton, R. Ren-Patterson, D.L. Murphy, Brain-derived neurotrophic factor, serotonin transporter, and depression: comment on Kaufman et al, *Biol Psychiatry* 61 (2007) 1112-1113; author reply 1113-1115.

[80] A.V. Kalueff, M. Wheaton, R. Ren-Patterson, D.L. Murphy, Brain-Derived Neurotrophic Factor, Serotonin Transporter, and Depression: Comment on Kaufman et al, *Biol Psychiatry* (2007).

[81] T. Kato, Molecular genetics of bipolar disorder and depression, *Psychiatry and clinical neurosciences* 61 (2007) 3-19.

[82] J. Kaufman, B.Z. Yang, H. Douglas-Palumberi, D. Grasso, D. Lipschitz, S. Houshyar, J.H. Krystal, J. Gelernter, Brain-derived neurotrophic factor-5-HTTLPR gene interactions and environmental modifiers of depression in children, *Biol Psychiatry* 59 (2006) 673-680.

[83] J.L. Kennedy, L.A. Farrer, N.C. Andreasen, R. Mayeux, P. St George-Hyslop, The genetics of adult-onset neuropsychiatric disease: complexities and conundra?, *Science* 302 (2003) 822-826.

[84] S.G. Kernie, D.J. Liebl, L.F. Parada, BDNF regulates eating behavior and locomotor activity in mice, *The EMBO journal* 19 (2000) 1290-1300.

[85] H. Kim do, H. Li, K.Y. Yoo, B.H. Lee, I.K. Hwang, M.H. Won, Effects of fluoxetine on ischemic cells and expressions in BDNF and some antioxidants in the gerbil hippocampal CA1 region induced by transient ischemia, *Experimental neurology* 204 (2007) 748-758.

[86] J.M. Kim, R. Stewart, S.W. Kim, S.J. Yang, I.S. Shin, Y.H. Kim, J.S. Yoon, Interactions Between Life Stressors and Susceptibility Genes (5-HTTLPR and BDNF) on Depression in Korean Elders, *Biol Psychiatry* (2007).

[87] H. Koizumi, K. Hashimoto, M. Iyo, Dietary restriction changes behaviours in brain-derived neurotrophic factor heterozygous mice: role of serotonergic system, *Eur J Neurosci* 24 (2006) 2335-2344.

[88] E. Koponen, T. Rantamaki, V. Voikar, T. Saarelainen, E. MacDonald, E. Castren, Enhanced BDNF signaling is associated with an antidepressant-like behavioral response and changes in brain monoamines, *Cell Mol Neurobiol* 25 (2005) 973-980.

[89] M. Korte, T. Bonhoeffer, Activity-dependent synaptic plasticity: a new face of action for neurotrophins, *Mol Psychiatry* 2 (1997) 197-199.

[90] M. Korte, P. Carroll, E. Wolf, G. Brem, H. Thoenen, T. Bonhoeffer, Hippocampal long-term potentiation is impaired in mice lacking brain-derived neurotrophic factor, *Proc Natl Acad Sci U S A* 92 (1995) 8856-8860.

[91] S.D. Kuipers, C.R. Bramham, Brain-derived neurotrophic factor mechanisms and function in adult synaptic plasticity: new insights and implications for therapy, *Current opinion in drug discovery & development* 9 (2006) 580-586.

[92] S.D. Kuipers, A. Trentani, J.A. Den Boer, G.J. Ter Horst, Molecular correlates of impaired prefrontal plasticity in response to chronic stress, *J Neurochem* 85 (2003) 1312-1323.

[93] L.E. Kuo, K. Abe, Z. Zukowska, Stress, NPY and vascular remodeling: Implications for stress-related diseases, *Peptides* 28 (2007) 435-440.

[94] L.E. Kuo, J.B. Kitlinska, J.U. Tilan, L. Li, S.B. Baker, M.D. Johnson, E.W. Lee, M.S. Burnett, S.T. Fricke, R. Kvetnansky, H. Herzog, Z. Zukowska, Neuropeptide Y acts directly in the periphery on fat tissue and mediates stress-induced obesity and metabolic syndrome, *Nature medicine* 13 (2007) 803-811.

[95] M.H. Larsen, H. Rosenbrock, F. Sams-Dodd, J.D. Mikkelsen, Expression of brain derived neurotrophic factor, activity-regulated cytoskeleton protein mRNA, and enhancement of adult hippocampal neurogenesis in rats after sub-chronic and chronic treatment with the triple monoamine re-uptake inhibitor tesofensine, *Eur J Pharmacol* 555 (2007) 115-121.

[96] J.M. Lauder, Ontogeny of the serotonergic system in the rat: serotonin as a developmental signal, *Ann N Y Acad Sci* 600 (1990) 297-313; discussion 314.

[97] K.P. Lesch, Genetic alterations of the murine serotonergic gene pathway: the neurodevelopmental basis of anxiety, *Handb Exp Pharmacol* (2005) 71-112.

[98] K.P. Lesch, Serotonergic gene inactivation in mice: models for anxiety and aggression?, *Novartis Found Symp* 268 (2005) 111-140; discussion 140-116, 167-170.

[99] K.P. Lesch, R. Mossner, Inactivation of 5HT transport in mice: modeling altered 5HT homeostasis implicated in emotional dysfunction, affective disorders, and somatic syndromes, *Handb Exp Pharmacol* (2006) 417-456.

[100] K.P. Lesch, Y. Zeng, A. Reif, L. Gutknecht, Anxiety-related traits in mice with modified genes of the serotonergic pathway, *Eur J Pharmacol* 480 (2003) 185-204.

[101] D.F. Levinson, The genetics of depression: a review, Biol Psychiatry 60 (2006) 84-92.

[102] Q. Li, Cellular and molecular alterations in mice with deficient and reduced serotonin transporters, *Mol Neurobiol* 34 (2006) 51-66.

[103] Q. Li, C.H. Wichems, L. Ma, L.D. Van de Kar, F. Garcia, D.L. Murphy, Brain region-specific alterations of 5-HT2A and 5-HT2C receptors in serotonin transporter knockout mice, *J Neurochem* 84 (2003) 1256-1265.

[104] A. Lira, M. Zhou, N. Castanon, M.S. Ansorge, J.A. Gordon, J.H. Francis, M. Bradley-Moore, J. Lira, M.D. Underwood, V. Arango, H.F. Kung, M.A. Hofer, R. Hen, J.A. Gingrich, Altered depression-related behaviors and functional changes in the dorsal raphe nucleus of serotonin transporter-deficient mice, *Biol Psychiatry* 54 (2003) 960-971.

[105] I. Lucki, The spectrum of behaviors influenced by serotonin, *Biol Psychiatry* 44 (1998) 151-162.

[106] B.A. Luellen, L.E. Bianco, L.M. Schneider, A.M. Andrews, Reduced brain-derived neurotrophic factor is associated with a loss of serotonergic innervation in the hippocampus of aging mice, *Genes Brain Behav* 6 (2007) 482-490.

[107] W.E. Lyons, L.A. Mamounas, G.A. Ricaurte, V. Coppola, S.W. Reid, S.H. Bora, C. Wihler, V.E. Koliatsos, L. Tessarollo, Brain-derived neurotrophic factor-deficient mice develop aggressiveness and hyperphagia in conjunction with brain serotonergic abnormalities, *Proc Natl Acad Sci U S A* 96 (1999) 15239-15244.

[108] G.M. MacQueen, K. Ramakrishnan, S.D. Croll, J.A. Siuciak, G. Yu, L.T. Young, M. Fahnestock, Performance of heterozygous brain-derived neurotrophic factor knockout mice on behavioral analogues of anxiety, nociception, and depression, *Behav Neurosci* 115 (2001) 1145-1153.

[109] L.A. Mamounas, C.A. Altar, M.E. Blue, D.R. Kaplan, L. Tessarollo, W.E. Lyons, BDNF promotes the regenerative sprouting, but not survival, of injured serotonergic axons in the adult rat brain, *J Neurosci* 20 (2000) 771-782.

[110] L.A. Mamounas, M.E. Blue, J.A. Siuciak, C.A. Altar, Brain-derived neurotrophic factor promotes the survival and sprouting of serotonergic axons in rat brain, *J Neurosci* 15 (1995) 7929-7939.

[111] M.P. Mattson, S. Maudsley, B. Martin, BDNF and 5-HT: a dynamic duo in age-related neuronal plasticity and neurodegenerative disorders, *Trends in neurosciences* 27 (2004) 589-594.

[112] M.P. Mattson, S. Maudsley, B. Martin, A neural signaling triumvirate that influences ageing and age-related disease: insulin/IGF-1, BDNF and serotonin, *Ageing research reviews* 3 (2004) 445-464.

[113] A.J. Mayorga, I. Lucki, Limitations on the use of the C57BL/6 mouse in the tail suspension test, *Psychopharmacology* (Berl) 155 (2001) 110-112.

[114] A.K. McAllister, Subplate neurons: a missing link among neurotrophins, activity, and ocular dominance plasticity?, *Proc Natl Acad Sci U S A* 96 (1999) 13600-13602.

[115] J.H. Meyer, Imaging the serotonin transporter during major depressive disorder and antidepressant treatment, *J Psychiatry Neurosci* 32 (2007) 86-102.

[116] L. Minichiello, M. Korte, D. Wolfer, R. Kuhn, K. Unsicker, V. Cestari, C. Rossi-Arnaud, H.P. Lipp, T. Bonhoeffer, R. Klein, Essential role for TrkB receptors in hippocampus-mediated learning, *Neuron* 24 (1999) 401-414.

[117] A. Montkowski, F. Holsboer, Intact spatial learning and memory in transgenic mice with reduced BDNF, *Neuroreport* 8 (1997) 779-782.

[118] R. Mossner, S. Daniel, D. Albert, A. Heils, O. Okladnova, A. Schmitt, K.P. Lesch, Serotonin transporter function is modulated by brain-derived neurotrophic factor (BDNF) but not nerve growth factor (NGF), *Neurochem Int* 36 (2000) 197-202.

[119] MPD, Mouse Phenome Database, http://www.jax.org/phenome (2007).

[120] D.L. Murphy, A. Lerner, G. Rudnick, K.P. Lesch, Serotonin transporter: gene, genetic disorders, and pharmacogenetics, *Mol Interv* 4 (2004) 109-123.

[121] D.L. Murphy, Q. Li, S. Engel, C. Wichems, A. Andrews, K.P. Lesch, G. Uhl, Genetic perspectives on the serotonin transporter, *Brain Res Bull* 56 (2001) 487-494.

[122] D.L. Murphy, G.R. Uhl, A. Holmes, R. Ren-Patterson, F.S. Hall, I. Sora, S. Detera-Wadleigh, K.P. Lesch, Experimental gene interaction studies with SERT mutant mice as models for human polygenic and epistatic traits and disorders, *Genes Brain Behav* 2 (2003) 350-364.

[123] M. Narita, K. Aoki, M. Takagi, Y. Yajima, T. Suzuki, Implication of brain-derived neurotrophic factor in the release of dopamine and dopamine-related behaviors induced by methamphetamine, *Neuroscience* 119 (2003) 767-775.

[124] M. Nibuya, E.J. Nestler, R.S. Duman, Chronic antidepressant administration increases the expression of cAMP response element binding protein (CREB) in rat hippocampus, *J Neurosci* 16 (1996) 2365-2372.

[125] K. Nishimura, K. Nakamura, A. Anitha, K. Yamada, M. Tsujii, Y. Iwayama, E. Hattori, T. Toyota, N. Takei, T. Miyachi, Y. Iwata, K. Suzuki, H. Matsuzaki, M. Kawai, Y. Sekine, K. Tsuchiya, G. Sugihara, S. Suda, Y. Ouchi, T. Sugiyama, T. Yoshikawa, N. Mori, Genetic analyses of the brain-derived neurotrophic factor (BDNF) gene in autism, *Biochemical and biophysical research communications* 356 (2007) 200-206.

[126] M.J. Owens, C.B. Nemeroff, Role of serotonin in the pathophysiology of depression: focus on the serotonin transporter, *Clinical chemistry* 40 (1994) 288-295.

[127] N. Ozaki, D. Goldman, W.H. Kaye, K. Plotnicov, B.D. Greenberg, J. Lappalainen, G. Rudnick, D.L. Murphy, Serotonin transporter missense mutation associated with a complex neuropsychiatric phenotype, *Mol Psychiatry* 8 (2003) 933-936.

[128] T. Panagiotaropoulos, A. Papaioannou, S. Pondiki, A. Prokopiou, F. Stylianopoulou, K. Gerozissis, Effect of neonatal handling and sex on basal and chronic stress-induced corticosterone and leptin secretion, *Neuroendocrinology* 79 (2004) 109-118.

[129] L. Pezawas, A.L. Goldman, V. B.A., V.S. Mattay, G. Chen, J.H. Callicott, B.S. Kolachana, R.E. Straub, M.F. Egan, A. Meyer-Lindenberg, D.R. Weinberger, Interactions of SERT & BDNF: A Complex Genetic Model of Depression. *ECNP*, 2006.

[130] A. Piccinni, D. Marazziti, M. Catena, L. Domenici, A. Del Debbio, C. Bianchi, C. Mannari, C. Martini, E. Da Pozzo, E. Schiavi, A. Mariotti, I. Roncaglia, A. Palla, G. Consoli, L. Giovannini, G. Massimetti, L. Dell'osso, Plasma and serum brain-derived neurotrophic factor (BDNF) in depressed patients during 1 year of antidepressant treatments, *J Affect Disord* (2007).

[131] L.D. Pozzo-Miller, W. Gottschalk, L. Zhang, K. McDermott, J. Du, R. Gopalakrishnan, C. Oho, Z.H. Sheng, B. Lu, Impairments in high-frequency transmission, synaptic vesicle docking, and synaptic protein distribution in the hippocampus of BDNF knockout mice, *J Neurosci* 19 (1999) 4972-4983.

[132] T. Rantamaki, P. Hendolin, A. Kankaanpaa, J. Mijatovic, P. Piepponen, E. Domenici, M.V. Chao, P.T. Mannisto, E. Castren, Pharmacologically Diverse Anti- depressants Rapidly Activate Brain-Derived Neurotrophic Factor Receptor TrkB and Induce Phospholipase-Cgamma Signaling Pathways in Mouse Brain, *Neuropsychopharmacology* (2007).

[133] R.F. Ren-Patterson, L.W. Cochran, A. Holmes, K.P. Lesch, B. Lu, D.L. Murphy, Gender-dependent modulation of brain monoamines and anxiety-like behaviors in mice with genetic serotonin transporter and BDNF deficiencies, *Cell Mol Neurobiol* 26 (2006) 755-780.

[134] R.F. Ren-Patterson, L.W. Cochran, A. Holmes, S. Sherrill, S.J. Huang, T. Tolliver, K.P. Lesch, B. Lu, D.L. Murphy, Loss of brain-derived neurotrophic factor gene allele exacerbates brain monoamine deficiencies and increases stress abnormalities of serotonin transporter knockout mice, *J Neurosci Res* 79 (2005) 756-771.

[135] R.F. Ren-Patterson, D.K. Kim, X. Zheng, S. Sherrill, S.J. Huang, T. Tolliver, D.L. Murphy, Serotonergic-like progenitor cells propagated from neural stem cells in vitro: survival with SERT protein expression following implantation into brains of mice lacking SERT, *Faseb J* 19 (2005) 1537-1539.

[136] M. Rios, G. Fan, C. Fekete, J. Kelly, B. Bates, R. Kuehn, R.M. Lechan, R. Jaenisch, Conditional deletion of brain-derived neurotrophic factor in the postnatal brain leads to obesity and hyperactivity, *Molecular endocrinology* (Baltimore, Md 15 (2001) 1748-1757.

[137] M. Rios, E.K. Lambe, R. Liu, S. Teillon, J. Liu, S. Akbarian, S. Roffler-Tarlov, R. Jaenisch, G.K. Aghajanian, Severe deficits in 5-HT2A -mediated neurotransmission in BDNF conditional mutant mice, *J Neurobiol* 66 (2006) 408-420.

[138] R. Rosmond, M.F. Dallman, P. Bjorntorp, Stress-related cortisol secretion in men: relationships with abdominal obesity and endocrine, metabolic and hemodynamic abnormalities, *The Journal of clinical endocrinology and metabolism* 83 (1998) 1853-1859.

[139] C. Rossi, A. Angelucci, L. Costantin, C. Braschi, M. Mazzantini, F. Babbini, M.E. Fabbri, L. Tessarollo, L. Maffei, N. Berardi, M. Caleo, Brain-derived neurotrophic factor (BDNF) is required for the enhancement of hippocampal neurogenesis following environmental enrichment, *Eur J Neurosci* 24 (2006) 1850-1856.

[140] G. Rudnick, Serotonin transporters--structure and function, *The Journal of membrane biology* 213 (2006) 101-110.

[141] G. Rudnick, Structure/function relationships in serotonin transporter: new insights from the structure of a bacterial transporter, *Handb Exp Pharmacol* (2006) 59-73.

[142] D. Rudrauf, P. Venault, C. Cohen-Salmon, A. Berthoz, R. Jouvent, G. Chapouthier, A new method for the assessment of spatial orientation and spatial anxiety in mice, *Brain Res Brain Res Protoc* 13 (2004) 159-165.

[143] P. Rumajogee, D. Verge, N. Hanoun, M.J. Brisorgueil, R. Hen, K.P. Lesch, M. Hamon, M.C. Miquel, Adaption of the serotoninergic neuronal phenotype in the absence of 5-HT autoreceptors or the 5-HT transporter: involvement of BDNF and cAMP, *Eur J Neurosci* 19 (2004) 937-944.

[144] D. Senkowski, M. Linden, D. Zubragel, T. Bar, J. Gallinat, Evidence for disturbed cortical signal processing and altered serotonergic neurotransmission in generalized anxiety disorder, *Biol Psychiatry* 53 (2003) 304-314.

[145] A. Serretti, R. Calati, L. Mandelli, D. De Ronchi, Serotonin transporter gene variants and behavior: a comprehensive review, *Curr Drug Targets* 7 (2006) 1659-1669.

[146] I.S. Shiah, L.N. Yatham, Serotonin in mania and in the mechanism of action of mood stabilizers: a review of clinical studies, *Bipolar disorders* 2 (2000) 77-92.

[147] J.A. Siuciak, C. Boylan, M. Fritsche, C.A. Altar, R.M. Lindsay, BDNF increases monoaminergic activity in rat brain following intracerebroventricular or intraparenchymal administration, *Brain Res* 710 (1996) 11-20.

[148] J.A. Siuciak, M.S. Clark, H.B. Rind, S.R. Whittemore, A.F. Russo, BDNF induction of tryptophan hydroxylase mRNA levels in the rat brain, *J Neurosci Res* 52 (1998) 149-158.

[149] J.M. Solano, L. Jacobson, Glucocorticoids reverse leptin effects on food intake and body fat in mice without increasing NPY mRNA, *The American journal of physiology* 277 (1999) E708-716.

[150] J.H. Son, H.S. Chun, T.H. Joh, S. Cho, B. Conti, J.W. Lee, Neuroprotection and neuronal differentiation studies using substantia nigra dopaminergic cells derived from transgenic mouse embryos, *J Neurosci* 19 (1999) 10-20.

[151] T. Strekalova, N. Gorenkova, E. Schunk, O. Dolgov, D. Bartsch, Selective effects of citalopram in a mouse model of stress-induced anhedonia with a control for chronic stress, *Behav Pharmacol* 17 (2006) 271-287.

[152] T. Strekalova, R. Spanagel, D. Bartsch, F.A. Henn, P. Gass, Stress-induced anhedonia in mice is associated with deficits in forced swimming and exploration, *Neuropsychopharmacology* 29 (2004) 2007-2017.

[153] T. Strekalova, R. Spanagel, O. Dolgov, D. Bartsch, Stress-induced hyperlocomotion as a confounding factor in anxiety and depression models in mice, *Behav Pharmacol* 16 (2005) 171-180.

[154] K.J. Sufka, M.W. Feltenstein, J.E. Warnick, E.O. Acevedo, H.E. Webb, C.M. Cartwright, Modeling the anxiety-depression continuum hypothesis in domestic fowl chicks, *Behav Pharmacol* 17 (2006) 681-689.

[155] M.E. Szapacs, T.A. Mathews, L. Tessarollo, W. Ernest Lyons, L.A. Mamounas, A.M. Andrews, Exploring the relationship between serotonin and brain-derived neurotrophic factor: analysis of BDNF protein and extraneuronal 5-HT in mice with reduced serotonin transporter or BDNF expression, *J Neurosci Methods* 140 (2004) 81-92.

[156] M.E. Szapacs, A.L. Numis, A.M. Andrews, Late onset loss of hippocampal 5-HT and NE is accompanied by increases in BDNF protein expression in mice co-expressing mutant APP and PS1, *Neurobiology of disease* 16 (2004) 572-580.

[157] O.A. Tjurmina, I. Armando, J.M. Saavedra, D.S. Goldstein, D.L. Murphy, Exaggerated adrenomedullary response to immobilization in mice with targeted disruption of the serotonin transporter gene, *Endocrinology* 143 (2002) 4520-4526.

[158] J.R. Tonra, M. Ono, X. Liu, K. Garcia, C. Jackson, G.D. Yancopoulos, S.J. Wiegand, V. Wong, Brain-derived neurotrophic factor improves blood glucose control and alleviates fasting hyperglycemia in C57BLKS-Lepr(db)/lepr(db) mice, *Diabetes* 48 (1999) 588-594.

[159] S.J. Tsai, The P11, tPA/plasminogen system and brain-derived neurotrophic factor: Implications for the pathogenesis of major depression and the therapeutic mechanism of antidepressants, *Med Hypotheses* 68 (2007) 180-183.

[160] K.L. Tucker, M. Meyer, Y.A. Barde, Neurotrophins are required for nerve growth during development, Nat Neurosci 4 (2001) 29-37.

[161] K. Turlejski, Evolutionary ancient roles of serotonin: long-lasting regulation of activity and development, *Acta Neurobiol Exp* (Wars) 56 (1996) 619-636.

[162] P. Van Meer, J. Raber, Mouse behavioural analysis in systems biology, *Biochem J* 389 (2005) 593-610.

[163] D.E. Vetter, C. Li, L. Zhao, A. Contarino, M.C. Liberman, G.W. Smith, Y. Marchuk, G.F. Koob, S.F. Heinemann, W. Vale, K.F. Lee, Urocortin-deficient mice show hearing impairment and increased anxiety-like behavior, *Nat Genet* 31 (2002) 363-369.

[164] T. Vitalis, O. Cases, S. Passemard, J. Callebert, J.G. Parnavelas, Embryonic depletion of serotonin affects cortical development, *Eur J Neurosci* 26 (2007) 331-344.

[165] T. Vitalis, J.G. Parnavelas, The role of serotonin in early cortical development, *Dev Neurosci* 25 (2003) 245-256.

[166] P.M. Whitaker-Azmitia, The discovery of serotonin and its role in neuroscience, *Neuropsychopharmacology* 21 (1999) 2S-8S.

[167] P.M. Whitaker-Azmitia, Role of serotonin and other neurotransmitter receptors in brain development: basis for developmental pharmacology, *Pharmacological reviews* 43 (1991) 553-561.

[168] P.M. Whitaker-Azmitia, Serotonin and brain development: role in human developmental diseases, *Brain Res Bull* 56 (2001) 479-485.

[169] P.M. Whitaker-Azmitia, M. Druse, P. Walker, J.M. Lauder, Serotonin as a developmental signal, *Behav Brain Res* 73 (1996) 19-29.

[170] L.A. White, M.J. Eaton, M.C. Castro, K.J. Klose, M.Y. Globus, G. Shaw, S.R. Whittemore, Distinct regulatory pathways control neurofilament expression and neurotransmitter synthesis in immortalized serotonergic neurons, *J Neurosci* 14 (1994) 6744-6753.

[171] M. Wichers, G. Kenis, N. Jacobs, R. Mengelers, C. Derom, R. Vlietinck, J. van Os, The BDNF Val(66)Met x 5-HTTLPR x child adversity interaction and depressive symptoms: An attempt at replication, *Am J Med Genet B Neuropsychiatr Genet* (2007).

[172] S.A. Willis-Owen, J. Flint, The genetic basis of emotional behaviour in mice, *Eur J Hum Genet* 14 (2006) 721-728.

[173] S.A. Willis-Owen, J. Flint, Identifying the genetic determinants of emotionality in humans; insights from rodents, *Neurosci Biobehav Rev* 31 (2007) 115-124.

[174] D.P. Wolfer, W.E. Crusio, H.P. Lipp, Knockout mice: simple solutions to the problems of genetic background and flanking genes, *Trends in neurosciences* 25 (2002) 336-340.

[175] D.P. Wolfer, U. Muller, M. Stagliar, H.P. Lipp, Assessing the effects of the 129/Sv genetic background on swimming navigation learning in transgenic mutants: a study using mice with a modified beta-amyloid precursor protein gene, *Brain Res* 771 (1997) 1-13.

[176] K. Yamada, M. Mizuno, T. Nabeshima, Role for brain-derived neurotrophic factor in learning and memory, *Life Sci* 70 (2002) 735-744.

[177] L.N. Yatham, P.F. Liddle, I.S. Shiah, G. Scarrow, R.W. Lam, M.J. Adam, A.P. Zis, T.J. Ruth, Brain serotonin2 receptors in major depression: a positron emission tomography study, *Arch Gen Psychiatry* 57 (2000) 850-858.

[178] K. Yoshida, H. Higuchi, M. Kamata, H. Takahashi, K. Inoue, T. Suzuki, K. Itoh, N. Ozaki, The G196A polymorphism of the brain-derived neurotrophic factor gene and the antidepressant effect of milnacipran and fluvoxamine, *J Psychopharmacol* (2006).

[179] R. Yoshimura, M. Mitoma, A. Sugita, H. Hori, T. Okamoto, W. Umene, N. Ueda, J. Nakamura, Effects of paroxetine or milnacipran on serum brain-derived neurotrophic factor in depressed patients, *Prog Neuropsychopharmacol Biol Psychiatry* 31 (2007) 1034-1037.

[180] H. Zhang, F. Ozbay, J. Lappalainen, H.R. Kranzler, C.H. van Dyck, D.S. Charney, L.H. Price, S. Southwick, B.Z. Yang, A. Rasmussen, J. Gelernter, Brain derived neurotrophic factor (BDNF) gene variants and Alzheimer's disease, affective disorders, posttraumatic stress disorder, schizophrenia, and substance dependence, *Am J Med Genet B Neuropsychiatr Genet* 141 (2006) 387-393.

[181] S. Zhao, J. Edwards, J. Carroll, L. Wiedholz, R.A. Millstein, C. Jaing, D.L. Murphy, T.H. Lanthorn, A. Holmes, Insertion mutation at the C-terminus of the serotonin transporter disrupts brain serotonin function and emotion-related behaviors in mice, *Neuroscience* 140 (2006) 321-334.

[182] F.C. Zhou, K.P. Lesch, D.L. Murphy, Serotonin uptake into dopamine neurons via dopamine transporters: a compensatory alternative, *Brain Res* 942 (2002) 109-119.

[183] C.B. Zhu, J.A. Steiner, J.L. Munn, L.C. Daws, W.A. Hewlett, R.D. Blakely, Rapid stimulation of presynaptic serotonin transport by a3 adenosine receptors, *J Pharmacol Exp Ther* 322 (2007) 332-340.

In: Behavioral Models in Stress Research
Editors: A.V. Kalueff and J.L. LaPorte, pp. 185-209
ISBN: 978-1-60456-361-0

Chapter 8

IMPAIRED FUNCTION OF PHOSPHODIESTERASE 4B IN MUTATED DISC1 MICE LEADS TO DEVELOPMENT OF PRONOUNCED DEPRESSION-LIKE BEHAVIOR WITH MILD EXPRESSION OF SCHIZOPHRENIA-LIKE PHENOTYPE

Tatiana V. Lipina[1*], Steven J. Clapcote[2], Kirsty Millar[2], Yoichi Gondo[3], R. Mark Henkelman[4,5], David J. Porteous[2] and John C. Roder [1,5]

[1] Samuel Lunenfeld Research Institute at Mount Sinai Hospital, Toronto, ON, Canada.
[2] Medical Genetics Section, University of Edinburgh Centre for Molecular Medicine, Edinburgh, United Kingdom.
[3] RIKEN Genomic Sciences Center, Yokohama, Japan.
[4] Hospital for Sick Children, Mouse Imaging Centre, Toronto, ON, Canada.
[5] University of Toronto, Departments of Medical Biophysics and Molecular & Medical Genetics, Toronto, ON, Canada

DEPRESSION AND SCHIZOPHRENIA: OVERLAPPING GENES

Mental disorders affect human population around the world [45]. Genetic factors are known to play an important role in influencing susceptibility to many most common disorders [39]. The functional psychoses are the most severe adult-onset psychiatric disorders and for over 100 years have been divided into two main diagnostic categories: schizophrenia and bipolar disorder. Schizophrenia is characterized by psychotic features (delusions and hallucinations), disorganization, dysfunction in normal affective responses, and altered cognitive functioning. Bipolar disorder is characterized by disturbances in mood ranging from

* E-mail address: lipina@mshri.on.ca. Phone: 1 (416) 586-4800 ext 8242. Samuel Lunenfeld Research Institute, Mount Sinai Hospital, 600 University Avenue, Toronto, Ontario, M5G 1X5, Canada

extreme elation (mania) to severe depression often accompanied by psychotic features and cognitive changes. However, there has been a long tradition of dissent against the validity of such classification of functional psychosis [10, 21, 61] and the veracity of the so-called Kraepelinian dichotomy have been increasingly challenged by emerging data from many fields of psychiatric research [19, 64, 45]. The most convincing challenges have come in recent years from genetic studies.

Most current genetic studies of schizophrenia and bipolar disorder have been based on the assumption of independence, with individual studies typically focusing on only one or the other disorder. Cases with a mix of mood and psychotic features, while common, have been ignored or subsumed into some broader category of either schizophrenia or bipolar disorder. However, family studies have shown statistically significant evidence that bipolar disorder occurs at increased rates in relatives of probands with schizophrenia [62] and that schizophrenia occurs at increased frequency in relatives of probands with bipolar disorder [63]. Twin study demonstrated a clear overlap in genetic susceptibility to syndromically defined mania and schizophrenia [14]. Genetic linkage studies have identified some chromosome regions that show convergent or overlapping regions of interest in both schizophrenia and bipolar disorder—including regions of 13q, 22q, 6q, and 18 [5, 7] for significant co-occurrence. The hypothesis that loci exist that influence susceptibility across the schizophrenia-bipolar divide receives further support. A genome scan using families selected on the basis of a member with DSM-IV schizoaffective disorder, and bipolar type, demonstrated significance at 1q42 and suggestive linkage at 22q11 with linkage evidence being contributed equally by ''schizophrenia'' families (i.e., where other members had predominantly schizophrenia) and ''bipolar families'' (i.e., where other members had predominantly bipolar disorder) [26]. More direct evidence has come from reports implicating variation in the same genes as influencing susceptibility to both schizophrenia and bipolar disorder. There are a number of genes (DAOA, dysbindin, COMT, BDNF, Disc1 and NRG1) that contribute susceptibility across the Kraepelinian divide to schizophrenia, bipolar disorder and schizoaffective disorder [5, 25, 59], where Disc1 rank as gene most equally contributing to both prototypical Kraepelinian illnesses [20].

The Role of Disc1 and PDE4B in Depression and Schizophrenia

Disc1, which was originally identified at one breakpoint of a chromosomal t(1;11) (q42.1;q14.3) translocation that cosegregates in a large Scottish family with major mental illness, including schizophrenia, bipolar disorder, and major depression [42, 8]. The 1q42 region [52] and mutations in Disc1 have also been associated with psychiatric conditions in independent Finnish, European and American populations [52]. Disc1 has also been reported to associate with impaired memory, reduced gray matter density and volume, and abnormal hippocampal volume and function [11, 13, 27, 29]. A consistent pattern has emerged from Disc1 expression studies in the human, primate, and rodent brain, with evidence for developmental regulation and high protein levels in subregions, particularly the hippocampus, implicated in the pathogenesis of psychiatric disorders [4, 41, 49, 57].

Disc1 functions as a molecular scaffold, interacting with multiple proteins required for neuronal migration, neurite outgrowth, signal transduction, cyclic adenosine monophosphate (cAMP) signaling, cytoskeletal modulation, and translational regulation [52, 55, 12]. Among the known Disc1 binding partners, cAMP-hydrolyzing phosphodiestirase 4B (PDE4B) is particularly important.

Phosphodiesterase (PDE) is any enzyme that catalyzes the hydrolysis of phosphodiester bonds, for instance a bond in a molecule of cAMP. PDE-4 enzymes are the multi-gene family, specifically hydrolyze cAMP, a key second messenger inside cells, thus providing a pivotal means of regulating cAMP levels. In the central nervous system PDE4 is expressed in neurons of the cerebral cortex and hippocampus, hypothalamus and striatum, dopaminergic neurons of the substantia nigra and in astrocytes [15, 23]. Previous studies show that schizophrenics have decreased levels of intracellular cAMP [44]. cAMP-dependent protein kinase levels are also altered in schizophrenia [60] as well as PDEs are increased in cortex and temporal lobes of patients with schizophrenia, that reduce intracellular cAMP. PDE4B gene was found to be disrupted by a translocation in two related individuals with psychosis [43]. Recent study on Scottish population revealed significant association between single nucleotide polymorphisms (SNPs) of PDE4B and schizophrenia in females [51]. Rolipram, a PDE4 inhibitor, demonstrated antipsychotic capacities in animal models of schizophrenia [35, 38]. However, historically PDE4 inhibitors have been studied for development new treatment for depression two decades ago [68]. Indeed, rolipram has antidepressant effects through prevention of cAMP hydrolysis [9, 32] and showed efficacy in preclinical models of depression, and is thought to be an integral component in the mechanisms of action for various types of antidepressants [48]. Millar et al [43] suggested that Disc1 acts as a molecular scaffold, forming a complex with PDE4B in specific subcellular compartments and sequestering it in a low activity form until it is needed to switch off cAMP signaling. Hence, both Disc1 and PDE4B are important genes involved in the development of schizophrenia and depression through cAMP-dependent mechanisms.

We therefore sought to establish a mutation in the mouse Disc1 gene, which can affect interactions with PDE4B and then use cognate measures of behavior, brain anatomy, neurophysiology, and pharmacological response to support and extend the clinical findings. The method of choice was N-nitroso-N-ethylurea (ENU) mutagenesis, which induces locus-specific point mutations with a high rate [18]. We found different impacts of two such mutations, one producing depressive-like phenotype and the other resulting in a schizophrenia-like phenotype in mice with a C57BL/6J genetic background [16]. Here, we will focus particularly on one ENU-induced mutation producing a marked depressive-like phenotype with mild cognitive deficit.

Identification of Missense Mutations in Disc1

Disc1 isoforms are encoded by 13 major exons, of which Exon 2 at 955 bp is the longest. Exon 2 is present in all known splice isoforms and encodes most of the protein head domain, which we have shown previously in the human to interact with PDE4B [43]. Association between schizophrenia, schizoaffective disorder, and aspects of working memory in the human have been reported for Exon 2-spanning haplotypes [13, 29, 37; Figure 1A] We screened Exon 2 of Disc1 in 1686 F1 progeny of ENU-mutagenized C57BL/6JJcl males and untreated DBA/2JJcl females, and detected one independently-derived missense mutations. Mutant transcript Disc1Rgsc1393 has a 127A/T transversion (Figure 1B) resulting in amino acid exchange Q31L. The Exon 2 sequences of the C57BL/6JJcl and DBA/2JJcl parental strains are identical (data not shown), suggesting that both mutations arose as a result of ENU administration. Heterozygous N2 backcross progeny of the founder 31L heterozygous (DBA/2JJcl x C57BL/6JJcl) F1 males and wild-type C57BL/6JJcl females were backcrossed through the male and female lines to C57BL/6J for four generations (N3–N6) before heterozygotes with a predominantly C57BL/6J genetic background (average of 98.4375% at N6) were intercrossed to generate homozygous (31L) for phenotypic testing, which were viable and grossly indistinguishable from their wild-type littermates.

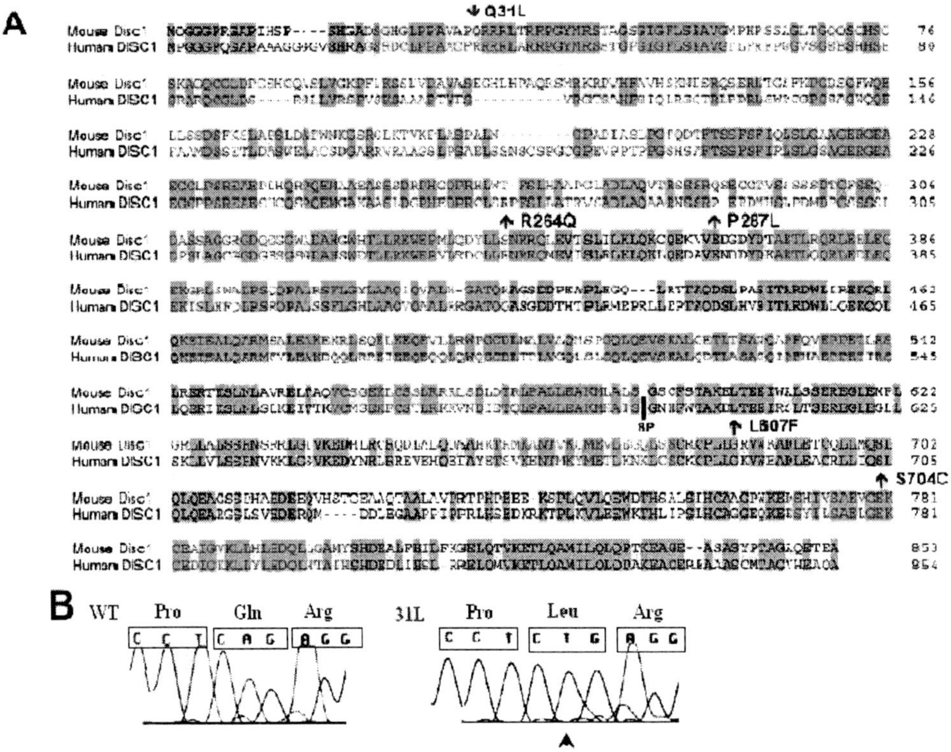

Figure 1. Missense mutations in mouse Disc1. (A) Alignment of the predicted disrupted in schizophrenia 1 protein sequences of mouse Disc1 (ENSMUSP00000074147) and human DISC1 (ENSP0000295051) using the ClustalW1.81 program. Residues identical between mouse and human have a gray background. The codon mutated in Disc1 mouse line is shown in red text and indicated by a red arrow. Nonsynonymous SNPs in human DISC1 associated with increased risk for major depression [27], bipolar disorder [37], schizophrenia [11, 13], or schizoaffective disorder are shown in green text and indicated by green arrows. The breakpoint of the translocation that segregates with psychiatric illness in a large Scottish family [42] is indicated by a vertical bar (j) labeled BP. Dashes indicate alignment gaps. Numbers to the right of the alignment show the amino acid position of the corresponding aligned sequence. Exons are indicated by alternating black and blue text. (B) DNA sequence chromatogram showing point mutation in Disc1 Exon 2. Transversion 127A/T is predicted to convert residue 31 from CAG glutamine (Gln) to CTG leucine (Leu). Adapted from [16].

Mice were used at 12-16 weeks old for all experiments. In order to detect earliest age for behavioral abnormalities, independent cohort of mice was tested at 8 weeks old in forced swim test and prepulse inhibition as the quick and robust measures for depression and schizophrenia in animals [22, 24]. 31L/+ (Rgsc1393) mice are available from the RIKEN BioResource Center (www.brc.riken.jp/lab/animal/en/gscmouse.shtml).

Depressive-like Phenotype in 31L Disc1 Mutant Mice

No differences in locomotion and exploratory activity in the novel open field were observed between wild-type mice and 31L mutants (Figure 2), mice of both genotypes similarly expressed horizontal and vertical activity. In the elevated plus maze, a test of anxiety [48], the duration of time on the open arms, as the main parameter of anxiety, was not different between genotypes (Table 1).

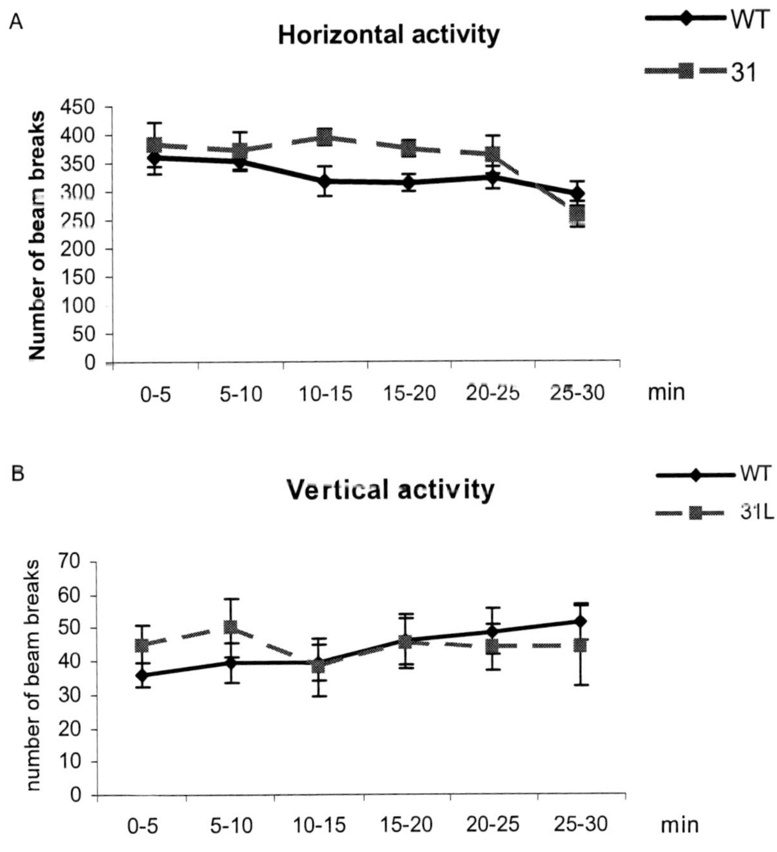

Figure 2. (A) Horizontal and (B) Vertical activity in the open field. Mean number of beam breaks (±SEM) in 5 min bins by 31L ($n = 6$), and wild-type (WT; $n = 14$) mice. There was no effect of genotype ($p > 0.05$), but there was significant effect of time interval [$F(5, 120) = 5.8$, $p = 0.0003$]. Adapted from [16].

In the forced swim test, a well-established paradigm to detect depressive-like behavior in rodents [22], the duration of floating immobile on the surface of the water was increased in 31L mutants (Figure 3A). The depression-like behavior was stable and detectable at an early age, since this phenomenon was observed at 8 and 12 weeks old Disc1-31L mutants. Moreover, in order to distinguish if 31L mutation indeed causing effect of behavioral despair rather than effect of fatigue, in a 2 day forced swim test procedure, a new cohort of mice have been tested. All mice had greater immobility on day 2 than on day 1 (Figure 3B), suggesting that all genotypes were exhibiting learned helplessness, but 31L mutants displayed higher immobility than wild-type animals on both days (Figure 3B), suggesting an effect of despair rather than fatigue. As antidepressant drugs have been shown to decrease the duration of immobility [22], this measure is used as an index of behavioral despair. We found that acute administration of bupropion reduced immobility in both wild-type and 31L mice (Figure 3C). Chronic treatment with rolipram at lower dose (0.5 mg/kg) was able to produce antidepressant-like effect on wild-type mice, without any effect on 31L mutants (Figure 3C). However, rolipram at a higher dose decreased immobility in 31L mutant mice, also with pronounced effect on wild-type animals, indicating lower sensitivity of Disc1 mutants to PDE4B inhibitor.

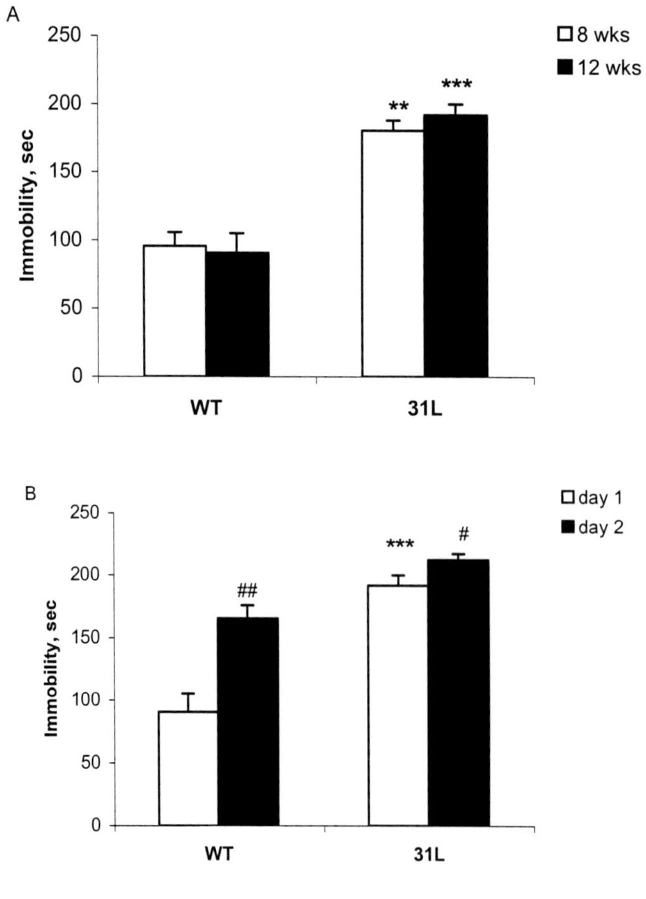

Figure 3. Continued on next page.

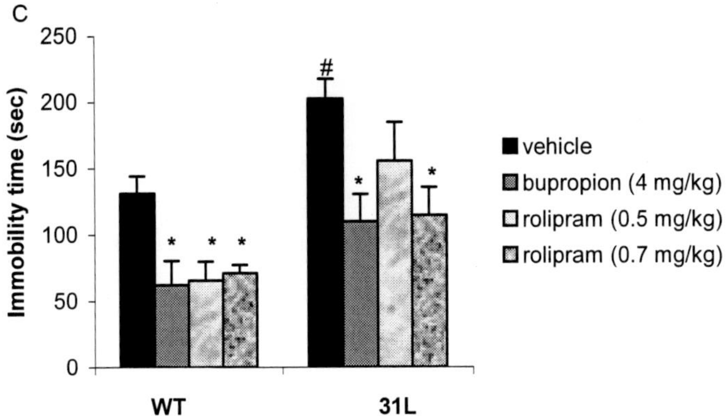

Figure 3. (A) Forced swim test. Mean duration of immobility (±SEM) of 31L at 8 weeks old (n = 8), 31L at 12 weeks old (n = 10), wild type (WT) at 8 weeks old (n = 10) and at 12 weeks old (n = 27) mice. There was a significant effect of age [F(1, 53) = 6.8, p= 0.03] and genotype [F(1,53) = 7.6, p = 0.005]. Immobility time was longer in 31L mice. **p < 0.01, ***p < 0.001 versus WT mice. (B) Two-day forced swim test procedure. Mean time of immobility (± SEM) of 31L (n = 15) and WT mice (n = 8) mice. There were significant effects of genotype [F(1,20) = 40.2, p = 4 x 10^{-7}], test day [F(1,20) = 21.5, p = 2.2 x 10^{-5}] and genotype x test day interaction [F(1,20) = 4.5, p = 0.01] on immobility time. Mice of both genotypes had higher immobility times on the second day, but 31L mutants showed increased immobility duration on both test days (p < 0.001). #p < 0.05, ##p < 0.01 compared with Day 1 within each genotype; ***p < 0.001 compared with WT mice. (C) Pharmacological responses in the forced swim test. Mean effects of bupropion (acute administration; 4 mg/kg, 30 min, i.p.) and rolipram (daily administration for 8 days; 0.5 mg/kg, and 0.75 mg/kg, 30 min, i.p.) on duration of immobility (±SEM) in 31L (n = 7-10) and WT (n = 6-9) mice. There were significant effects of drug [F(3,47) = 6.31, p = 0.01] and genotype [F(1,47) = 15.3, p = 0.004]. Bupropion, and rolipram at the high dose, significantly decreased immobility duration in 31L mice (p = 0.01 and p = 0.05). However, all drugs had antidepressant effects on WT mice. *p < 0.05, **p < 0.01 versus vehicle-treated mice within each genotype. ##p < 0.01 versus vehicle-treated WT mice. Adapted from [16].

Table 1. Behavior in the elevated plus-maze test of Disc1 mice

Behavioral parameters	WT n = 27	31L n = 13
Open arms time, %	2.8 ± 0.5	3.9 ± 0.8
Central platform time, %	14.0± 1.3	10.2 ± 1.8
Enclosed arms time, %	83.4 ± 1.5	85.8 ± 2.5
Open arms entries, %	8.5 ± 1.2	12.2 ± 2.4
Central platform entries, %	36.2 ± 1.5	31.6 ± 2.8
Enclosed arms entries, %	55.2 ± 1.8	59.2 ± 3.0
Total entries	27.7 ±1.4	24.4 ± 2.6
Head-dips	5.4 ± 0.5	4.2 ± 0.8
Passages	4.1 ± 0.5	4.6 ± 0.7
Open arm end exploration	1.2 ± 0.3	1.5 ± 0.3

Next, we investigated whether the depressive-like behavior of 31L mice in the forced swim test was also exhibited in the less-aversive social interaction [47] and reward responsiveness [53] tests, deficits in which are analogous to key symptoms of depression [1].

In the sociability phase of the social interaction test, we found that wild-type mice spent more time and making more entries in a chamber containing an unfamiliar mouse inside a cylinder than in the opposite chamber containing an empty cylinder, whereas 31L mutants did not (Figure 4A-B). In the social novelty phase of the test, wild-type mice demonstrated a shift in preference, spending more time and entering more often in a chamber containing a new unfamiliar mouse (''stranger 2'') than in the chamber containing the original unfamiliar mouse from the sociability phase (''stranger 1''), whereas again 31L mutants did not (Figure 4A-B). The lower social interaction of 31L mutants was not a result of diminished olfaction, as we found that the time required to find buried food in an olfaction test was not different between genotypes (Figure 4C) or decreased exploratory activity during the test (Figure 4D). We tested reward responsiveness by measuring sucrose consumption. Low responsiveness to rewards, such as a sweet taste, is comparable to anhedonia, the core symptom of the melancholic subtype of depression [67]. When presented with two drinking bottles, one containing water and the other containing 10% sucrose, 31L mice consumed significantly less of the sucrose solution relative to water than wild-type animals on the last 3 days of the 4-day test (Figure 5A), which is consistent with a lower responsiveness to rewards. We found no evidence for a difference in taste sensitivity in 31L mice (Figure 5B).

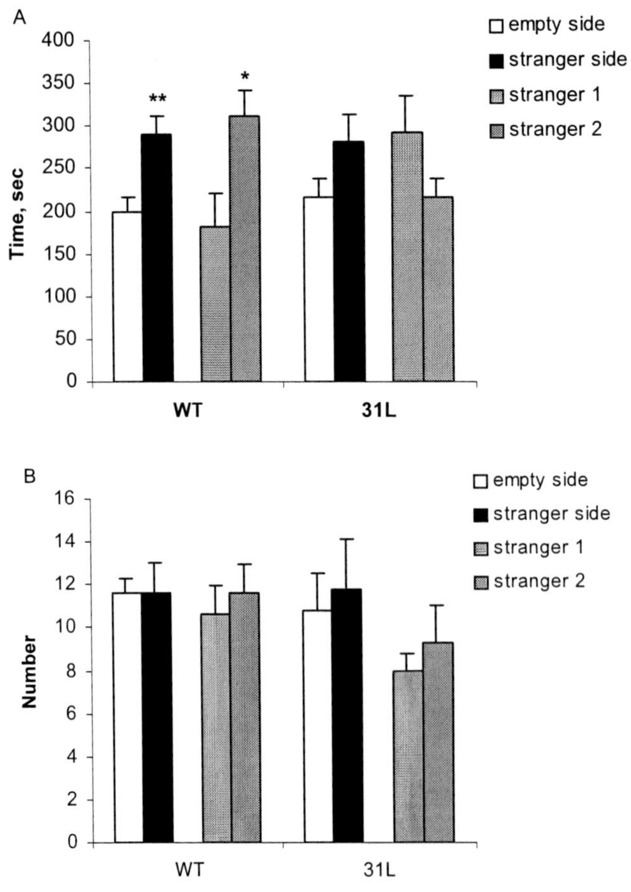

Figure 4. Continued on next page.

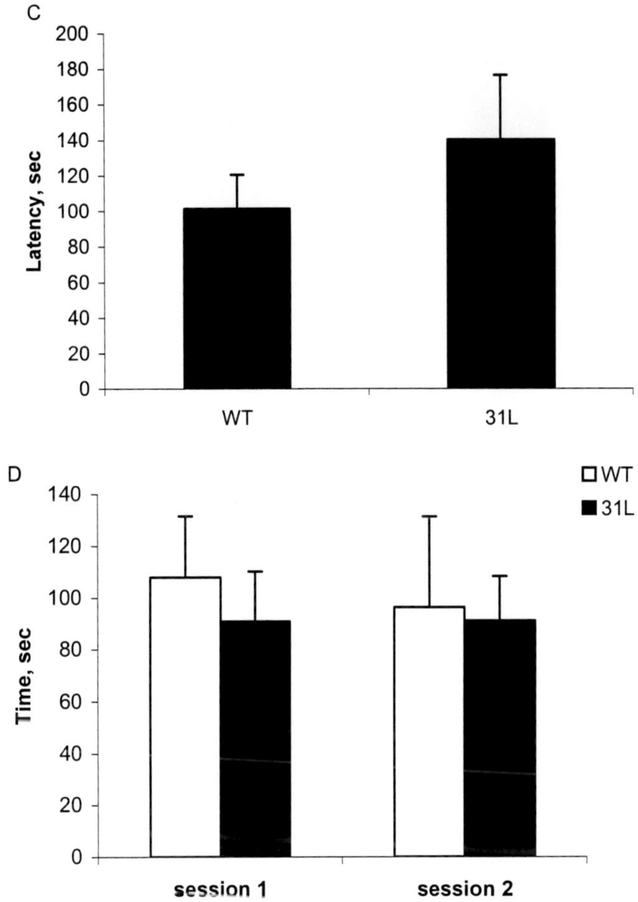

Figure 4. Sociability and Social Novelty. (A) Mean duration of time spent and (B) number of entries (±SEM) to the chamber with the stranger ("stranger side") than in the opposite chamber ("empty side") and in the chamber with the unfamiliar mouse from the sociability phase ("stranger 1") and in the opposite chamber with a new unfamiliar mouse ("stranger 2"). Unlike wild-type (WT) mice ($n = 6$), 31L ($n = 6$) male mice failed to demonstrate a preference for social proximity and preference for social novelty by spending significantly more time in the stranger side or chamber with stranger 2, rather than the more familiar stranger 1, without differences in number of entries (B). There were significant effects of genotype [$F(1, 11) = 7.8$, $p = 3.7 \times 10^{-5}$], presence of "stranger 1" [$F(1, 22) = 4.5 \times 10^{-4}$] and presence of new partner (stranger 2) [$F(1, 11) = 6.2$, $p = 0.02$]. *$p < 0.05$, **$p < 0.01$ versus empty side or versus "stranger 1" within each genotype. (C) Olfactory function. Mean latency (sec, ± SEM) of 31L ($n = 7$) and WT ($n = 8$) mice to find buried food. There was no difference between genotypes in this measure of olfactory function [$F(4, 28) = 1.21$, $p > 0.05$].(D) Exploratory activity. Mean duration of non-social, exploratory activity (sec, ± SEM) of 31L ($n = 7$) and WT ($n = 8$) mice during the testing period. There was no difference between genotypes in this measure of exploratory activity [$F(1, 13) = 0.98$, $p > 0.05$]. Adapted from [16].

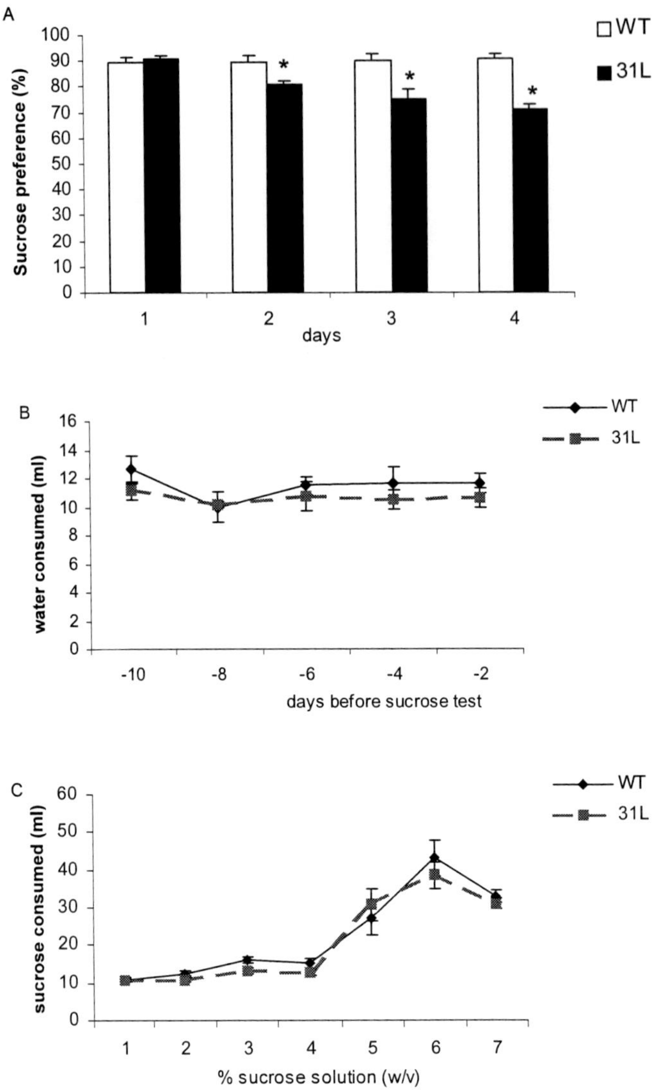

Figure 5. Sucrose consumption test. (A) Preference for a 10% sucrose solution over water is shown as the mean percentage (±SEM) of sucrose solution ingested relative to the total amount of liquid consumed. There were significant effects of genotype [$F(2,17) = 28.43$, $p = 0.0001$], test day [$F(3,51) = 5.39$, $p = 0.003$], and genotype x test day interaction [$F(6, 51) = 8.35$, $p = 53 \times 10^{-6}$] on sucrose preference. 31L ($n = 6$) mice consumed significantly less of the sucrose solution relative to water than wild-type (WT) ($n = 8$) mice on day 2 ($p < 0.01$), day 3 ($p < 0.05$), and day 4 ($p < 0.01$), consistent with a lower responsiveness to rewards. *$p < 0.05$, **$p < 0.01$ versus WT mice. (B, C) Sucrose sensitivity. (B) Mean water consumption (± SEM) of 31L ($n = 6$) and WT ($n = 6$) mice measured 10 days before the sucrose sensitivity test to establish baseline water intake. There was no difference between genotypes in consumption of water ($p > 0.05$). (C) Mean intake of various concentrations of sucrose solution (± SEM) by 31L ($n = 6$) and WT ($n = 6$) mice. There was a significant effect of sucrose concentration on intake of sucrose solution [$F(6, 60) = 32.1$, $p = 5 \times 10^{-17}$], but no effect of genotype was detected ($p > 0.05$). Adapted from [16].

Schizophrenic-Like Phenotype in 31L Disc1 Mutant Mice

Deficits in attention and information processing are considered a central feature of schizophrenia, which lead to stimulus overload, cognitive fragmentation, and thought disorders [50]. Prepulse inhibition and latent inhibition are the most common methods to quantify information-processing deficits in schizophrenia with a reasonable amount of face, predictive, and construct validity [24], and they can be used in both human and animal experiments [2]. Prepulse inhibition is the degree to which the acoustic startle response is reduced when the startle-eliciting stimulus is preceded by a brief low-intensity stimulus that does not elicit a startle response. We found that 31L mutants had mild, but detectably lower prepulse inhibition than wild-type mice (Figure 6A) at all three prepulses at the age of 12 weeks old, but no prepulse inhibition deficit was detected at 8 weeks old. There were no differences between genotypes in the acoustic startle response at both ages (Figure 6B). The typical antipsychotic haloperidol, a dopamine D2 receptor antagonist, and the atypical antipsychotic clozapine, an antagonist of both D2 and 5-HT2 receptors, had no effect improving prepulse inhibition deficit in 31L mutants (Figure 6C). Whereas the antidepressant bupropion, a dopamine and norepinephrine reuptake inhibitor, abolished the mild prepulse inhibition deficit of 31L mice (Figure 6C). The PDE4 inhibitor rolipram at the lower dose had no effect to facilitate prepulse inhibition in 31L mutants, whereas at the higher dose rolipram was able to restore mild prepulse inhibition deficit in this Disc1 31L mutant mice (Figure 6C). None of the drugs affected the acoustic startle response (Table 2).

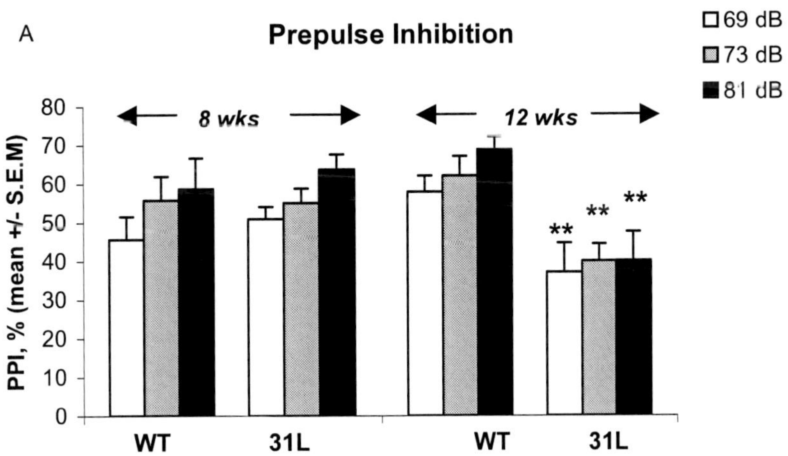

Figure 6. Continued on next page.

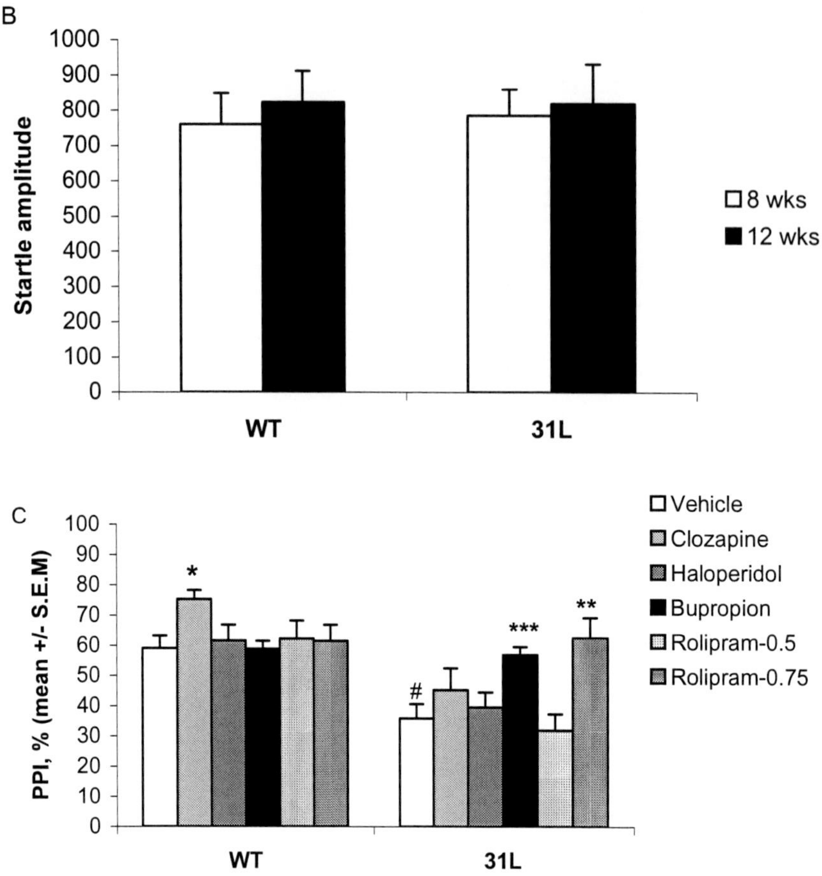

Figure 6. (A) Prepulse inhibition of acoustic startle response at 8 weeks old and 12 weeks old mice. Prepulse inhibition assay using a combination of startle (120 dB) and three prepulse levels (69 dB, 73 dB, and 81 dB) in 31L (n = 9, 24) and wild-type (WT) (n = 11, 21) mice. Prepulse inhibition is expressed as the mean percent reduction (± SEM) in startle amplitude at all three prepulses. Higher y axis values represent greater percent prepulse inhibition. There were significant effects of genotype [$F(1,63) = 12.6$, $p = 0.005$], prepulse intensity [$F(2, 128) = 6.3$, $p = 0.004$]. **$p < 0.01$ versus WT mice.(B) Mean acoustic startle response (ASR, ± SEM) to the startle stimulus (120 dB, 40 ms) without a prepulse. There was no effect of genotype [$F(1, 63) = 0.87$, $p > 0.05$] on startle expression. (C) Pharmacological responses in prepulse inhibition assay. Mean effects of clozapine (3 mg/kg), haloperidol (0.4 mg/kg), bupropion (4 mg/kg), rolipram (0.5 mg/kg) and rolipram (0.75 mg/kg) on average of three prepulses (mean PPI) (± SEM) in 31L (n = 10–16), and WT (n = 11–16) mice. There were significant effects of genotype [$F(1, 165) = 28.3$, $p = 10^{-6}$], drug [$F(5,165) = 8.9$, $p = 0.0002$], and genotype-drug interaction [$F(5,165) = 6.2$, $p = 4.8 \times 10^{-5}$]. *$p < 0.05$; **$p < 0.01$; ***$p < 0.001$ versus vehicle-treated mice within each genotype. #$p < 0.05$ versus vehicle-treated WT mice. Adapted from [16].

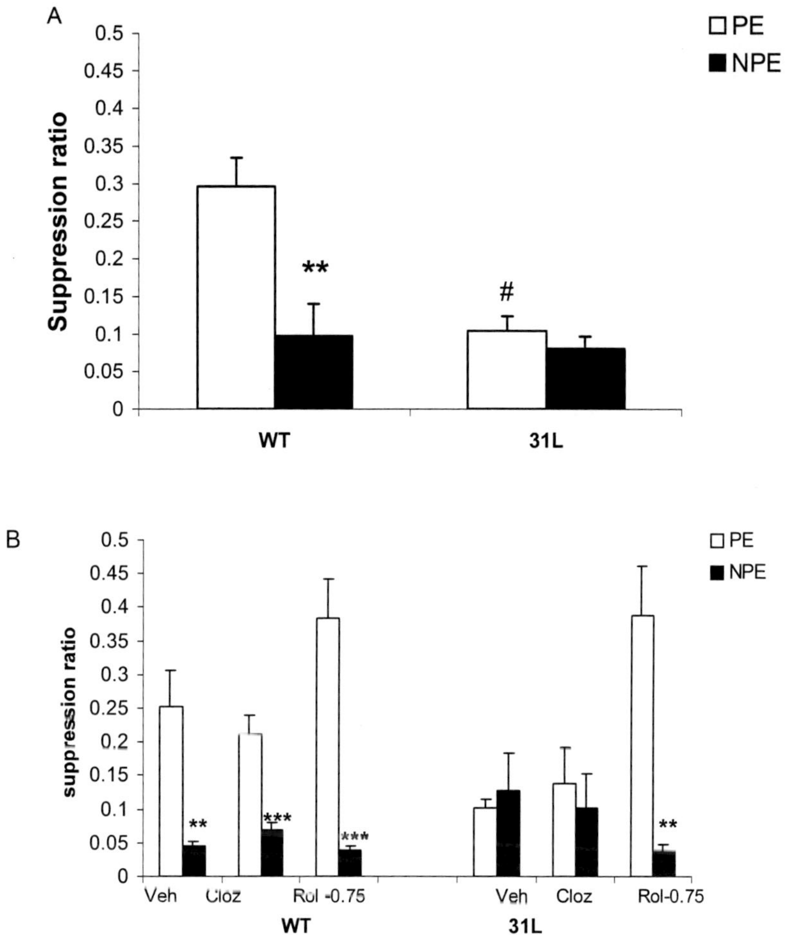

Figure 7. Latent inhibition of fear conditioning. (A) Mean suppression ratios (±SEM) of pre-exposed (PE) and non-pre-exposed (NPE) 31L (n = 9 PE, n = 8 NPE), and wild-type (WT; n = 13 PE, n = 8 NPE) mice to the conditioned stimulus. Latent inhibition is indicated by a lower suppression of drinking (higher suppression ratio) in PE compared to NPE mice, and was exhibited in WT mice. There were significant effects of genotype [F(1,35) = 4.6 p = 0.01] and pre-exposure [F(1,35) = 9.2, p = 0.005]. There was no difference between genotypes in the time to complete licks 50–75 (period A; before conditioned stimulus onset). **p < 0.01 versus PE mice within each genotype. # p < 0.001 versus PE-WT mice. (B) Pharmacological response in latent inhibition assay. Mean suppression ratios (±SEM) of vehicle-, clozapine-treated (3mg/kg) and rolipram (0.75 mg/kg) (PE, NPE) mice: 31L (n= 6-8) and WT (n = 6-10) mice. Only rolipram (0.75 mg/kg) rescued the latent inhibition deficit of 31L mice. Both drugs were administered intraperitoneally 30 min before both the pre-exposure and conditioning stages. There were significant effects of genotype [F(1,73) = 5.8, p = 0.01], drug [F(2,73) = 9.2, p = 0.001], pre-exposure [F(1, 73) = 41.2, p = 0.0001], and genotype x drug x pre-exposure interactions [F(2,73) = 6.2, p = 0.005]. **p < 0.01; ***p<0.001 versus PE mice within each genotype and drug treatment. Adapted from [16].

Table 2. Startle magnitude of Disc1 mice administered vehicle, clozapine (3 mg/kg), haloperidol (0.4 mg/kg), bupropion (4 mg/kg), rolipram (0.5 mg/kg) and rolipram (0.75 mg/kg)

Drug	WT	31L
Vehicle	812.3 ± 58.6 (n = 16)	805.6 ± 39.3 (n = 16)
Clozapine	765.5 ± 70.9 (n = 8)	690.3 ± 83.9 (n = 9)
Haloperidol	843.0 ± 44.0 (n = 8)	636.5 ± 102.7 (n = 8)
Bupropion	707.6 ± 66.8 (n = 7)	708.6 ± 93.8 (n = 9)
Rolipram-0.5	753.5 ± 85.2 (n = 7)	700.1 ± 92.2 (n = 9)
Rolipram-0.75	975 ± 91.4 (n = 8)	980.3 ± 88.4 (n = 6)

Latent inhibition is the phenomenon by which prior exposure to a nonrewarding conditioned stimulus decreases the salience of the conditioned stimulus when it is later paired with an unconditioned stimulus [66]. The drinking performance of mutant mice was normal, as genotypes were not different in time to complete licks 50–75 before conditioned stimulus onset during testing (overall A period = 10.2 sec). Latent inhibition was disrupted in 31L mice pre-exposed to the conditioned stimulus (Figure 7A). A strong association between conditioned and unconditioned stimulus was exhibited by non-pre-exposed mice of both genotypes, suggesting that the latent inhibition disruption was not due to a gross cognitive deficit but a specific information-processing deficit. The administration of clozapine had no effect on the disrupted latent inhibition in 31L mice (Figure 7B), however rolipram at the higher dose (as the effective dose based on prepulse inhibition and forced swim test data), was able to restore impaired latent inhibition in 31L mutant mice.

We assessed working memory using a discrete paired trial variable-delay T-maze task [3]. When this task was previously applied to Disc1 mutation carriers, they performed normally during training but showed a deficiency across all choice delay intervals in the working memory test [36]. We found that wild-type and 31L mice required the same training period to reach 70% correct responses on 3 consecutive days (Figure 8A). In the working memory test, 31L mutants had fewer correct responses at shorter (5 s and 10 s) but not longer (30 s) choice delay intervals (Figure 8B).

To sum up, the suggested depression-like phenotype detected in forced swim test already at 8 weeks of age, was supported by reduced sociability and social novelty and reduced reward responsiveness, behavioral paradigms that model anhedonia [17], a key depression symptom characterized by markedly diminished interest or pleasure in everyday activities [1]. The depression-like phenotype in 31L Disc1 mutants was treatable by antidepressant bupropion and rolipram at the higher dose.

Schizophrenia-like behavior in prepulse inhibition was mild, but significant in 31L mutants at the older age (12 weeks old). We found disrupted latent inhibition and modest deficit in working memory compare to another Disc1 mutant line (100P), described by Clapcote with colleagues [16]. Moreover, pharmacological analysis found that 31L mutants were sensitive particularly to antidepressant, which reduced high immobility in forced swim test, and even mild deficit in prepulse inhibition, however antipsychotics had no effect on prepulse inhibition impairments and latent inhibition deficit of 31L mice.

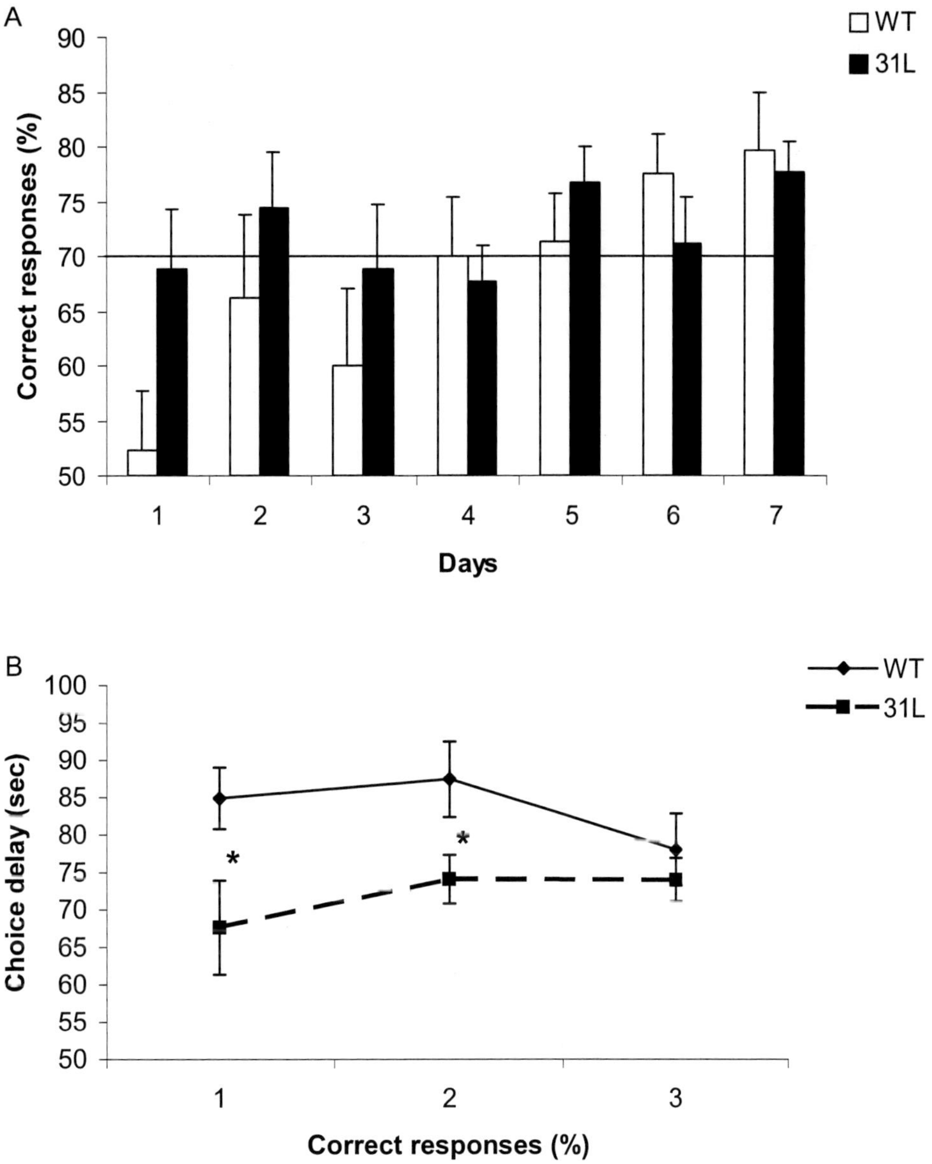

Figure 8. T-maze alternation training. (A) Mean percentage of correct responses (± SEM) during the discrete paired-trial delayed alternation training by 31L (n = 9), and wild-type (WT; n = 8) mice. There was no effect of genotype [$F(1,15) = 1.1$, $p > 0.05$] on correct responses. Note that the *y* axis starts at 50% correct responses, which represents the baseline response accuracy expected by chance. (B) Working memory performance with different delay intervals. Mean percentage of correct responses (± SEM) by 31L (n = 9), and WT (n = 8) mice in the working memory test with choice delay intervals of 5 s, 15 s, and 30 s. There was a significant effect of genotype [$F(1,15) = 6.3$, $p = 0.004$] on correct responses. *$p < 0.05$ versus WT mice. Adapted from [16].

Hence, we suggest here that depressive phenotype might be a dominant pattern over the mild and resistant to antipsychotic schizophrenia phenotype in our Disc1 genetic model. Next, we statistically estimated the effect of sample sizes between independent experiments among depressive (immobility in the forced swim test) and schizophrenia (prepulse inhibition) phenotypes. We found no effect of experimental group size on immobility of 31L mutants in the forced swim test (Table 3), however ANOVA detected significant effect size on prepulse inhibition, indicating on higher variation for that parameter. In addition, as expected, we also revealed no effect of the age on immobility in the forced swim test [$F(1,60) = 0.01$, $p > 0.05$], but found a significant effect of the age on prepulse inhibition [$F(1,35) = 9.5$, $p < 0.01$]. This data support our suggestion that depressive phenotype is more pronounced and stable than schizophrenia-like behavior.

Table 3. Immobility in the forced swim test and magnitude of prepulse inhibition of startle response in 31L mice among independent experiments

	Experiment 1	Experiment 2	Experiment 3	Experiment 4	ANOVA
Forced Swim Test					
Immobility, sec	211.3 ± 9.6 (n = 30)	198.1 ± 8.1 (n = 15)	196.5 ± 10.9 (n = 14)		F (2, 56) = 1.3; p = 0.28
Prepulse Inhibition (PPI)					
PPI at 69 dB, %	37.2 ± 3.7 (n = 33)	35.1 ± 9.2 (n = 7)	36.8 ± 6.2 (n = 13)	26.1 ± 6.6 (n = 16)	F (3, 65) = 5.32; p = 0.002
PPI at 73 dB, %	42.9 ± 3.9	38.6 ± 8.9	38.2 ± 5.6	27.8 ± 6.4	F (2, 130) = 24.9; $p = 10^{-4}$
PPI at 81 dB, %	48.1 ± 3.8	39.3 ± 5.4	38.8 ± 7.1	29.5 ± 4.6	

ANOVA detected significant effect of experimental sample size on prepulse inhibition of 31L mutants and significant effect of prepulse intensity, whereas no effect sizes on immobility in the forced swim test was found. Data presented as mean ± standard error.

Reduced Brain Volume in 31L Disc1 Mutant Mice

Structural and functional imaging studies have established that brain anatomical abnormalities are a consistent feature of schizophrenia [55]. Recent evidence suggests that this is also true in depression [40] and bipolar affective disorder [28]. We therefore examined 31L mutants for evidence of structural differences in the brain. Magnetic resonance imaging (MRI) revealed reductions in overall brain volume of 6% in 31L mice (Figure 9A-C). These overall reductions were accompanied by tissue contraction in the cortex, entorhinal cortex, thalamus, and cerebellum of 100P and 31L mutants.

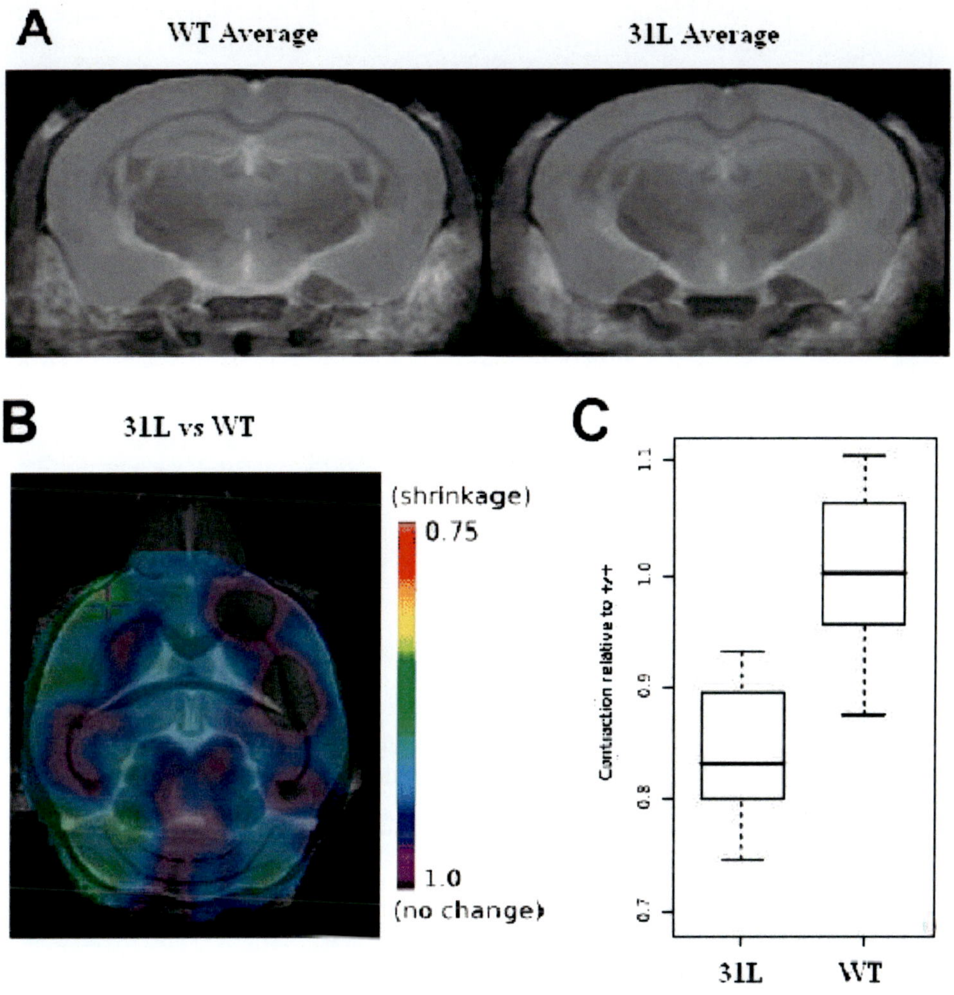

Figure 9. (A) Composite mean brains of the wild-type (WT; n = 6) and 31L (n = 6) groups after nonlinear registration. The overall volume of the 31L mutant brains was lower by 6% (U = 10, p = 0.05). (B) Relative contraction of brain regions. Differences in Jacobian determinants in the 31L (p = 0.046) compared to the WT. The contraction around the cortex is noticeable while the white matter tracts remain unchanged. (C) Box plot of tissue contraction (normalized to the mean Jacobian of the WT group) of a single voxel in the cortex. Adapted from [16].

DISC1 Protein Levels and Subcellular Distribution of DISC1 and PDE4B in 31L Disc1 Mutant Brain

To assess the effects of the 31L mutations on DISC1 protein levels in the brain, we measured DISC1 immunoreactivity in brain extracts from 31L and wild-type mice. An antibody that recognizes an epitope at the C-terminal region of DISC1, unaffected by both missense mutations, detected major bands of ~71, 75, and 105 kDa, plus several minor bands, in all genotypes [16]. The three major bands match comparably sized bands previously detected in the adult mouse brain [32, 33, 57] using different DISC1 C-terminal antibodies, the ~105 kDa band likely representing full-length DISC1. This suggests that 31L mutant line expressed all of the major DISC1 isoforms. Densitometry analysis of the ~71, 75, and 105 kDa signals revealed no significant differences between genotypes.

To examine effects of the 31L mutations on the subcellular distribution of DISC1, we prepared subcellular fractions from wild-type and 31L mouse brains. Immunoblotting of synaptosomes and postsynaptic density (PSD) fractions revealed that the ~71 and 75 kDa DISC1 isoforms were predominantly expressed in purified synaptosomes, PSD1, and PSD2 fractions, but not enriched within the core PSD fraction, PSD3 [43]. We also investigated the distribution of PDE4B in these fractions, since we have previously demonstrated a robust interaction between PDE4B and DISC1 in human cells [43]. PDE4B was associated with synaptosomes and all PSD fractions [16]. In addition, we analyzed P3 (microsomes and light membranes) and ER-G (a mixture of ER, Golgi, and plasma membranes) fractions, and immunoblotting analysis detected the ~71 and 75 kDa DISC1 isoforms in both fractions. PDE4B was also abundant in both fractions [16]. There was no detectable difference in distribution of DISC1 or PDE4B in any of the fractions tested from wild-type and 31L mice.

Reduced PDE4B Binding to Mutant DISC1

The N-terminal head domain of DISC1 is important for interaction with PDE4B, and we previously demonstrated that amino acids 220–283 are involved in binding [43]. However, a peptide array profiling approach has since demonstrated that multiple regions of human DISC1 act as contact sites for PDE4B, including additional sites within the head domain that encompass the positions of the mouse 31L mutations (unpublished data). We therefore investigated the possibility that this missense mutation influence binding between DISC1 and PDE4B. When DISC1 and PDE4B were exogenously coexpressed in HEK293 cells, we found that 31L mutation significantly reduced binding between DISC1 and PDE4B (Figure 10A). However, the degree of binding was variable, suggesting that mutant DISC1 binding to PDE4B is influenced by yet unknown cellular factors.

Lower PDE4B Activity in 31L Disc1 Mutant Brain

We previously reported that the head domain of Disc1 binds to the UCR2 domain of PDE4B in a cAMP-dependent fashion [43] and proposed a model in which Disc1 binds to an unphosphorylated low-activity form of PDE4B that is released as a PKA-phosphorylated high-activity form in response to cAMP upregulation, thus providing a negative-feedback mechanism to modulate cAMP levels [43]. Given this dynamic interaction and the decreased response of 31L mutants to the PDE4 inhibitor rolipram, we determined the relative amounts of DISC1 and PDE4B protein and PDE4B catalytic activity in 31L and wild-type brain extracts. Intriguingly, although we saw no difference in the amount of PDE4B protein between Disc1 mutant and wild-type brains (Figure 10C), we did see a striking 50% reduction in PDE4B activity in 31L brains (Figure 10B). This reduction in PDE4B activity is consistent with the resistance of 31L mutants to treatment with rolipram (Figure 3B, 6C) and efficacy of the higher dose rolipram (Figure 3B, 6C, 7B).

Rolipram is a potent, highly selective inhibitor of PDE4, the enzyme that specifically catalyzes the hydrolysis of cAMP [32], a second messenger in most neuromodulatory systems in the brain. Thus, the administration of rolipram increases cAMP levels in brain tissue, which is thought to have behavioral effects via the consequent upregulation of cAMP signaling in the hippocampus and cerebral cortex [56, 65]. However, the 50% reduction in cAMP hydrolyzing activity of brain PDE4B, coupled with the rolipram-insensitivity of 31L mutants, suggests that increased cAMP levels are likely a dominant contributory factor to the depressive-like behavior of 31L mice.

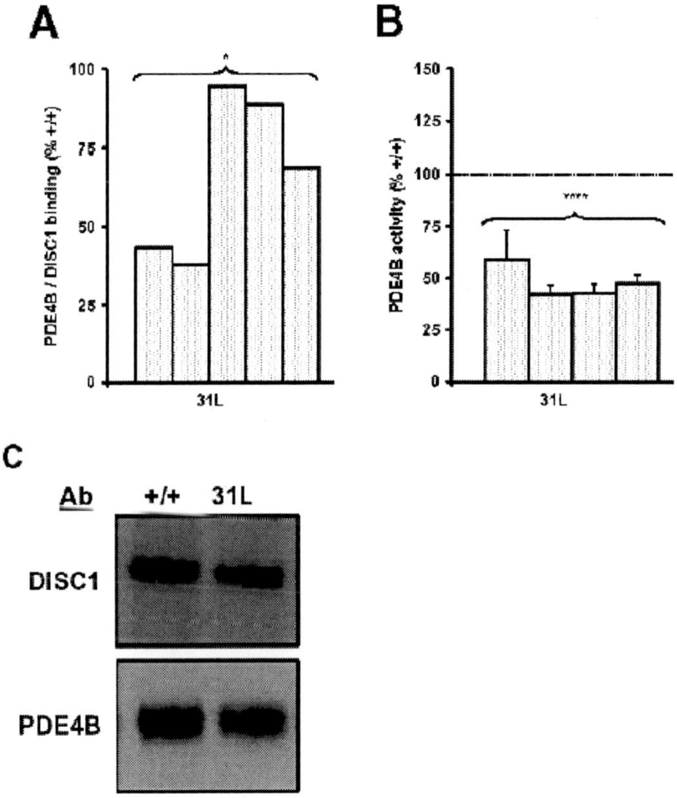

Figure 10. Reduced PDE4B binding to mutant Disc1. (A) Association between exogenously expressed PDE4B isoforms and mutant Disc1 expressed as percentage of WT mice. Five independent binding assays were performed, using PDE4B1 (B1) or PDE4B3 (B3). Numbers within bars refer to each independent experiment. Binding of PDE4B isoforms to 31L mutant Disc1 was significantly reduced * $p < 0.05$. (B) Lower PDE4B activity in 31L mutant brain. Activity of PDE4B in the brains of 31L ($n = 4$) mice. Activity measurements are means (±SEM) for three separate experiments. Results are expressed as a percentage of WT ($n = 4$) mice, to minimize inter-experiment variation. Mean PDE4B activity in 31L brain tissue was reduced by 47.98% ± 6.70% ($p < 0.00001$). ****$p < 0.00001$.(C) Typical blot of PDE4B1 immunoprecipitations probed with anti-Disc1 (upper) and pan-PDE4B (lower) antibodies. The bands shown correspond to the predicted 100 kDa DISC1 and 80 kDa PDE4B1 forms. Adapted from [16].

Disc1 abundance and subcellular distribution are indistinguishable from normal in mutant brains, indicating that altered Disc1 function, rather than expression, gives rise to the depression-like and schizophrenic-like phenotypes of 31L mutant mice. Indeed, we have demonstrated that the 31L mutations reduce PDE4B binding to Disc1 and additionally results in reduced PDE4B activity, which is not attributable to altered PDE4B expression, suggesting that altered Disc1/PDE4B function is fundamentally related to the phenotypes of 31L mutant mice. However, these observations are at odds with our previous data implying that decreased binding of PDE4B to Disc1 is likely to result in increased phosphodiesterase activity [43]. This discrepancy is perhaps partially due to our incomplete understanding of the biochemistry of Disc1/PDE4B binding. Indeed, we now appreciate that the interaction between these two proteins is far more complex than we previously anticipated, since multiple points of contact between human Disc1 and PDE4B have recently been identified (unpublished data). Such multipoint attachment allows Disc1/PDE4B complexes to show isoform-specific responses to elevated cAMP levels (unpublished data). This emerging complexity indicates that the differential impact of the Disc1 mutations upon PDE4B binding, and possibly also activity, is likely to be crucially related to their position within distinct contact sites for PDE4B. Alternatively, the reduced PDE4B activity of 31L mutants may not be related to the reduced Disc1 binding but may instead involve the extracellular-regulated kinase pathway that is known to directly regulate PDE4B catalytic activity by phosphorylation of its catalytic unit [6]. Intriguingly, it was recently demonstrated that extracellular-regulated kinase activation is modulated by Disc1 expression levels in cultured neurons [27, 58]; thus, mutant DISC1 may influence extracellular-regulated kinase activity leading indirectly to reduced PDE4B activity in the 31L mutant. Although we can only speculate upon the molecular causes of the altered PDE4B binding and activity in 31L mutants, we have shown that both Disc1 and PDE4B are abundant in synaptosomes and postsynaptic density fractions, and consequently it is possible that depressive-schizophrenia-like phenotypes of the 31L mutant mice may arise in part from altered synaptic cAMP signaling.

Conclusion

In conclusion, we have shown that missense mutation in mouse Disc1 elicit physiological, pharmacological, neuroanatomical, and behavioral phenotypes, which when taken together are strikingly consistent with the emerging picture from clinical and basic studies of Disc1 as a common genetic and biologically plausible risk factor for major mental illnesses. This mouse model supports both the neurodevelopmental role for Disc1 [34] and the proposed cAMP signaling role through modulation of PDE4 activity [52, 43]. Various types of disturbances of DISC1, such as deletion in a coding exon [36], missense mutation [16], and expression of a dominant-negative mutant [31], lead to common behavioral changes in some paradigms, but not in others [31]. Comparative analysis between the Disc1 mouse models will provide a better understanding of the Disc1 functions in the brain. A combination of depressive and schizophrenia-like phenotypes of the 31L mutant mice, lend further credence to the growing recognition that schizophrenia and bipolar disorder share, at least in part, common genetic etiologies and thus underlie molecular mechanisms [19, 20]. Our results in the mouse thus emphasize the importance of replicating, validating, and resolving inconsistencies within the current picture of various Disc1 alleles and haplotypes associating with distinct clinical phenotypes and, indeed, normal variation in cognitive function and neurodevelopment [29, 52]. This study demonstrates that a *Disc1* missense mutation in mice give rise to depression-like and schizophrenia-like phenotypes, thus supporting the role of this gene in major mental illnesses.

REFERENCES

[1] American Psychiatric Association: *Diagnostic and Statistical Manual of Mental Disorders*, Fourth Edition; Washington, DC: American Psychiatric Association 1994

[2] Arguello, P.A., and Gogos, J.A: Modeling madness in mice: one piece at a time. *Neuron* 2006; 52, 179–196.

[3] Aultman, J.M., and Moghaddam, B: Distinct contributions of glutamate and dopamine receptors to temporal aspects of rodent working memory using a clinically relevant task. *Psychopharmacology* (Berl.) 2001; 153, 353–364.

[4] Austin, C.P., Ky, B., Ma, L., Morris, J.A., and Shughrue, P.J: Expression of Disrupted-In-Schizophrenia-1, a schizophrenia-associated gene, is prominent in the mouse hippocampus throughout brain development. *Neuroscience* 2004; 124, 3–10.

[5] Badner JA, Gershon ES: Meta-analysis of whole-genome linkage scans of bipolar disorder and schizophrenia. *Mol Psychiatry* 2002; 7:405–411.

[6] Baillie, G.S., MacKenzie, S.J., McPhee, I., and Houslay, M.D. (2000). Sub-family selective actions in the ability of Erk2 MAP kinase to phosphorylate and regulate the activity of PDE4 cyclic AMP-specific phosphodiesterases. *Br. J. Pharmacol.* 131, 811–819.

[7] Berrettini W: Evidence for shared susceptibility in bipolar disorder and schizophrenia. *Am J Med Genet*. 2003; 123C:59–64.

[8] Blackwood, D.H., Fordyce, A., Walker, M.T., St Clair, D.M., Porteous, D.J., and Muir, W.J: Schizophrenia and affective disorders—cosegregation with a translocation at chromosome 1q42 that directly disrupts brain-expressed genes: clinical and P300 findings in a family. *Am. J. Hum. Genet* 2001; 69, 428–433.

[9] Bonbon D, Breulet M, Gerard-Vandenhove M, Guiot-Goffioul F, Plomteux G, Sastre-y-Hernandez M, Schratzer M, Troisfontaines B, van Frenckeill R, Watchel H: Is phosphodiestirase inhibition a new mechanism of antidepressant action? A double blind double-dummy study between rolipram and desipramine in hospitalized major and/or endogenious depressives. *Eur Arch Psychiatry Neurol Sci* 1988; 238: 2-6.

[10] Brockington IF, Kendell RE, Wainwright S, Hillier VF, Walker J: The distinction between the affective psychoses and schizophrenia. *Br J Psychiatry* 1979; 135:243–248.

[11] Callicott, J.H., Straub, R.E., Pezawas, L., Egan, M.F., Mattay, V.S., Hariri, A.R., Verchinski, B.A., Meyer-Lindenberg, A., Balkissoon, R., Kolachana, B., et al: Variation in DISC1 affects hippocampal structure and function and increases risk for schizophrenia. *Proc. Natl. Acad. Sci. USA* 2005; 102, 8627–8632.

[12] Camargo, L.M., Collura, V., Rain, J.C., Mizuguchi, K., Hermjakob, H., Kerrien, S., Bonnert, P., Whiting, P.J., and Brandon, N.J. Disrupted in schizophrenia 1 interactome: evidence for the close connectivity of risk genes and a potential synaptic basis for schizophrenia. *Mol. Psychiatry* 2007; 12, 74–86.

[13] Cannon, T.D., Hennah, W., van Erp, T.G.M., Thompson, P.M., Lonnqvist, J., Huttunen, M., Gasperoni, T., Tuulio-Henriksson, A., Pirkola, T., Toga, A.W., et al: Association of DISC1/TRAX haplotypes with schizophrenia, reduced prefrontal gray matter, and impaired short- and long-term memory. *Arch. Gen. Psychiatry* 2005; 62, 1205–1213.

[14] Cardno AG, Rijsdijk FV, Sham PC, et al: A twin study of genetic relationships between psychotic symptoms. *Am J Psychiatry* 2002; 159:539–545.

[15] Cherry JA, Davis RL: Cyclic AMP phosphodiestirase are localized in regions of the mouse brain associated with reinforcement, movement, and affect. *J Comp. Neurol.* 1999; 407, 287-301.

[16] Clapcot SJ, Lipina TV, Millar KJ, Mackie S, Christie S, Ogawa F, Lerch JP, Trimble K, Uchiyama M, Sakuraba Y, Kaneda H, Shiroishi T, Houslay MD, Henkelman RM, Sled JG, Gondo Y, Porteous DJ, Roder JC: Behavioral phenotypes of Disc1 missense mutations in mice. *Neuron* 2007; 54(3): 387-402.

[17] Cryan, JF, and Holmes, A: The ascent of mouse: advances in modeling human depression and anxiety. *Nat. Rev. Drug Discov* 2005; 4, 775–790.

[18] Coghill, E.L., Hugill, A., Parkinson, N., Davison, C., Glenister, P., Clements, S., Hunter, J., Cox, R.D., and Brown, S.D: A gene-driven approach to the identification of ENU mutants in the mouse. *Nat. Genet* 2002; 30, 255–256.

[19] Craddock N, Owen MJ: The beginning of the end for the Kraepelinian dichotomy. *Br J Psychiatry* 2005; 186:364–366.

[20] Craddock N, O'Donovan MC, Owen MJ: Genes for Schizophrenia and bipolar disorder? Implications for psychiatric nosology. *Schizophr Bull* 2006; 32(1):9-16.

[21] Crow TJ: The continuum of psychosis and its genetic origins. The sixty-fifth Maudsley lecture. *Br J Psychiatry* 1990; 156:788–797.

[22] David, D.J., Renard, C.E., Jolliet, P., Hascoet, M., and Bourin, M: Antidepressant-like effects in various mice strains in the forced swimming test. *Psychopharmacology* (Berl.) 2003;166: 373–382.

[23] Engels P, Abdel'Al S, Hulley P, Lubbert H: Brain distribution of four rat homologous of the Drosophila dunce cAMP phosphodiesterase. *J. Neurosci. Res.* 1995; 41, 169-178.

[24] Geyer, M.A., and Ellenbroek, B: Animal behavior models of the mechanisms underlying antipsychotic atypicality. *Prog. Neuropsychopharmacol. Biol. Psychiatry* 2003; 27, 1071–1079.

[25] Green E, Raybould R, McGregor S, Gordon-Smith K, Heron J, Hyde S, et al. The operation of the schizophrenia susceptibility gene, Neuregulin 1 (NRG1) across traditional diagnostic boundaries to increase risk for bipolar disorder. *Archives of General Psychiatry*. 2005; 62, 642–648.

[26] Hamshere ML, Williams NM, Segurado R, et al: Evidence for linkage (LOD=3.54) at IQ42 close to DISC1in a genome scan of functional psychosis pedigrees including at least one member with schizoaffective disorder of bipolar type [abstract]. *American Journal of Medical Genetics* 2004; 130B(1):127.

[27] Hashimoto, R., Numakawa, T., Ohnishi, T., Kumamaru, E., Yagasaki, Y., Ishimoto, T., Mori, T., Nemoto, K., Adachi, N., Izumi, A., et al: Impact of the DISC1 Ser704Cys polymorphism on risk for major depression, brain morphology, and ERK signaling. *Hum. Mol. Genet* 2006; 15, 3024–3033.

[28] Hasler, G., Drevets, WC., Gould, TD, Gottesman, II, and Manji, HK: Toward constructing an endophenotype strategy for bipolar disorders. *Biol. Psychiatry* 2006; 60, 93–105.

[29] Hennah, W., Tuulio-Henriksson, A., Paunio, T., Ekelund, J., Varilo, T., Partonen, T., Cannon, T.D., Lonnqvist, J., and Peltonen, L: A haplotype within the DISC1 gene is associated with visual memory functions in families with a high density of schizophrenia. *Mol. Psychiatry* 2005; 10, 1097–1103.

[30] Hennah, W., Thomson, P., Peltonen, L., and Porteous, D: Beyond schizophrenia: the role of DISC1 in major mental illness. *Schizophr. Bull* 2006; 32, 409–416.

[31] Hikida T, Jaaro-Peled H, Seshadri S, Oishi K, Hookway C, Kong S, Wu D, Xue R, Andradé M, Tankou S, Mori S, Gallagher M, Ishizuka K, Pletnikov M, Kida S, Sawa A: Dominant-negative DISC1 transgenic mice display schizophrenia-associated phenotypes detected by measures translatable to humans. *Proc Natl Acad Sci USA* 2007, in press.

[32] Houslay, M.D., Schafer, P., and Zhang, K: Phosphodiesterase-4 as a therapeutic target. *Drug Discov. Today* 2005; 10:1503–1519.

[33] Ishizuka, K., Paek, M., Kamiya, A., and Sawa, A: A review of disrupted-in-schizophrenia-1 (disc1): neurodevelopment, cognition, and mental conditions. *Biol. Psychiatry* 5 2006; 9, 1189–1197.

[34] Kamiya, A., Kubo, K., Tomoda, T., Takaki, M., Youn, R., Ozeki, Y., Sawamura, N., Park, U., Kudo, C., Okawa, M., et al; A schizophrenia- associated mutation of DISC1 perturbs cerebral cortex development. *Nat. Cell Biol* 2005; 7, 1167–1178.

[35] Kanes SJ. Tokarczyk J, Siegel SJ, Bilker W, Abel T, Kelly MP: Rolipram: a specific phosphodiesterase 4 inhibitor with potential antipsychotic activity. *Neurosci.* 2007; 144: 239-246.

[36] Koike, H., Arguello, P.A., Kvajo, M., Karayiorgou, M., and Gogos, J.A: Disc1 is mutated in the 129S6/SvEv strain and modulates working memory in mice. *Proc. Natl. Acad. Sci. USA* 2006; 103, 3693–3697.

[37] Maeda, K., Nwulia, E., Chang, J., Balkissoon, R., Ishizuka, K., Chen, H., Zandi, P., McInnis, M.G., and Sawa, A: Differential expression of Disrupted-in-Schizophrenia (DISC1) in bipolar disorder. Biol.Psychiatry 2006; 60, 929–935.

[38] Maxwell CR, Kanes SJ, Abel T, Siegel SJ: Phosphodiesterase inhibitors: a novel mechanism for receptor-independent antipsychotic medications. *Neurosci,* 2004; 129: 101-107.

[39] McGiffin P, Owen M, Gottesman II, eds.: *Psychiatric genetics and genomics*, Oxford: Oxford University Press, 2002.

[40] Meyer-Lindenberg, A, and Weinberger, DR: Intermediate phenotypes and genetic mechanisms of psychiatric disorders. *Nat. Rev. Neurosci.* 2006; 7, 818–827.

[41] Miyoshi, K., Honda, A., Baba, K., Taniguchi, M., Oono, K., Fujita, T., Kuroda, S., Katayama, T., and Tohyama, M: Disrupted-In-Schizophrenia 1, a candidate gene for schizophrenia, participates in neurite outgrowth. *Mol. Psychiatry* 2003; 8, 685–694.

[42] Millar, J.K., Wilson-Annan, J.C., Anderson, S., Christie, S., Taylor, M.S., Semple, C.A.M., Devon, R.S., St Clair, D.M., Muir, W.J., Blackwood, D.H.R., and Porteous, D.J: Disruption of two novel genes by a translocation co-segregating with schizophrenia. *Hum. Mol* 2000; Genet. 9, 1415–1423.

[43] Millar, J.K., Pickard, B.S., Mackie, S., James, R., Christie, S., Buchanan, S.R., Malloy, M.P., Chubb, J.E., Huston, E., Baillie, G.S., et al: DISC1 and PDE4B are interacting genetic factors in schizophrenia that regulate cAMP signaling. *Science* 2005; 310, 1187–1191.

[44] Muly C: Signal transduction abnormalities in schizophrenia: the cAMP system. *Psychopharmacol Bull* 2002; 135: 1783 – 1793.

[45] Murray RM, Sham P, Van Os J, Zanelli J, Cannon M, McDonald C: A developmental model for similarities and dissimilarities between schizophrenia and bipolar disorder. *Schizophr Res* 2004; 71:405–416.

[46] Murray CJL, Lopez AD, eds.: *The global burden of disease: a comprehensive assessment of mortality, injures, and risk factors in 1990 and projected to 2020*: Boston: Harvard School of public Health and the World Health Organization.

[47] Moy, S.S., Nadler, J.J., Perez, A., Barbaro, R.P., Johns, J.M., Magnuson, T.R., Piven, J., and Crawley, J.N: Sociability and preference for social novelty in five inbred strains: an approach to assess autistic-like behavior in mice. *Genes Brain Behav* 2004; 3, 287–302.

[48] O'Donnell JM, Zhang HT: Antidepressant effects of inhibitors of cAMP phosphodieserase (PDE4). *Trends Pharmacol Sci* 2004; 25, 158-163.

[49] Ozeki, Y., Tomoda, T., Kleiderlein, J., Kamiya, A., Bord, L., Fujii, K., Okawa, M., Yamada, N., Hatten, M.E., Snyder, S.H., et al: Disrupted-in-Schizophrenia-1 (DISC1): mutant truncation prevents binding to NudE-like (NUDEL) and inhibits neurite outgrowth. *Proc. Natl. Acad. Sci. USA* 2003; 100, 289–294.

[50] Perry, W., and Braff, D.L: Information-processing deficits and thought disorder in schizophrenia. *Am. J. Psychiatry* 1994; 151, 363–367.

[51] Pickard BS, Thomson PA, Christoforou A, Christoforou A, Evans KL, Morris SW, Porteous DJ, Blackwood DH, Muir WJ: The PDE4B gene confers sex-specific protection against schizophrenia. *Psychiatr Genet* 2007; 17(3): 129-133.

[52] Porteous DJ, Thomson P, Brandon NJ, Millar JK: The genetics and biology of Disc1 – an emerging role in psychosis and cognition. *Biol Psychiatry* 2006; 60: 123-131.

[53] Pothion, S., Bizot, J.C., Trovero, F., and Belzung, C: Strain differences in sucrose preference and in the consequences of unpredictable chronic mild stress. Behav. Brain Res. 2004; 155, 135–146.

[54] Rogers, R.J., and Cole, J.C: Anxiolytic-like effect of (S)-WAY 100135, a 5-HT1A receptor antagonist, in the murine elevated plusmaze test. *Eur. J. Pharmacol.* 1994; 261, 321–325.

[55] Ross, C.A., Margolis, R.L., Reading, S.A.J., Pletnikov, M., and Coyle, J.T: Neurobiology of schizophrenia. *Neuron* 2006; 52, 139–153.

[56] Scheider, H.H: Brain cAMP response to phosphodiesterase inhibitors in rats killed by microwave irradiation or decapitation. *Biochem. Pharmacol.* 1984; 33, 1690–1693.

[57] Schurov, I.L., Handford, E.J., Brandon, N.J., and Whiting, P.J: Expression of disrupted in schizophrenia 1 (DISC1) protein in the adult and developing mouse brain indicates its role in neurodevelopment. *Mol. Psychiatry* 2004; 9, 1100–1110.

[58] Shinoda, T., Taya, S., Tsuboi, D., Hikita, T., Matsuzawa, R., Kuroda, S., Iwamatsu, A., and Kaibuchi, K: DISC1 regulates neurotrophininduced axon elongation via interaction with Grb2. *J. Neurosci.* 2007; 27, 4–14.

[59] Stefansson H, Sigurdsson E, Steinthorsdottir V, et al. Neuregulin 1 and susceptibility to schizophrenia. *Am J Hum Genet*. 2002; 71,877–892.

[60] Tardito D, Tura GB, Bocchio L, Bignotti S, Pioli R, Racagni G, Perez J: Abnormal levels of cAMP-dependent protein kinase regulatory subunits in platelets from schizophrenic patients. *Neuropsychopharmacolgy* (Berl) 2000; 23: 216-219.

[61] Taylor MA: Are schizophrenia and affective disorder related? A selective literature review. *Am J Psychiatry* 1992; 149: 22–32.

[62] Tsuang MT, Winokur G, Crowe RR: Morbidity risks of schizophrenia and affective disorders among first degree relatives of patients with schizophrenia, mania, depression and surgical conditions. *Br J Psychiatry* 1980; 137:497–504.

[63] Valles V, Van Os J, Guillamat R, et al: Increased morbid risk for schizophrenia in families of in-patients with bipolar illness. *Schizophr Res*. 2000; 42:83–90.

[64] Van Os J, Gilvarry C, Bale R, Van Horn E, Tattan T, White I, Murray R: A comparison of the utility of dimensional and categorical representations of psychosis. UK700 Group. *Psychol Med* 1999; 29:595–606.

[65] Wachtel, H: Potential antidepressant activity of rolipram and other selective cyclic adenosine 30,50-monophosphate phosphodiesterase inhibitors. *Neuropharmacology* 1983; 22, 267–272.

[66] Weiner, I: The ''two-headed'' latent inhibition model of schizophrenia: modeling positive and negative symptoms and their treatment. *Psychopharmacology (Berl.)* 2003; 169, 257–297.

[67] Willner, P: Validity, reliability and utility of the chronic stress model of depression: a 10-year review and evaluation. *Psychopharmacology (Berl.)* 1997; 134, 319–329

[68] Zeller E, Stief HJ, Pflug B, Sastre-y-Hernández M: Results of a phase II study of the antidepressant effect of rolipram. *Pharmacopsychiatry* 1984; 17: 188-190.

INDEX

A

B

C

D

E

F

G

J

K

L

N

O

P

Q

R

S

T

U

V

W

X

Y

Z